WATER RESOURCE PLANNING, DEVELOPMENT AND MANAGEMENT

ADVANCES IN HYDROGEOCHEMISTRY RESEARCH

WATER RESOURCE PLANNING, DEVELOPMENT AND MANAGEMENT

Additional books and e-books in this series can be found on Nova's website under the Series tab.

WATER RESOURCE PLANNING, DEVELOPMENT AND MANAGEMENT

ADVANCES IN HYDROGEOCHEMISTRY RESEARCH

S. SAEID ESLAMIAN

AND

FAEZEH ESLAMIAN

EDITORS

nova
science publishers
New York

Library of Congress Cataloging-in-Publication Data
ISBN: 978-1-53617-829-6
Library of Congress Control Number: 2020936328

Published by Nova Science Publishers, Inc. † New York

CONTENTS

PREFACE

Hydrogeochemistry is of great application in reconstructing the previous groundwater circulation under ice sheets due to meltwater recharging the subsurface having a geochemical signature that could be discriminated from the other sources of water produced under warmer climates. Many hydrogeological factors could affect the geochemical composition of groundwater. These include the climate conditions, the lithology of the substrate and its hydraulic conductivity, the weathering, the length of the flow path, the penetration depth, the groundwater flow velocity and the residence time of groundwater.

In recent years, hydrogeochemistry advances in analytical technology have allowed hydrogeochemical surveys to become a very powerful exploration technique. The new research works in this area are very broad and extensive. In this mini book, it is impossible to bring all of these novel techniques. The current Advances in Hydrogeochimistry Research book is limited to some diverse examples in different regions of the earth. As an introduction, chemical interaction between surface water and groundwater has been explained and discussed. A hydrogeochemical perspective of the wetland systems is then drawn. It has been combined with the biological remediation. Contaminant fate and transport in both groundwater and soil are also studied using different image analysis techniques. A case study in an Arab region of the world is comprehensively undertaken and the impact of groundwater pollution and quality on regional economics has been investigated. The concepts, classifications and applications of the groundwater modeling for solute transport simulation have been defined in the saturated porous media of the earth. An African analysis was done on assessing the microbiological quality of potable groundwater from the selected protected and unprotected wells in Zimbabwe. 36Cl for dating groundwater was contributed on the Guarani Aquifer System, located beneath the surface of Argentina, Brazil, Paraguay and Uruguay and is the second largest known aquifer system in the world and is attributed as an important source of freshwater. For semi-arid and arid zones of Iran, geospatial techniques in hydrogeochemistry have been used by intersecting the layers and statistical analysis. Then, a hydrochemical analysis of the bottled mineral waters has been comparatively studied. In another part of Africa, a semiarid part of North Eastern Nigeria, physicochemical characteristics of groundwater are widely determined. It has been also shown that the hydraulic fracking associated with the coalbed methane production has a great impact on water resources. Finally, various case studies have been done on the environmental state of freshwater, natural resources and mine waters in the some parts of Serbia, Spain and Central Asia. In addition, the implications on regional water resources management have been indicated.

This book is intended primarily as a textbook at the graduate/research level and as a guide for field engineers, enabling them to remain current with scientific developments. As a simple prerequisite, the readers should have basic background knowledge in groundwater engineering and a general understanding of environmental and geological processes.

In: Advances in Hydrogeochemistry Research
Editors: S. Saeid Eslamian and Faezeh Eslamian
ISBN: 978-1-53617-829-6

Chapter 1

CHEMICAL INTERACTION BETWEEN GROUNDWATER AND SURFACE WATERS

Saeid Eslamian[1,2] and Mousa Maleki[3,*]
[1]Department of Water Engineering, Collage of Agriculture,
Isfahan University of Technology, 8415683111, Isfahan, Iran
[2]Center of Excellence in Risk Management and Natural Hazards,
Isfahan University of Technology, 8415683111, Isfahan, Iran
[3]Civil Engineering Department, Islamic Azad University, Shahrekord Branach,
Shahrekord, Iran

ABSTRACT

The interactions between groundwater and surface water are complex and dynamic and are known to show considerable spatial and temporal variability. Generally hydrological studies that investigate these interactions are conducted at weekly to yearly timescales and inadvertently lose information contained at the neglected shorter timescales. To perceive interactions in relation to biotic factors, geology, climate, and landform, a sound hydrogeoecological framework is required. Within the upper few centimeters of sediments beneath nearly all surface-water bodies, biogeochemical processes (hyporheic zone) have a remarkable effect on the water interchange chemistry, and here is where most of the recent research has been focusing.

Keywords: surface water, water interaction, geochemical, groundwater

[*] Corresponding Author's Email: mousa.maleki.7@gmail.com.

INTRODUCTION

Groundwater and surface water have the same origin within the closed system of the totality of waters existing on Earth. These are estimated to total approximately $1.5 \times 10^9\ m^3$ and are nearly all concentrated in one layer about 20 km thick, of which 10 km lies above sea level and 10 km beneath it. The waters above sea level take the form of water vapor in the lower atmosphere and of circulating freshwater both on the land surface and underground. Those below sea levels are mainly ocean saltwaters, accounting for 95% of the total waters of the Earth. The remaining 5% is freshwater: of this, 80% is stabilized in solid form as ice caps, mountain glaciers, and snowy mantles. Nearly all remaining waters in a liquid state (approximately 99%) are groundwaters, and very little of these are in circulation on the land surface or are accumulating in freshwater lakes. The negligible volume of circulating water is a representative of extraordinarily dynamic and vital phase in the hydrologic cycle. These processes are responsible for the circulation of the water masses between the oceans and the continents, together with the atmospheric transport of water vapor.

In the hydrologic cycle only 8% of the water evaporated from the oceans is transferred onto continents and an equal amount comes back to the oceans, partly in vapor form (2%) but mostly through the runoff of surface waters and through slow groundwater movements (Cavazza and Pagliara, 2017).

The motion and quality characteristics of surface and groundwater vary greatly, often leading to very different effects and conditions. It is because of these differences that the interactions and exchanges between surface and groundwater are of remarkable interest. Much of this interest is connected with their hydrological and hydrogeological implications, and with the potential value of surface and groundwater resources.

The fundamental differences between surface and groundwater can be characterized as follows:

1. Most surface water is in a solid state (snow and ice); other surface waters constitute the runoff from hillsides and the water in rivers, springs, and lakes.
2. Surface water movement is basically a function of the Earth's gravitational field. Within lakes there are fields of motion caused by variations of temperature or by the action of the wind. The fields of motion in groundwater are essentially regulated by the gravitational field in saturated two-phase flow (solid–liquid). In unsaturated porous media (where three phases - solid, liquid and gas, and vapor - are present) the influence of molecular forces becomes very important, and the superficial tension forces (thermal and vapor forces) often prevail over gravitational ones.

3. Groundwater moves at extremely low speeds, and in highly permeable aquifers. By contrast, surface waters flow on hillsides, and in streams and rivers, at higher speeds.
4. Surface waters have a free surface at which the piezometric line coincides with the water level; groundwater is often held under pressure in artesian aquifers, and here the piezometric line is above the level of the stratum.
5. Groundwater generally contains more dissolved chemical elements than surface water. In groundwater the concentration of such elements increases with the persistence of waters in the soil, the solubility of the rocks traversed by the waters, and the prevailing temperature, pressure, and pH. The oligo-mineral waters have low quantities of dissolved elements because they circulate, over relatively short timespans, in insoluble siliceous rocks. The chemical elements dissolved in groundwater are therefore of natural origin, and are largely governed by the rocks within which it flows. Human made pollution may result in the presence of chemical compounds from other sources. Surface water in normal conditions has less concentrations of chemical elements, representating its short-term contact with riverbeds and hillslope rocks. However, surface waters hold concentrations of pollutants, due to outflow caused by anthropogenic activities, and also contain suspended solid materials due to the hydrodynamic forces within its flow. (Cavazza and Pagliara, 2017).
6. As a consequence of (5), surfase water is vulnerable to pollution processes and very quickly disperses the polluting element by dilution, or by means of chemical and biological processes in which dissolved oxygen has a fundamental role. By contrast, groundwaters are less vulnerable, depending on how distant the recharge areas are and on the permeability of the geological deposits that separate the aquifer from the surface. The phreatic strata are therefore more vulnerable than artesian aquifers, due to their direct contact with the surface waters that infiltrate and percolate in the soil. Pollution of an aquifer may be discovered long after its contact with the contaminant source. This delay is a function of the water's travel time, and of the diffusion, dispersion, and transportation of the solute. These detection problems aggravate the pollution issue, since pollution may already have assumed dangerous proportions by the time it is discovered. Any remediation process is therefore lengthy, involving wholesale replacement of waters in the aquifer after elimination of the polluting source.
7. Temperatures can vary much more rapidly in surface water bodies than in groundwater. The temperature of surface waters varies with the temperature of the air, and thus between day and night and with the seasons. Groundwater temperatures depend on the thermal gradient inside the aquifer. Variations are therefore slow; many extensive aquifers cannot be influenced by seasonal variations, and temperatures within them remain practically constant. The

temperature of very deep groundwater increases with depth due to the presence of geothermal flow. Common of these forms are thermo-mineral waters, characterized by quite high temperatures and high levels of dissolved salts. (Cavazza and Pagliara, 2017).

Anthropogenic use of the surface water is easy and direct when it is possible to derive the necessary amount from the watercourse using gravity. When the waters are loaded with sediments, extraction becomes expensive and complex however: surface water use can be difficult when treatment is needed, or when the water required is available at a great distance or only in certain seasons, calling for long aqueducts and reservoirs. The anthropogenic use of groundwater is easier in these respects because spring or well capacities vary less with the seasons.

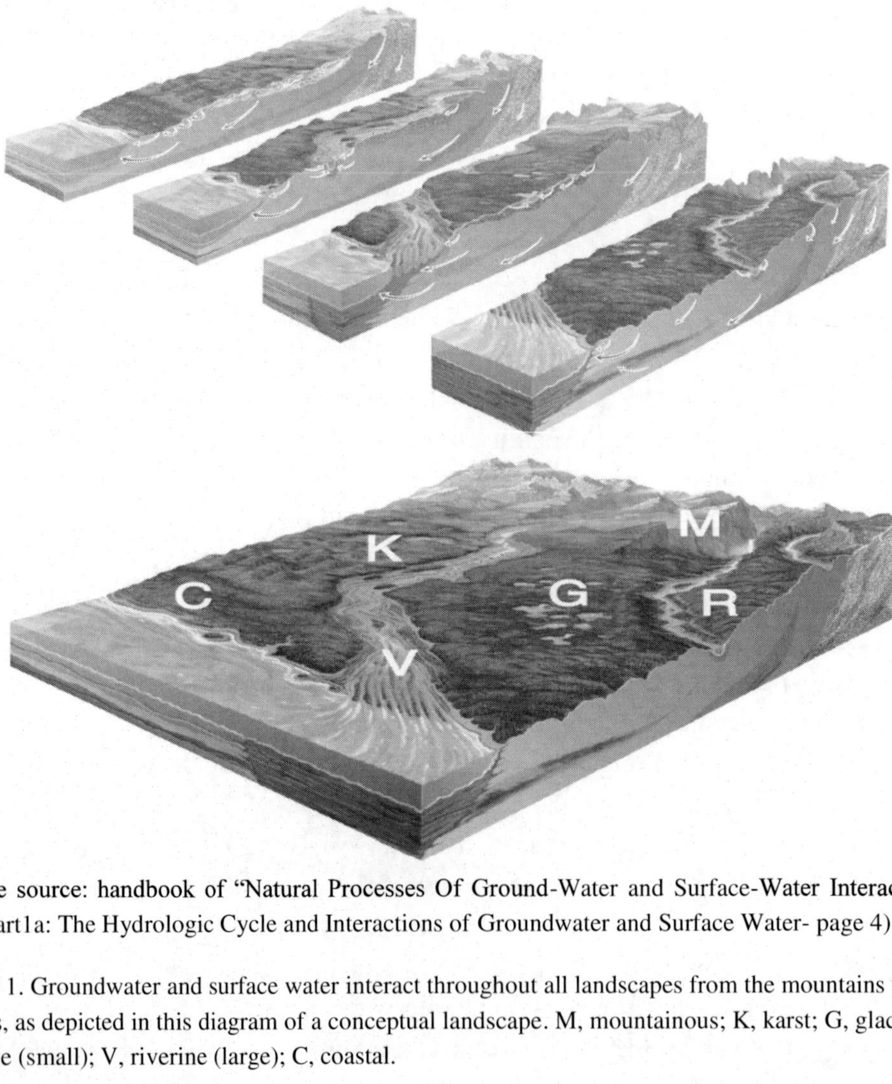

(Figure source: handbook of "Natural Processes Of Ground-Water and Surface-Water Interaction" – Part1a: The Hydrologic Cycle and Interactions of Groundwater and Surface Water- page 4).

Figure 1. Groundwater and surface water interact throughout all landscapes from the mountains to the oceans, as depicted in this diagram of a conceptual landscape. M, mountainous; K, karst; G, glacial; R, riverine (small); V, riverine (large); C, coastal.

Extracting water from wells involves energy consumption, and this influences operating costs. However, groundwater use does not involve water treatment of the kind needed for surface water. The conditions of motion in surface water and groundwater are generally different. The motion of surface water is usually governed by open-channel hydraulics, and the flow is almost always turbulent. The motion of groundwater in the aquifers may result from pressure flow (artesian aquifers) or free surface flow (phreatic aquifers), but is governed in most cases by the laminar motions set out in Darcy's law (Cavazza and Pagliara, 2017).

Water movement in the atmosphere and on the land surface is relatively easy to visualize, but the movement of groundwater is not. As depited in Figure 1, groundwater moves along flow paths of varying lengths from areas of recharge to areas of discharge. The generalized flow paths in Figure 1 start at the water table, continue through the groundwater system, and terminate at the stream or at the pumped well. The source of water to the water table (ground-water recharge) is infiltration of precipitation through the unsaturated zone.

In the highest, unlimited aquifer, flow paths near the stream can be tens to hundreds of feet in length and have corresponding traveltimes of days to a few years. The longest and deepest flow paths in Figure 1 may be thousands of feet to tens of miles in length, and traveltimes may range from decades to millennia. In general, shallow groundwater is more susceptible to contamination from human sources and activities because of its close proximity to the land surface. As a result, shallow, local patterns of groundwater flow near surface water are emphasized in this Circular.

SURFACE WATERS TYPES AND THEIR INTERACTIONS WITH GROUNDWATER

Surface water is water on the surface of continents such as in a river, lake, or wetland. It can be contrasted with groundwater and atmospheric water.

Non-saline surface water uses are replenished by precipitation and by recruitment from ground-water. It is lost through evaporation, seepage into the ground where it becomes ground-water, used by plants for transpiration, extracted by mankind for agriculture, living, industry, etc. or discharged to the sea where it becomes saline.

Groundwater-surface water interaction is important for both physical and chemical fluxes in the hydrologic cycle. Groundwater and surface water systems have commonly been studied separately, making it difficult to quantify the fluxes. Nonetheless, there is increasing recognition of the importance of understanding how this interface influences hydrologic budgets, geochemical cycles, ecosystems, and global change.

Surface-water/groundwater interactions include the exchange of water, and the chemicals that may be present in the water, which can lead to issues with water supply and water quality. Groundwater can be a major contributor to streams, lakes, and wetlands while surface water can contribute recharge to groundwater. The USGS Texas Water Science Center (TXWSC) uses geophysics and other techniques to study surface-water/groundwater interaction across the state. Visit the links below for more information on our data and science.

Streams interact with groundwater in all types of landscapes. The interaction takes place in three basic ways:

- streams gain water from inflow of groundwater through the streambed (gaining stream),
- streams lose water to groundwater by outflow through the streambed (losing stream), or
- they do both, gaining in some reaches and losing in other reaches.

For groundwater to discharge into a stream channel, the altitude of the water table in the vicinity of the stream must be higher than the altitude of the stream-water surface. Conversely, for surface water to seep to groundwater, the altitude of the water table in the vicinity of the stream must be lower than the altitude of the stream-water surface. Surface-water depletion is directly associated with chronic lowering of groundwater levels. The amount of water that is gained by a stream from groundwater is called baseflow. Several different methods of analyzing hydrographs have been used by hydrologists to determine the baseflow component of streamflow.

The movement of water between groundwater and surface water provides a major pathway for chemical transfer between terrestrial and aquatic systems. As chemicals are transferred between groundwater and surface water, the supply of carbon, oxygen, nutrients such as nitrogen and phosphorus, and other chemical constituents that enhance biogeochemical processes on both sides of the interface can be affected. This transfer can affect the biological and chemical characteristics of aquatic systems downstream.

The Impact of River Percolation on Chemistry of Shallow Groundwater in a Reclaimed Water Irrigation Area

Water shortages in arid and semiarid regions become increasingly prominent, which promotes wastewater irrigation worldwide (Bouri et al., 2008; USEPA, 2012; FAO, 2012; Chopra and Pathak, 2015). Agricultural wastewater reuse reduces the use of chemical fertilizers (Rosenqvist and Dawson, 2005; Teklehaimanot et al., 2015) and increases production (Xu et al., 2010; Wu et al., 2010.). But, wastewater sewage has a relatively high sodium adsorption ratio (SAR), and exchange between sodium and calcium ions may increase salinity during infiltration (Jain et al., 2005; Kass et al., 2005; Candela et al., 2007; Bouri et al., 2008). The infiltration of sewage irrigation may cause groundwater contamination by nitrogen (Jiang et al., 1997; Mitra et al., 2000; Kass et al., 2005; Yin et al., 2012) and heavy metals (Weng and Chen, 2000; Rattan et al., 2005). However, other studies have presented opposite results, indicating that irrigation with sewage containing high concentrations of heavy metals does not pollute groundwater (Mooers and Alexander 1994; Stuart and Miline, 2001; Yadav et al., 2002; Domínguez-Mariani, 2004).

Scholars have stated that groundwater contamination by sewage irrigation depends on groundwater vulnerability (Liu et al., 2006; Quayle et al., 2010.), and groundwater pollution risk is assessed by matching land use using DRASTIC model (Shirazi et al., 2013). The stable isotope method was used for quantitative evaluation of groundwater recharge sources in a sewage irrigation district (Domínguez-Mariani, 2004; Kass et al., 2005; Chen et. al, 2006; Li et al., 2012). There are few *in situ* investigations focusing on the impacts of long-term continuous infiltration of river water on groundwater chemistry in reclaimed water irrigation districts. The main objectives of the study were as follows: (1) to evaluate recharge sources of different aquifers under reclaimed water infiltration by stable hydrogen and oxygen isotopes; (2) to identify the impacts of reclaimed water infiltration on water quality of different aquifers, including of salinity, nitrogen and heavy metals; (3) to investigate groundwater vulnerability and removal efficiency of major types of pollutants such as nitrogen, phosphorus and heavy metals, in sand-clay alternating hydrogeological conditions.

Hydrogeological conditions characterized by sand-clay alternation result in different oxidizing and reducing environments, which facilitates the nitrification and denitrification reactions to promote high nitrogen removal efficiencies (up to 95%). During reconstruction of the Fengheying gate, the clearance of riverbed silt reduced the ammonia nitrogen removal ability and caused preferential cross-layer flow recharge at the WC3 well. Heavy metal contents of all sampling wells were 1–2 orders of magnitude lower than permissible limits owing to low contamination input from recharge water and high vulnerability of the aquitard.

INTERACTION OF RESERVOIR DAM AND GROUNDWATER

Among surface-water/groundwater interactions, probably the most distinct and complex are those between artificial lakes—referred to also as "reservoirs", following the Chapman (1996) terminology—and groundwater. Artificial lakes are designed to store surface water, so their beds are expected to have low permeability and low seepage.

However, they are affected by natural and particularly strong artificial (due to lake regulation) driving forces, which imply large and fast changing lake-water levels (lake-stages), implying large impact on groundwater dynamics. That impact may result, for example, in flooding of an area adjacent to a reservoir also affecting agricultural productivity or contamination of that area (Wildi 2010).

For all these reasons, the management of reservoirs and adjacent areas requires appropriate methodologies and models well accounting for surface-water/groundwater interactions. The simplest and most direct way of studying a lake interaction with groundwater, is by investigating exchange of seepage between that lake water and groundwater by point field measurements using methods such as seepage meters, thermic profiling, dye tracing, piezometers and potentio-manometers (Lee 1977; Ong and Zlotnik 2011; Otz et al., 2003; Winter et al., 1988) or a combination of methods (Anibas et al., 2009; Owor et al., 2011; Su et al., 2016). However, such methods are vulnerable to heterogeneity and therefore difficult for spatial integration. Hydrochemical analysis and mass balance methods (Sacks et al., 1998; Stauffer 1985) are better in that respect but require a conservative tracer not influenced by land-use practices such as for example stable isotopes (Brindha et al. 2014; Kanduč et al. 2014; Krabbenhoft et al. 1990; Sacks et al. 2014); additionally, convenient semianalytical water balance models (Ghosh et al. 2015; Rudnick et al. 2015) involve quite a number of assumptions simplifying spatial heterogeneity.

The most complete and reliable assessment of lake–groundwater interaction is by increasingly sophisticated physically based distributed and integrated hydrological models, particularly if based on solid characterization of hydrogeological conditions of a lake and its adjacent area and time series data, as proposed in this artificial lake– groundwater interaction study. The general scarcity of physically based distributed models that simulate interactions of lakes with groundwater is surprising, but even more surprising is that, till now, to the authors' knowledge, there has not been any such study published on the interaction of artificial lakes with groundwater.

Some of the few available examples of artificial lake studies, include the following: Liuzzo et al. (2015), who used a lumped conceptual model (TOPDM) to characterize the effect of climate change on water resources availability in the Belice River Basin (Italy), in which artificial Lake Garcia is located; Fowe et al. (2015), who used a genetic algorithm model to couple Boura reservoir (Burkina Faso) with the adjacent groundwater system and to optimize water allocation; and Chhuon et al. (2016), who coupled the semi-distributed hydrologic SWAT model with the MODSIM decision support system to investigate hydrological consequences of a future upland reservoir on the Prek Te River, Cambodia.

Simulations of interactions of natural lakes with groundwater have been carried out for decades, mainly by the widely used MODFLOW (McDonald and Harbaugh, 1988) type of solutions, also applied in this study. The simplest MODFLOW lake implementations include representing the lake as a head (stage) dependent sink or source through the River

Package (Leblanc et al., 2007), General Head Boundary (GHB) Package (Mylopoulos et al., 2007) or Reservoir Package (Fenske et al., 1996).

The disadvantage of these approaches is that the simulated lake stages constrain lakebed seepages, thus a prior knowledge of their seepage amounts is required (Merritt and Konikow, 2000) instead of calculating them internally. Another approach that requires cells of a model grid representing lake volume is called the "High-K" method (Lee, 1996), where extremely high hydraulic conductivity and a storage coefficient equal to 1 are used to simulate lake cells. The main disadvantages of the "High-K" method are that it is difficult to represent accurately the connections between streams and a lake (Merritt and Konikow 2000) and that the method is prone to numerical instability (Yihdego and Becht, 2013).

Surface Water–Groundwater Interaction and the Influence of Ion Exchange Reactions on River Chemistry for Domestic and Irrigation Purposes

Many rivers show an inverse relationship between discharge and total dissolved solids (TDS) load. This relationship is often interpreted to result from mixing of groundwater-fed base flow by more dilute waters. However, the observation that river waters often have fixed ion ratios through the year is inconsistent with simple mixing and suggests that water-rock interaction controls the major ion chemistry. It has commonly been assumed that after correction for atmospheric input, weathering stoichiometry is the dominant control on river chemistry. Based on this assumption, numerous authors have used the mass flux from rivers to calculate rock and mineral weathering rates (e.g., Meybeck, 1987; Holland, 1978).

However, Drever (1988) suggested that equilibrium cationex change reactions between Ca/Na-smectite and Ca/Mg-smectite may play an important role in controlling the major ion chemistry of some rivers. This suggestion is supported by Cerling et al. (1989), who indicated that Na-Ca ion exchange is a dominant control on the chemistry of rivers draining shale basins. Similarly, in examining numerous rivers draining basaltic terrains, Bluth and Kump (1994) noted that cation activity ratios always plot on trends paralleling smectite exchange-reaction boundaries. In their words (p. 2347), "if river chemistry depended only on bedrock composition and weathering stochiometry, we would expect a much greater variation among the cation activity ratios." Bluth and Kump (1994) concluded that the major cation ratios are likely controlled by smectite exchange reactions. It is important to evaluate the importance of cation exchange reactions in controlling river chemistry, particularly when the mass flux of rivers is used to calculate mineral weathering rates.

In order to examine surface-water–groundwater interaction, we consider groundwater data Ca2+ and Mg2+ activity ratios for groundwaters are, along with the reaction boundaries calculated herein. As with the Bow River data, Ca2+ and Mg2+ activity ratios define a slope of 1, and plot between boundaries for reactions 1 and 2. Three groups of waters are observed: (1) waters that cluster around the Na-K exchange boundary and the metastable extension, (2) waters that cluster around the Na-smectite óillite + K-feldspar boundary, and (3) waters that cluster around the albite óK-feldspar boundary. These data may be interpreted in terms of a classical evolution of waters during progressive weathering of rock-forming minerals (e.g., Drever, 1988). Alternatively, groups 1 and 3 may represent waters in equilibrium with different lithologies and group 2 may represent resulting mixtures.

There is a general trend of group 1 waters being associated with wells completed in either feldspathic sand stones of the Brazeau Formation or till, and group 3 waters being associated with the Wapiabi shale. It appeared that ion-exchange reactions influence groundwaters, and that groundwater evolves in rivers to have Ca/Mg and Na/K ratios that are nearly constant. Given this, caution is necessary when using river chemistry as a proxy for the stoichiometry of primary mineral weathering in a basin.

Hydrogeochemical Characteristics and Transformation Relationships between Surface Water and Groundwater in Rivers

The regularity of the water cycle and the conversion between surface water and groundwater is the basis for the study of pollutant transport in river-groundwater systems (Lu et al., 2016a; Shang et al., 2017). Different water bodies have different hydrogeochemical characteristics and isotopic signatures because of their different sources of recharge, environments and circulation conditions. Hence, hydrogeochemical characteristics are the ideal tracers for tracking water circulation processes. Analyses of hydrogeochemical and isotopic characteristics of rivers and groundwater can effectively reveal the relationship in the transformation of river water to groundwater. Descriptive statistics, graphic analysis and multivariate statistical analysis can be used to analyze the hydrogeochemical characteristics (Lu et al., 2015). Graphic methods include the Durov diagram, Stiff diagram, Piper diagram and Gibbs diagram. Common multivariate statistical analysis methods include factor analysis (Keesari et al., 2016), principal component analysis (Chattopadhyay and Singh, 2103) and cluster analysis (Zhang et al., 2012). Stable isotopes of δD and δ18O are ideal tracers for tracking various hydrogeochemical processes (Liu et al., 2014). Dogramaci et al. (2012) studied the hydrogeochemical and isotopic characteristics of the Hamersley Basin in northwestern Australia and provided a theoretical basis for the sustainable development of local water resource utilization. The descriptive statistical method, the Piper diagram and the main ion component proportion coefficient and factor analysis method were used to study hydrogeochemical characteristics of groundwater in the Sara Wusu aquifer system in the Ordos Basin (Yang et al., 2016). Fuzzy mathematics and multivariate statistical methods were used to study the quality characteristics of surface water and groundwater in the Songnen plain (Zhang et al., 2012). Zeng et al. (2013) studied the spatial distribution of hydrogeochemical and isotopic characteristics of different water bodies, including spring water, river water and lake water, in different parts of Tajikistan and discussed their origins and environmental significance. Although there are many studies related to the chemical and isotopic characteristics of groundwater and surface water, the relationship between surface water and groundwater transformation is still a prevalent topic in hydrology and water resource studies (Wang et al., 2016; Lu et al., 2016b), hydrogeochemistry, biogeochemistry, and ecohydrology. Hydrogen and oxygen stable isotopes (δ18O and δD) and electrical conductivity (EC) were used to study the mutual relationship among precipitation, river water and groundwater in Taiwan Douliushan (Peng et al., 2014).

Multivariate statistical analysis methods and isotope analysis methods were used to study the hydraulic linkage between surface water and groundwater and their temporal and spatial variation in the Condamine River in Australia (Martinez et al., 2015). Hydrogen and oxygen isotopes were used to study the relationship of recharge and discharge between the various water bodies on the Portuguese island of Madeira, from which a hydrogeological conceptual model of Madeira Island was established (Prada et al., 2016). By analyzing the hydrogeochemical characteristics of surface water and groundwater in the Heihe River Basin, Nie et al. (2005) identified the transformation relationship between groundwater and surface water in the main stream of Heihe River. Hydrogen and oxygen isotopes and water chemistry were used to investigate the relationship between surface water and groundwater of the Second Songhua River, and the end element method was used to quantitatively calculate a conversion proportion between surface water and groundwater (Zhang et al., 2014).

The surface water of Weihe River is seriously polluted and has become a major pollution source for nearby shallow groundwater. This seriously affects the exploitation, utilization and protection of groundwater resources and endangers the ecological safety and the health of the residents (Lu et al., 2016c). The surface water and groundwater were sampled in several typical sections of the Weihe River Basin to study the conversion relationship between surface water and groundwater based on the hydrogeochemical and isotopic characteristics and to provide a basis for groundwater protection, restoration and management.

Table 1. Major and trace elements concentrations ($\mu mol/l$) and $87_{Sr}/86_{Sr}$ ratios of the dissolved load collected during the daily survey

Sample name	Date	W(m^3/s)	pH[a]	T (°C)	Conductivity (µS/cm)	HCO_3^-	Cl^-	SO_4^{2-}	NO_3^-	Ca^{2+}	Na^+	K^+	Mg^{2+}	Rb	Sr	$^{87}Sr/^{86}Sr$
jlol	05/16/1995	310	8.28	14.8	235	1860	392.1	190.2	142.9	970.5	491.3	79.2	209.6	0.05	1.44	nd
jlo2	05/17/1995	339	8.21	14.9	227	1590	383.4	177.1	143.9	857.8	473.5	85.4	194.6	0.05	1.30	nd
jlo3	05/18/1995	351	7.79	14.3	221	1560	355.8	165.9	147.3	860.0	416.5	79.2	174.2	0.05	1.15	nd
jlo4	05/19/1995	389	8.05	13.4	229	1700	331.8	169.0	163.5	734.3	407.8	81.8	163.3	0.15	3.99	nd
jlo5	05/20/1995	433	7.51	14.2	222	1770	305.1	159.1	161.1	724.0	360.0	72.1	148.8	0.12	3.13	0.711253
jlo6	05/21/1995	475	7.47	15.7	214	1620	301.1	157.1	154.5	875.8	338.7	75.9	161.3	0.20	5.27	nd
jlo7	05/22/1995	460	7.68	17.0	203	1760	262.5	135.7	115.8	811.0	331.3	63.8	152.5	0.05	1.17	nd
jlo8	05/23/1995	372	7.40	19.4	202	1630	279.2	140.0	121.6	807.8	340.9	66.4	152.9	0.05	1.15	nd
jlo9	05/24/1995	341	7.56	17.8	211	1620	305.1	152.7	130.0	838.5	360.9	70.8	164.2	0.05	1.20	0.711336
Jlo10	05/25/1995	317	7.55	18.2	214	1730	286.2	142.2	114.8	875.3	382.9	72.8	174.2	0.05	1.24	nd
Jlo11	05/26/1995	308	7.50	19.1	217	1680	312.4	157.8	123.7	933.8	423.9	74.4	187.5	0.05	1.31	nd
jlo12	05/28/1995	272	7.55	21.5	222	1840	344.5	165.9	108.9	761.3	370.4	61.5	161.7	0.05	1.36	0.711262
jlo13	05/31/1995	220	7.50	18.6	244	1860	376.3	177.4	114.8	834.8	420.4	64.6	177.9	nd	nd	nd
jlo14	06/03/1995	254	7.46	17.7	234	1880	352.7	173.9	116.8	1245.8	587.0	103.3	285.8	0.06	1.43	nd
jlo15	06/05/1995	352	7.73	18.1	209	1400	354.6	160.3	95.2	992.0	590.0	87.9	245.4	0.05	1.18	nd
jlo16	05/12/1995	197	7.43	17.7	228	1730	372.7	178.1	81.6	1144.5	629.1	98.5	290.4	0.05	1.40	0.711372
jlo17	06/17/1995	170	7.32	19.1	233	1710	415.8	191.4	89.5	1114.5	699.6	103.6	302.9	0.06	1.50	nd
jlo18	06/23/1995	128	7.29	18.0	225	1530	440.6	202.4	66.3	928.5	736.5	121.0	310.0	0.13	2.85	0.711388
jlo19	06/25/1995	171	7.41	21.5	215	1490	429.9	202.1	43.1	954.3	755.7	117.7	306.7	0.11	2.48	nd
jlo20	06/29/1995	222	7.28	23.3	211	1390	436.9	194.1	51.5	835.5	728.3	135.9	316.3	0.07	1.35	0.711379
jlo21	07/01/1995	99	7.10	24.1	222	1360	468.2	207.3	40.2	545.0	584.3	76.9	303.3	0.10	1.69	nd

nd = non-determined values

d.l = detection limit.

Surface Water–Groundwater Interactions in an Alluvial Plain-Chemical and Isotopic Systematics

Wetlands serve as water-storage areas that reduce flooding by retaining water excess and releasing it to streams, and their importance for the future of our living world is becoming increasingly recognized. Not only are wetlands areas of exceptional biodiversity, but they also play a key role in the conservation and management of fresh-water resources (Whittaker and Likens, 1975; Holland et al., 1990; Mitsch and Gosselink, 1993; Hunt et al., 1998). As wetlands are complex and sensitive ecosystems, a slight change can have a strong impact not only on the wetlands themselves, but also on large areas of the enclosing catchment (Fustec and Frochot, 1996). The hydrology and water quality of rivers, lakes and wetlands are strongly controlled by the exchanges of water and solutes with adjacent river banks and uplands, and detailed studies dedicated to evaluating mechanisms of water and solute transfer between the wetland and surrounding upland have been carried out in different environments (Kehew et al., 1998; Hayashi et al., 1998; Hunt et al., 1998). In France, where wetlands have been listed as being highly sensitive areas from the biological and water-resources standpoints, a National Research Program dedicated to wetlands (the Programme National de Recherche surles Zones Humides [PNRZH]) is currently underway. Previous geochemical investigations of the Loire were focused on the middle part of the river near Orleans and Tours, and characterized the temporal variations in the dissolved and suspended loads with regard to natural and anthropogenic inputs (Ne´grel and Grosbois, 1999; Ne´grel et al., 2000; Grosbois).

Using Isotopes or Tracers for Investigation of the Interactions between Groundwater and Surface Water

The groundwater-surface water interactions are pretty common in large inland basins, playing an absolutely necessary role in maintaining the ecosystem of river basins (Kalbus et al., 2006). It is crucial to understand the relationship between groundwater and surface water for sustainable utilization and management of the precious water resources especially in the arid and semi-arid areas. Numerous researches have been conducted on the groundwater-surface water interactions (Kalbus et al., 2006; Unland et al., 2013), generally focusing on areas such as sources identification, hydrologic dynamics of interchanges, and exchange characteristics of water quality and quantity, etc. (Stellato et al., 2008; Xu et al., 2017; Liao et al., 2018). Various methods and techniques have been developed, making great contributions to promoting advances in above studies.

The most frequently used approaches include the environmental isotopic tracers hydrogen and oxygen (Lambs, 2004; Chen et al., 2006; Xu et al., 2017), strontium (Negrel et al., 2003; Paces and Wurster, 2014), radium (Kraemer, 2005) and radon (Cook et al., 2003; Wu et al., 2004; Martinez et al., 2015; Su et al., 2015), chemical components (such as anions and cations) (Cook et al., 2003), electrical conductivity (Carey and Quinton, 2005; Kalbus et al., 2006), heat tracing (Becker et al., 2004; Anderson, 2005; Xu et al., 2017) by distributed temperature sensing (DTS) (Westhoff et al., 2010), hydrologic parameters monitoring by piezometer (Kelly and Murdoch, 2003) and seepage meters (Landon et al., 2001), as well as numerical modeling or simulations (Liang et al., 2003; Scibek et al., 2007; Fleckenstein et al., 2010).

Conclusions have been drawn that limitations and uncertainties of different methods and the measurement scale effects should be taken into considerations when selecting appropriate methods and materials (Kalbus et al., 2006). Therefore, the multi-scale approaches combining different techniques that can considerably reduce uncertainties are highly recommended in the studies on groundwater and surface water interactions.

Radioactive isotope 222Rn with short half-life (3.83 days) is a natural conservative tracer widely used to investigate the exchange between groundwater and surface water (Cook et al., 2003; Santos et al., 2010; Dimova et al., 2013). The sources of 222Rn in groundwater include the recoil supply from the decay of 226Ra in the aquifer, production from 226Ra adsorbed and in solution. However, the latter were insignificant and often neglected (Carvalho et al., 2014). The 222Rn in groundwater is mainly removed by decay. For surface water without recharge from groundwater, the sources of 222Rn are production from 226Ra in water and sediment diffusion which were often insignificant (Wu et al., 2004; Zhang et al., 2015). The sinks of 222Rn in surface water include decay and atmospheric losses. The 222Rn concentration in groundwater is often 2–3 orders of magnitude higher than that in surface water (Dimova and Burnett, 2011). This significant difference enables researchers to identify the exchange between groundwater and surface water. Stable isotopes of hydrogen (deuterium, or D) and oxygen (18O) carry the information of water circulation, recharge and discharge that makes these two common stable isotopes quite useful to identify the sources of different types of water, which are widely used in hydrological and hydrogeological studies in recent decades (Clark and Fritz, 1997; Kendall and McDonnell, 1998; Gat, 2010; Gu et al., 2011; Qian et al., 2014; Xu et al., 2017).

Conclusion

The data presented in this study demonstrate the spatial and temporal variability of surface water and groundwater interactions. Furthermore, this study has shown that this interaction is spatially complex and displays a wide degree of temporal variability, on sub-daily as well as longer (monthly) timescales. In order to capture temporal variability of this scale in surface water and groundwater interaction, high frequency hydrochemical sampling and monitoring is required. This highlights the potential of using multivariate statistical methods and historic hydrochemical medians to provide a rapid and cost-effective preliminary screening method to identify areas of potential groundwater and surface water interaction. The use of such preliminary investigations is likely to reduce the chance of investigating areas of little interest and these locations can be further validated and investigated using a range of field techniques.

There is still much to be learned about the complex nature surrounding groundwater surface water interactions and the principal processes that influence this phenomenon across various temporal scales and climatic and geomorphic environments. Although we have demonstrated the need for high frequency datasets, long term datasets are still required to elucidate the role seasonal cycles and environmental change may have on such processes. The establishment of long term high resolution hydrological and chemical monitoring programmes will allow the scientific community to determine not only seasonal changes in this interaction but also sub-daily changes that were previously ignored.

REFERENCES

Bluth, G. J. S. & Kump, L. R. (1994). *Lithological and climatological controls on river chemistry: Geochimica et Cosmochimica Acta*, v. *58*, p. 2341–2359.

Bouri, S., Abida, H. & Khanfir, H. (2008). Impacts of wastewater irrigation in arid and semi arid regions: case of Sidi Abid region. Tunisia. *Environ Geol*, *53*, 1421–1432.

Candela, L., Fabregat, S., Josa, A., Suriol, J., Vigués, N. & Mas, J. (2007). Assessment of soil and groundwater impacts by treated urban wastewater reuse. A case study: Application in a golf course (Girona, Spain). *Science of the Total Environment*, *374*, 26-35.

Cavazza, S. & Pagliara, S. (2017). Groundwater and Surface Water Interactions. *GROUNDWATER*–Vol. I. https://www.eolss.net/ebooklib/ sc_cart.aspx?File=E2-09-01-03. University of Pisa, Italy.

Cerling, T. E., Pederson, B. L. & Von Damm, K. L. (1989). Sodium-calcium ion exchange in the weathering of shales: Implications for global weathering budgets: *Geology*, v. *17*, p. 552–554.

Chapman, D. (1996). *Water quality assessments: a guide to the use of biota, sediments and water in environmental monitoring*, 2nd edn. Spon, London.

Chattopadhyay, P. B. & Singh, V. S. (2013). Hydrochemical evidences: Vulnerability of atoll aquifers in Western Indian Ocean to climate change, *Global & Planetary Change*, *106*, 123-140.

Chen, J. Y., Tang, C. Y. & Yu, J. J. (2006). Use of 18O, 2H and 15N to identify nitrate contamination of groundwater in a wastewater irrigated field near the city of Shijiazhuang, China. *Journal of Hydrology*, *326*, 367-378.

Chopra, A. K. & Pathak, C. (2015). Accumulation of heavy metals in the vegetables grown in wastewater irrigated areas of Dehradun, India with reference to human health risk. *Environmental Monitoring and Assessment*, *187*(7), 445. doi: 10.1007/s10661-015-4648-6.

Dogramaci, S., Skrzypek, G., Dodson, W. & Grierson, P. F. (2012). Stable isotope and hydrochemical evolution of groundwater in the semi-arid Hamersley Basin of subtropical northwest Australia, *Journal of Hydrology*, *475* (26), 281-293.

Domínguez-Mariani, E., Carrillo-Chávez, A., Ortega, A. & Orozco-Esquivel, M. T. (2004). Wastewater reuse in Valsequillo agricultural area, Mexico: Environmental impact on groundwater. *Water, Air, and Soil Pollution*, *155*(1-4), 251-267.

Drever, J. I. (1988). *The geochemistry of natural waters* (second edition): New Jersey, Prentice Hall, 437 p.

Eslamian, S., Okhravi, S. & Eslamian, F. (2014). Groundwater–Surface Water Interactions, in *Handbook of Engineering Hydrology, Ch. 13, Vol. 1: Fundamentals and Applications*, Ed. By Eslamian, S., Francis and Taylor, CRC Group, USA, 251-279.

Eslamian, S. S. & Nekoueineghad, B. (2009). A Review on Interaction of Groundwater and Surface Water, *International Journal of Water*, *Special Issue on Groundwater and Surface Water Interaction (GSWI)*, Vol. *5*, No. 2, 82-99.

Fustec, E. & Frochot, B. (1996). *Les fonctions et valeurs des zones humides* [*The functions and values of wetlands*]. Laboratoire de Ge´ologie Applique´e, Univ. Paris-VI, Laboratoire d'Ecologie, Univ. Dijon, Agence de l'Eau Seine- Normandie, p. 134.

Gat, J. R. (2010). *Isotope Hydrology – A study of the Water Cycle*. Imperial College Press, London.

Grosbois, C., Ne´grel, Ph., Fouillac, C. & Grimaud, D. (2000). Chemical and isotopic characterization of the dissolved load of the Loire river. *Chem. Geol.*, *170*, 179–201.

Grosbois, C., Ne´grel, Ph., Fouillac, C. & Grimaud, D. (2001). An overview of dissolved and suspended matter fluxes in the Loire river basin: natural and anthropogenic inputs. *Aq. Geochem.*, *7*, 81–105.

Hayashi, M., van der Kamp, G. & Rudolph, D. (1998). Water and solute transfer between a prairie wetland and adjacent uplands; 1-water balance. *J. Hydrol.*, *207*, 42–55.

Holland, H. D. (1978). *The chemistry of the atmosphere and oceans*: New York, Wiley, 351 p.

Holland, M. M., Whigham, D. F. & Gopal, B. (1990). In: Naiman, R. J., De´camps, H. (Eds.), The characteristics of wetland ecotones, The Ecology and Management of Aquatic-Terrestrial Ecotones. *MAB-UNESCO*, vol. *4*., pp. 171–198.

Hunt, R., Bullen, T. D., Krabbenhoft, D. P. & Kendall, C. (1998). Using stable isotopes of water and strontium to investigate the hydrology of a natural and a constructed wetland. *Groundwater*, *36*, 434–443.

. H. (2005). Impact of post-machination distillery effluent irrigation on groundwater quality. *Environmental Monitoring and Assessment*, *110*, 243-255.

Jiang, C. L., Xia, Z. Q., Liu, L., et al. (1997). Variation trends of three types of nitrogen in the soil and groundwater after wastewater irrigation. *Advances in Water Science*, *8*(2), 184-188.

Kalbus, E., Reinstorf, F. & Schirmer, M. (2006). Measuring methods for groundwater – surface water interactions: a review. *Hydrol. Earth Syst. Sci. 10* (6), 873–887. https://doi.org/10.5194/hess-10-873-2006.

Kass, A., Gavrielia, I., Yechielia, Y., Vengoshc, A. & Starinsky, A. (2005). The impact of freshwater and wastewater irrigation on the chemistry of shallow groundwater: a case study from the Israeli Coastal Aquifer. *Journal of Hydrology*, *300*, 314-331.

Keesari, T., Ramakumar, K. L., Chidambaram, S., Pethperumal, S. & Thilagavathi, R. (2016). Understanding the hydrochemical behavior of groundwater and its suitability for drinking and agricultural purposes in Pondicherry area, South India – A step towards sustainable development, *Groundwater for Sustainable Development*, 2-3, 143-153.

Kehew, A. E., Passero, R. N., Kishanamurthy, R. V., Lovett, C. K. & Dayharsh, B. A. (1998). Hydrogeochemical interaction between a wetland and an unconfined glacial drift aquifer Southwestern Michigan. *Groundwater*, *36*, 849–856.

Krabbenhoft, D. P., Bowser, C. J., Anderson, M. P. & Valley, J. W. (1990). Estimating groundwater exchange with lakes: 1. the stable isotope mass balance method. *Water Resour Res*, *26*, 2445–2453. doi: 10.1029/WR026i010p02445.

Kraemer, T. F. (2005). Radium isotopes in Cayuga Lake, New York: Indicators of inflow and mixing processes. *Limnol. Oceanogr.*, *50* (1), 158–168. https://doi.org/10.4319/lo. 2005.50.1.0158.

Lambs, L. (2004). Interactions between groundwater and surface water at river banks and the confluence of rivers. *J. Hydrol.*, *288* (3–4), 312–326. https://doi.org/10.1016/j.jhydrol.2003.10.013.

Lee, T. M. (1996). Hydrogeologic controls on the groundwater interactions with an acidic lake in karst terrain, Lake Barco, Florida. *Water Resour Res*, *32*, 831–844.

Lissey, A. (1971). Depression-focused transient groundwater flow patterns in Manitoba. *Geol. Assoc. Can. Spec. Paper* 9, 333–341.

Li, H., Satoh, A., Taniguchi, M., Kitching, M., et al. (2012). Groundwater contamination due to irrigation of treated sewage effluent in the Werribee Delta. *Int. J. of Geomate*, *3*(1), 332-338.

Liu, C. L., Zhang, Y., Ye, H., et al. (2006). Experimental studies of the filtration capability of clayey soils in the vadose zone and its influence on the evaluation of groundwater vulnerability to pollution. *Acta Geoscientica Sinic*, *27*(4), 349-354.

Liu, F. S., Wang, Y. & Lan, X. Hu. (2014). Environmental isotopes and exchanges of surface water-groundwater system in the Zhangye basin of Heihe River watershed, *South-to-North Water Transfers and Water Science& Technology*, *12* (2), 92-96.

Lu, S. B., Zhang, X. L., Bao, H. J. & Skitmore, M. (2016a). Review of social water cycle research in a changing environment, *Renewable & Sustainable Energy Reviews*, *63*, 132-140.

Lu, S. B., Wang, J. H. & Pei, L. (2016b). Study on the Effects of Irrigation with Reclaimed Water on the Content and Distribution of Heavy Metals in Soil, *International Journal of Environmental Research & Public Health*, *13*(3), 298.

Lu, S. B., Bao, H. J. & Pan, H. L. (2016c). Urban water security evaluation based on similarity measure model of Vague sets, *International Journal of Hydrogen Energy*, *41*(35), 15944-15950.

Lu, S. B., Pei, L. & Bai, X. (2015). Study on method of domestic wastewater treatment through new-type multi-layer artificial wetland, *International Journal of Hydrogen Energy*, *40*(34), 11207-11214.

Martinez, J. L., Raiber, M. & Cox, M. E. (2015). Assessment of groundwater–surface water interaction using long-term hydrochemical data and isotope hydrology: Headwaters of the Condamine River, Southeast Queensland, Australia, *Science of the Total Environment*, *536*, 499-516.

Meybeck, M. (1987). Global chemical weathering of surficial rocks estimated from river dissolved loads: *American Journal of Science*, v. *287*, p. 401–428.

Mitra, A. & Gupta, S. K. (2000). Evaluation of groundwater pollution potential of sewage-irrigated vegetable growing areas of the eastern fringe of Calcutta city. *Schriftenreihe Des Vereins Fur Wasser-, Boden-Und Lufthygiene*, *105*, 261-267.

Mitsch, W. J. & Gosselink, J. G. (1993). *Wetlands*, second ed., Van Nostrand Reinhold Company, New York, p. 722.

Mooers, H. D. & Alexander, E. C. (1994). Contribution of spray irrigation of wastewater to groundwater contamination in the karst of Southeastern Minnesota, USA. *Applied Hydrogeology*, *2*(1), 34-44.

Ne´grel, Ph. (1999). Geochemical study in a granitic area, the Margeride, France: chemical element behavior and 87Sr/86Sr constraints. *Aq. Geochem.*, *5*, 125–165.

Ne´grel, Ph., Grosbois, C. & Kloppmann, W. (2000). The labile fraction of suspended matter in the Loire river (France): multi-element chemistry and isotopic (Rb-Sr and C-O) systematics. *Chem. Geol.*, *166*, 271–285.

Nie, Z., Chen, Z., Chen, X., Hao, M. & Zhang, G. (2005). The chemical information of the interaction of unconfined groundwater and surface water along the Heihe River, Northwestern China, *Journal of Jilin University (Earth Science Edition)*, *35*(1), 48-51.

Peng, T., Lu, W., Chen, K., Zhan, W. & Liu, T. (2014). Groundwater-recharge connectivity between a hills-and-plains' area of western Taiwan using water isotopes and electrical conductivity, *Journal of Hydrology*, *517*, 226-235.

Prada, S. J., Cruz, V. & Figueira, C. (2016). Using stable isotopes to characterize groundwater recharge sources in the volcanic island of Madeira, Portugal, *Journal of Hydrology*, *536*, 409-425.

Quayle, W. C., Jayawardane, N. & Arienzo, M. (2010). Impacts of winery wastewater irrigation on soil and groundwater at a winery land application site. 2010 *19th World Congress of Soil Science, Soil Solutions for a Changing World.* 1-6, Brisbane, Australia. Published on DVD.

Rattan, R. K., Datta, S. P., Chhonkar, P. K., Suribabu, K. & Singh, A. K. (2005). Long-term impact of irrigation with sewage effluents on heavy metal content in soils, crops and groundwater—a case study. *Agriculture, Ecosystems and Environment*, *109*, 310-322.

Rosenqvist, H. & Dawson, M. (2005). Economics of using wastewater irrigation of willow in Northern Ireland. *Biomass and Bioenergy*, *29*, 83-92.

Rudnick, S., Lewandowski, J. & Nützmann, G. (2015). Investigating groundwater–lake interactions by hydraulic heads and a water balance. *Groundwater*, *53*, 227–237. doi:10.1111/gwat.12208.

Shang, Y. Z., Lu, S. B., Li, X. F., Hei, P. F., Lei, X. H., Gong, J. G., Liu, J. H., Zhai, J. Q. & Wang, H. (2017). Balancing development of major coal bases with available water resources in China through 2020, *Applied Energy*, *194*, 735-750.

Shirazi, S. M., Imran, M. H., Akib, S., Yusop, Z. & Harun, Z. B. (2013). Groundwater vulnerability assessment in Melaka state of Malaysia using DRASTIC and GIS techniques. *Environmental Earth Sciences*, *70*, 2293-2304. doi 10.1007/s12665-013-2360-9.

Stuart, M. E. & Milne, C. J. (2001). The risk to groundwater from wastewater irrigation using highchromium tannery effluent. *Environmental Geochemistry and Health*, *23*, 247-251.

Teklehaimanot, G. Z., Kamika, I., Coetzee, M. A. A. & Momba, M. N. B. (2015). Seasonal variation of nutrient loads in treated wastewater effluents and receiving water bodies in Sedibeng and Soshanguve, South Africa. *Environmental Monitoring and Assessment*, *187*(9), 4794. doi: 10.1007/s10661-015-4794-x.

USEPA (U.S. Environmental Protection Agency), 2012. *Guidelines for Water Reuse*.

Wang, J. H., Lu, S. B. & Pei, L. (2016). Study on rules of dynamic variation of nitrogen in soil after reclaimed water drip irrigation, *International Journal of Hydrogen Energy*, *41*(35), 15938-15943.

Weng, H. X. & Chen, X. H. (2000). Impact of polluted canal water on adjacent soil and groundwater systems. *Environmental Geology*, *39* (8), 945-950.

Whittaker, R. H. & Likens, G. E. (1975). In: Lieth, H., Whittaker, R.H. (Eds.), *The biosphere and Man, Primary Productivity of the Biosphere*, Springer, New York, p. 175.

Wu, W. Y., Xu, C. P., Liu, H. L., Hao, Z. Y., Ma, F. S. & Ma, Z. J. (2010). Effect of reclaimed water irrigation on yield and quality of fruity vegetables. *Transactions of the CSAE*, *26*(1), 36-40.

Xu, C. P., Wu, W. Y., Liu, H. L., Shi, Y. W. & Yang, S. (2010). Effect of reclaimed water irrigation on the yield and quality of leaf vegetables. *Journal of Irrigation and Drainage*, *29*(5), 23-26.

Yadav, R. K., Goyal, B., Sharma, R. K., Dubey, S. K. & Minhas, P. S. (2002). Post-irrigation impact of domestic sewage effluent on composition of soils, crops and ground water—A case study. *Environment International*, *28*, 481- 486.

Yang, Q., Wang, L., Ma, H., Yu, K. & Martín, J. D. (2016). Hydrochemical characterization and pollution sources identification of groundwater in Salawusu aquifer system of Ordos Basin, China, *Environmental Pollution*, 216: 340-349.

Yin, S. Y., Wu, W. Y., Liu, H. L., et al. (2012). Spatial variability of groundwater nitrate-nitrogen and cause analysis of its pollution for irrigation area with reclaimed water. *Transactions of the Chinese Society of Agricultural Engineering (Transactions of the CSAE)*, *28*(18), 200-207.

Zeng, H. & Wu, J. (2013). Water isotopic and hydrochemical characteristics and causality in Tajikistan, *Advances in Water Science*, *24* (2), 272-279.

Zhang, B., Song, X., Zhang, Y., Han, D., Tang, C., Yu, Y. & Ma, Y. (2012). Hydrochemical characteristics and water quality assessment of surface water and groundwater in Songnen plain, Northeast China, *Water Research*, *46* (8), 2737-2748.

Zhang, B., Song, X., Zhang, Y., Han, D., Yang, L. & Tang, C. (2014). Relationship between surface water and groundwater in the second Songhua River basin, *Advances in Water Science*, *25* (3), 336-347.

In: Advances in Hydrogeochemistry Research
Editors: S. Saeid Eslamian and Faezeh Eslamian
ISBN: 978-1-53617-829-6

Chapter 2

BIOLOGICAL REMEDIATION USING WETLAND SYSTEMS: A HYDRO-GEOCHEMICAL PERSPECTIVE

Anthony E. Mader[1], Saeid Eslamian[2,3] and Anthony R. Turton[4,*]
[1]University of the Witwatersrand, Johannesburg, South Africa
[2]Department of Water Engineering, Collage of Agriculture, Isfahan University of Technology, 8415683111, Isfahan, Iran
[3]Center of Excellence in Risk Management and Natural Hazards, Isfahan University of Technology, 8415683111, Isfahan, Iran
[4]University of the Free State, Bloemfontein, South Africa

ABSTRACT

Wetlands, characterized by biological and hydro-geochemical features, play a pivotal role in the remediation of various inorganic and organic contaminants, produced by anthropogenic activities. The composition and concentration of contaminants, within the environment, differ based on their sources of production, such as mining (including the generation of Acid Mine Drainage – AMD) or the production and degradation (i.e., the release of degradation products) of plastics. The aim of this chapter is to discuss the hydro-geochemical factors influencing the biological remediation of inorganic and organic contaminants by (natural and constructed) wetland systems. These biological remediation (i.e., microbial bioremediation and phytoremediation) processes, including the sorption, precipitation, co-precipitation, methylation, degradation, stabilization, and extraction of contaminants from wetland systems, are mediated by wetland hydro-geochemical factors [pH, redox potential (Eh), salinity, oxygen availability, element composition and concentration, and organic matter content] which fluctuate through space and time. Due to their presence and ecotoxicological importance, this chapter focuses on the remediation of AMD- and plastic [including sorbed organics such as persistent organic pollutants - e.g., polycyclic aromatic hydrocarbons (PAH)]-contaminated water. The source of contamination will be described in order to contextualize this chapter and rationalize the

* Corresponding Author's Email: tony@anthonyturton.com.

need to remove these harmful contaminants from the environment whereas the hydro-geochemical and biological remediation processes involved in rendering these contaminants chemically inert will be discussed relative to the fate (e.g., bioavailability, transport, and toxicity) of contaminants in wetland systems. To the knowledge of the authors, no previously published literature has identified and/or described the ability/capacity of wetland systems in the remediation of plastic (and their sorbed constituents)-contaminated water. Thus, as this highlights a gap in knowledge and research, this chapter argues that the complex biological and hydro-geochemical factors, as well as the interactions thereof, present in wetland systems, may effectively render these plastic-contaminants chemically (and therefore, biologically) inert. This presents the case for engineered wetland systems, i.e., constructed wetlands (CW), which have the ability to control biological (e.g., selecting certain plant and microbial species) and hydro-geochemical (e.g., frequency and duration of saturation, selection of substrate, photoperiod, and temperature) factors optimal for the remediation of a targeted contaminant(s). This chapter concludes that wetland systems play a vital role in facilitating various biological and hydro-geochemical processes involved in the remediation of (in)organic contaminants and that the ability to optimize biological remediation pathways in CWs may benefit the scientific, and non-scientific, community by creating holistic solutions to solve complex environmental and socio-economic challenges associated with the contamination of water by AMD, plastics, and their sorbed constituents.

Keywords: AMD, biological remediation, plastic, wetlands

INTRODUCTION

In accordance with the National Water Act of South Africa (Act No. 36 of 1998), wetlands are defined as areas which transition between aquatic and terrestrial ecosystems, periodically saturated with shallow water, and/or areas where the water table surfaces are near the surface (Figure 1) (Ollis et al. 2013). Wetlands are characterized by their hydro-geochemical (e.g., hydrological factors and substrate form) and biological (e.g., presence of flood-tolerant plant species) factors (Tutu et al. 2008; Ollis et al. 2013; Prasad 2013; Tiner 2016). These factors enable wetlands to play a vital role in ecosystem functioning, with regards to provisioning and supporting services such as water purification, groundwater recharge, substrate and nutrient retention, flood and erosion control, habitat for biodiversity, and cultural services (Zedler and Kercher 2005).

These hydro-geochemical and biological processes, as well as the interaction thereof, influence the remediation efficacy of wetland systems relative to the removal of contaminants (Prasad 2013). This is evident in the hydrological regime of a wetland (e.g., duration and frequency of soil saturation) which determines the system's geochemical parameters such as the wetland substrate geomorphology, oxygen (O_2) supply, as well as nutrient retention and cycling processes (i.e., biogeochemistry) (Ollis et al. 2013). Soils saturated for long periods of time become anaerobic due to biological activities, such as decomposition, which utilizes available O_2 for such reactions to take place. This transition between reduced and oxidized states (depending on the hydrological regime) results in the mottling of soil, where minerals, such as iron (Fe) and manganese (Mn), are re-oxidized – a characteristic of seasonally/intermittently saturated wetland systems. In comparison, unsaturated soils are in an oxidized state. These factors influence the biological component of wetlands as plant species, and associated microbial communities (i.e., species composition), differ in their ability to tolerate varying durations of soil saturation (Figure 1) (Ollis et al. 2013) and thus the hydrological regime, facilitating hydro-geochemical and biological processes in such systems, is a limiting factor for contaminant remediation by wetland systems (Niering and Lowe 1985; Cronk and Fennessy 2016). It is also important to note that wetlands are complex systems in which various hydro-geochemical and biological factors, as well as the interplay thereof, fluctuate through space and time.

Figure 1. Wetlands (channeled valley bottom hydrogeomorphic units) associated with the uThukela (A) and Mvoti (B) rivers, located in KwaZulu-Natal, South Africa. Note, changes in the composition of plant species [e.g., helophytes/hydrophytes – flood (permanently inundated) – tolerant plant species], which influence microbial communities, are mediated by hydro-geochemical factors such as the wetland's hydrological regime (i.e., the frequency and duration of saturation - e.g., permanently, seasonally, or intermittently inundated).

In order to elucidate the remediation potential of wetlands, the sources of contamination directly impacting water quality and quantity, which have deleterious impacts on fauna (including humans) and flora growth and survival, must be described.

1. Sources of Contaminants in the Environment

1.1. Acid Mine Drainage (AMD)

In South Africa, mining has major short- and long-term environmental impacts, related to land and water (surface and groundwater) contamination, as well as socio-economic impacts such as human health due to potable water pollution (Glaister and Mudd 2010; Farrell et al. 2012). Contaminated land and water sources may contain numerous inorganic [salts, metals, and radionuclides such as uranium (U)] and organic (e.g., cyanide complexes) contaminants. The formation of acid mine drainage (AMD) (Figure 2) is a major challenge associated with mining activities The composition and concentration of these contaminants depend on various factors, such as host geology, accounting for the complex management issues associated with the remediation of AMD in different geological areas (Winde 2006; Tutu et al. 2008; Turton 2013).

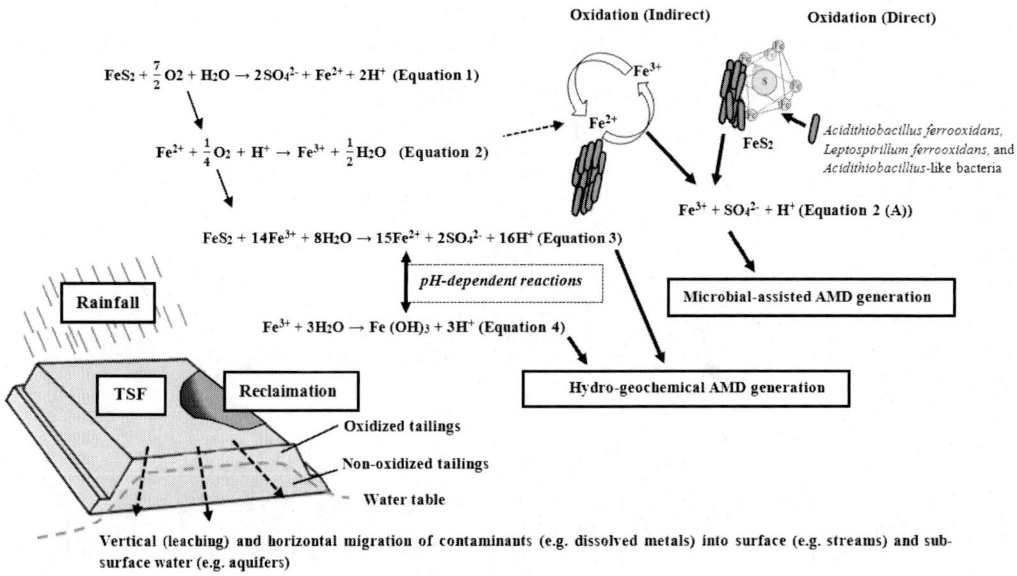

Figure 2. Acid mine drainage (AMD) is primarily a result of the oxidation of sulfide (S^{2-}) minerals, namely pyrite (FeS_2), present in tailings stored in tailing storage facilities (TSF) such as mine dumps. Metals in lower layers of TSFs, as well as in eroded tailings, become more soluble due to an elevated degree of weathering (e.g., hydrolysis, evaporation, and photolytic loss) and biotransformation processes, including the presence of sulfur-oxidizing *Acidothiobacillus* bacteria which catalyze AMD generation reactions. This is attributed to the reclamation of TSFs for residual precious metal(s) recovery. Pyrite is oxidized in the presence of H_2O (e.g., by the percolation of rainwater) resulting in dissolved Fe, sulfate (SO_4^{2-}) and hydrogen ions (H^+). This increases the total dissolved solids whilst decreasing the pH level due to acid production (Equation 1). Depending on the surrounding environment, namely the O_2 supply (bacterial aerobic activity (i.e., O_2 consumption)), ferrous Fe (Fe^{2+}) is oxidized to ferric Fe (Fe^{3+}) - which is a rate-determining reaction (Equation 2). Certain microbial species have the ability to catalyze AMD reactions thereby influencing the rate and degree of AMD generation (Equation 2 (A)) (Note, FeS_2 only depicts a section of the FeS_2 crystalline structure with the compound formulae [S=Fe=S]). Furthermore, additional FeS_2 may be oxidized under anaerobic conditions but is limited by the supply of FeS_2 and/or Fe^{3+} (Equation 3). Ferric Fe is hydrolyzed in the presence of H_2O and is subsequently precipitated out of solution when pH levels range between 2.3–3.5. When the pH level is greater than 3.5, Fe^{3+} is precipitated as $Fe(OH)_3$, an acid-generating reaction (Equation 4). The solution is further acidified due to the oxidation of FeS_2 and/or Fe^{3+} (depending on the rate-determining reaction – Equation 2). Adapted from Singer and Stumm (1970), Stumm and Morgan (1996), and Tutu et al. (2008). (Hallberg and Johnson 2005; Akcil et al. 2006).

These oxidation reactions (Figure 2) may result in the release of other heavy metals and metalloids including aluminum (Al), arsenic (As) cobalt (Co), chromium (Cr), copper (Cu), mercury (Hg), nickel (Ni), lead (Pb), and U (Naicker et al. 2003 Tutu et al. 2008; Turton 2015[b]). This is due to mineral sulfide sources being mined, such as arsenopyrite

(FeS_2.FeAs), chalcopyrite (Cu_2S), millerite (NiS), and/or galena (PbS), as well as the storage of such tailings (Sheridan et al. 2018). Due to the prevalence of gold (Au) mining in South Africa, U has been considered a contaminant of environmental concern due to its elevated presence in Witwatersrand complex Au-bearing ores (Turton 2015[a,b]). [Au-bearing Witwatersrand reefs contain 10-100 times more U compared with Au in which approximately 430 000 tonnes of U has been brought to the surface (Winde 2006)]. Uranium is a radioactive and highly reactive element and is thus rarely present in its elemental form in the environment (Turton 2013). Uranium, generally present as uraninite (UO_2) and triuranium octo-oxide (U_3O_8), is mobilized under certain hydro-geochemical conditions, including redox conditions at different pH levels, in aquatic ecosystems (Turton 2016). These dissolved metals and metalloids (e.g., As) may vertically migrate (i.e., leach) contaminating groundwater. Due to the mismanagement of TSFs during mine operation and closure, AMD will continue polluting the environment in years to come (Johnson and Hallberg 2005). Therefore, the presence of AMD in aquatic ecosystems is of great environmental importance due to the physicochemical properties (such as the bioavailability and mobility) of contaminants that dictate their degree of toxicity, within AMD, on fauna and flora health and survival. Thus, the use of wetland systems, incorporating various hydro-geochemical and biological factors, relative to the remediation of contaminants (namely heavy metals and salts at acidic pH levels) present within AMD will be discussed.

1.2. Plastics (and Their Sorbed Constituents)

Plastics comprise of inorganic or organic polymers of high molecular mass, which may be produced from natural materials or synthesized from hydrocarbons (Andrady and Neal 2009). Between 1950 to 2015, an estimated 6300 Mt (approximately, and on average, 96 923 076 tons/year) of plastic (primary and secondary (recycled)) has been produced of which, approximately 4977 Mt (79%) has accumulated in waste storage facilities (e.g., landfills) and natural environments. Of the total plastic produced, only 567 Mt (9%) is estimated to have been recycled whereas the remaining 756 Mt (12%) has been incinerated (Geyer et al. 2017). Of these produced plastics, frequently produced plastics including polyethylene (PE), polystyrene (PS), polyvinyl chloride (PVC), polyurethane (PUR), and polypropylene (PP) are utilized in various industries due to their differences in physicochemical properties. Polyethylene, produced as low-density PE (LDPE) or high-density PE (HDPE), is the most commonly produced type of plastic (29%) followed by PVC (17%) (Thompson et al. 2005). Due to this increased production, and consequently, its increased presence in the environment (especially aquatic ecosystems) after use, the remediation of PE will be focused on.

During plastic production, additives may be combined with a polymer in order to give plastic certain, desirable properties, e.g., phthalate is added to give plastic its flexibility. Although the aforementioned plastics are important environmental contaminants, abiotic (e.g., photochemical degradation and temperature) and biotic (e.g., microbial degradation) conditions degrade such plastics, increasing the release and transport of harmful degradation products in aquatic ecosystems (Andrady 2011). These degradation products include phthalates and adipates (which decompose into toxic monomers, such as bisphenol A (BPA)), or are transformed into toxic volatiles, such as dioxins, furans, Hg and polychlorinated biphenyl (PCBs) (Figure 3) (Verma et al. 2016), during incineration. Moreover, phthalate acid esters (PAEs), including di-*n*-butyl phthalate (DBP), are released into the environment during the production, use, and disposal of plastic polymers as DBP is not chemically bonded to certain plastic polymers (Net et al. 2015). Moreover, the high surface: volume ratio of plastics, and more specifically microplastics (< 5 mm), increase their ability to absorb persistent organic pollutants (POPs) such as polycyclic aromatic hydrocarbons (PAH), dichlorodiphenyltrichloroethane (DDT), polybrominated biphenyl (PBB), and PCBs (Do Sul and Costa 2014). The most toxicologically important POP is 2,3,7,8-tetrachlorodibenzo-p-dioxin (TCDD – known as Agent Orange), one of the most toxic compounds known to man (Figure 3) (McAdams and Aquino 1994). Due to the surface area: volume ratio of microplastics, the concentration of POPs absorbed/ adsorbed by microplastics may be orders of magnitude greater than POP concentrations in the surrounding environment, thereby increasing the environmental importance, and the need to remove plastics, such as PE, from aquatic systems (Teuten et al. 2007). Inorganic contaminants, including heavy metals such as cadmium (Cd), Co, Cu, Ni, Pb, silver (Ag), and Zn, have also been demonstrated to be adsorbed to plastics. This is caused by the change in surface charge within H_2O due to biological processes including the formation of biofilms (enhancing microbial activity, namely the precipitation and exudation of inorganic minerals – see section 2.2.1) and sorption of organic molecules varying in charge (Turner and Holmes 2015).

These additives and decomposition products, which may be present in plastics such as PE, have been implicated in developmental (e.g., endocrine disruption), reproductive, carcinogenic, and mutagenic toxicity (Matsumoto et al. 2008), having deleterious impacts on faunal health and survival. Negative impacts associated with the presence of plastics in aquatic environments include the bioaccumulation of microplastics, and more importantly, the toxic contaminants, which have been adsorbed/absorbed to the surface/ into the structure of the plastic, respectively, in the food chain (i.e., microplastics are ingested, transferred, and accumulated between different trophic levels) (Antunes et al. 2013; Farrell and Nelson 2013).

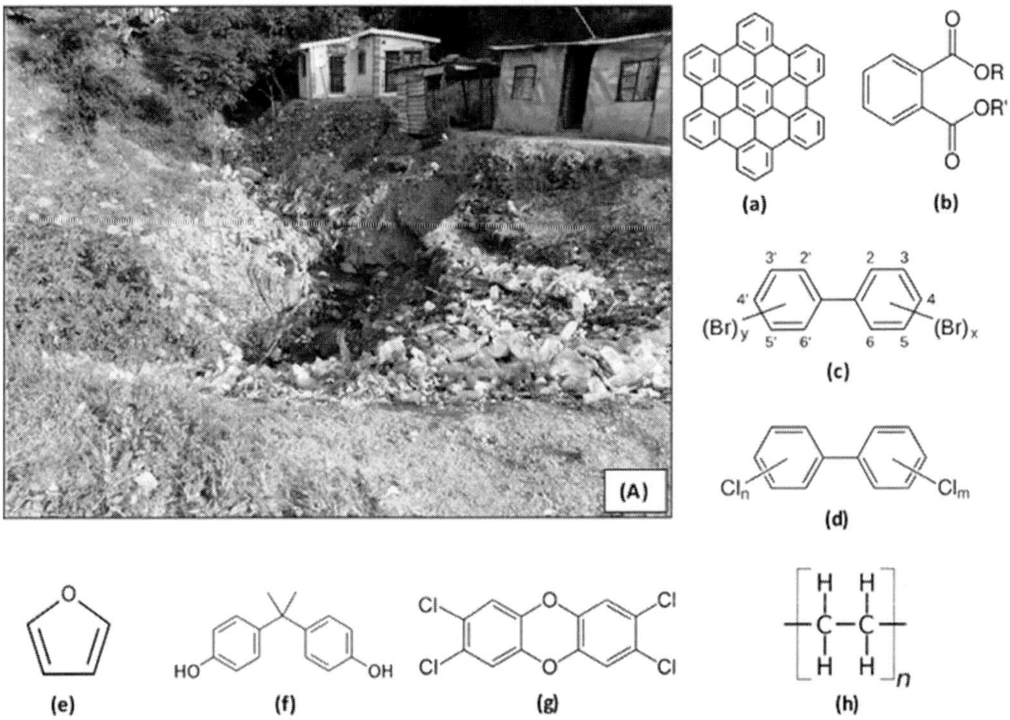

Figure 3. Chemical structures of some toxic persistent organic pollutants (POPs) sorbed to plastics which may be present in watercourses such as this tributary of the Umlazi River, located in KwaZulu-Natal, South Africa (A), *viz.* examples of (a): Polycyclic aromatic hydrocarbon (PAH) (b): Phthalates, (c): Polybrominated biphenyl (PBB), (d): Polychlorinated biphenyl (PCB), (e): Furan, (f): Bisphenol A (BPA), (g): Dioxin (2,3,7,8, Tetrachlorodibenzo-p-dioxin (TCDD) – TCDD is the reference dioxin structure for compounds with similar structures and dioxin-like toxicity) (h): Polyethylene (PE). Note, the structures of these compounds dictate their affinity for biomolecule binding, accounting for their elevated degree of toxicity in the environment.

Increasing research has been conducted on the presence and accumulation of plastics (as well as adsorbed/absorbed contaminants such as POPs) in freshwater ecosystems such as rivers, which are major contributors to the introduction of plastics into marine environments. Previously published studies have reported the assimilation of specific contaminants, such as PAHs, within wetland systems. Limited, previously published literature has focused on determining the efficacy of specific biological and hydro-geochemical processes involved in the remediation of these contaminants under certain conditions. Although such studies have been conducted, no previously published study, to the knowledge of the authors, has identified and described the use of wetland systems for the biological remediation of plastics and their sorbed constituents. This highlights a gap in knowledge and research in which the various (a)biotic variables (e.g., selection of specific plant and microbial species, such as bacterial communities isolated from wetland systems) tested under controlled conditions, naturally occur (and fluctuate through space and time) in wetland systems.

2. Remediation by Wetland Systems

Wetlands are complex systems, comprising of various biological and hydro-geochemical processes, as well as the interplay thereof, fluctuating through space-time. Wetlands possess the ability to reduce a range of contaminants including heavy metals [Cd, Co, Cr, Cu, Fe, Mn, Ni, Pb, and Zn], metalloids (e.g., As), radionuclides (e.g., U), and organics (ITRC 2009; Marschner 2011; Prasad 2013). *It must be noted that although this chapter focusses on biological (bioremediation and phytoremediation) remediation, hydro-geochemical processes mediate such processes (and vice-versa) and therefore, the interplay between these (a)biotic processes will be discussed for this reason. Thus, the following sections describe hydro-geochemical and biological processes involved in the remediation of various contaminants, present within AMD (viz. - heavy metals) - and plastic (viz. – PE and POPs) - contaminated water, by wetland systems.*

2.1. Hydro-geochemical Factors Influencing Contaminant Removal and Biological Remediation

Numerous hydro-geochemical factors influence the fate of contaminants (e.g., speciation of contaminants/bioavailability, toxicity, and transport) within the environment. These factors, including precipitation, co-precipitation, sorption, and methylation, are mediated by pH, dissolved oxygen (DO), competing species (i.e., competitive ion adsorption processes), presence of other elements/compounds (e.g., Fe and SO_4^{2-}), and carbon source(s) (Lizama et al. 2011). Contaminants, such as heavy metals, may be present as (i) readily available (i.e., bioavailable fractions which include dissolved or exchangeable forms – e.g., cation-exchange processes), (ii) potentially available (exchangeable, precipitated as metal-sulfides, co-precipitated with hydrous oxides, and/or complexed with organic matter (OM)), or (iii) unavailable (retained within the structure of certain minerals – e.g., isomorphic substitution resulting in the fixing of metals within the crystalline lattice of a mineral(s)) forms of the metal making up the total metal content within a wetland system.

Precipitation involves the interaction between dissolved species thereby forming a solid phase insoluble compound. Precipitation reactions result from redox reactions (i.e., oxidation/reduction reactions), changes in pH levels, and the presence of competitive ions/other elements and compounds influencing metal speciation within a solution (Henken and Hutchison 2009). In wetland systems, the source of phosphorus (P), as soluble inorganic phosphate (PO_4^{3-}), precipitates certain heavy metals, including Fe, Al, and Mn (at certain pH levels) out of solution forming insoluble, inorganic $FePO_4$ (Vollenweider 1975). Co-precipitation refers to the adsorption of a metal species on a developing precipitate. Iron has been suggested to be an agent for the co-precipitation of other contaminants, such as As, in aerobic areas of constructed wetlands (engineered systems incorporating various hydro-geochemical and biological factors, which can be controlled, present in natural wetlands) (Buddhawong et al. 2005).

Wetlands are sinks for various contaminants, which is attributed to the retention of contaminants in the substrate of wetlands via sorption processes, rendering the contaminant(s) chemically inert. Sorption, including both adsorption and absorption, processes result in the removal of contaminant species, present in different phases (e.g., gases or liquids) by their accumulation onto the surface of a solid substance (adsorption) and/or their assimilation into the chemical structure (absorption) of the solid substance (Henken and Hutchison 2009). Sorption processes, including ligand exchange-surface complexation interactions, may render contaminants chemically inert via the formation of insoluble complexes and their subsequent retention in wetland substrate (Wang and Mulligan 2006; Singhakant et al. 2009[a]). For example, U removal by hydro-geochemical and biological processes in wetland systems has been demonstrated to occur via sorption to organic carbon present in the substrate, Fe-oxides, and plant root binding sites (Gruau et al. 2000). The adsorption capacity of particular substrate (and biological) surfaces, present within wetland systems, are influenced by various hydro-geochemical factors including the physicochemical properties of the surface (e.g., the surface of substrate, mineral particles (such as metal (oxy)(hydr)oxides) and organic matter present within the wetland system), the contaminant species, pH, E_h, and the presence of other ions (i.e., competitive ion adsorption) (Stollenwerk 2003). According to the Irving-Williams series (see below), a metal may be displaced by another metal, upstream in the series, from a particular binding site as these metals share similar physicochemical properties (such as atomic radii and oxidative states), enabling certain metals to outcompete downstream metals for binding sites. This reduces the sorption of outcompeted metals thereby compromising their remediation and/or removal from the wetland system. It must be noted that the displacement of a particular metal(s) downstream of the series by an upstream element occurs under certain environmental (i.e., hydro-geochemical and biological) conditions.

Irving-Williams series: $Zn^{2+} < Cu^{+} > Cu^{2+} > Ni^{2+} > Co^{2+} > Fe^{2+} > Mn^{2+} > Mg^{2+} > Ca^{2+}$

The methylation (addition of methyl group – CH_3), followed by volatilization (Figure 4) of metalloids such as As and metals such as Hg, has been demonstrated in previous studies. The source of carbon in wetland systems (e.g., provided by OM and vegetation) stimulates SO_4^{2-} reduction and promotes SO_4^{2-}-reducing (SRB) and metal-oxidizing bacteria activity (Maurice et al. 2002; Sheoran and Sheoran 2006). The physicochemical properties of OM, such as the molecular weight, increases their affinity for heavy metal absorption and adsorption, resulting in the removal of elevated contaminant loads from solution (Maurice et al. 2002). As metal-S^{2-} precipitation serves as an effective process in the removal of metals, strong reducing environmental conditions (causing metals to be in a soluble, ionic form), contributing to metal precipitation processes. Stable metal-S^{2-} compounds may undergo further oxidation reactions in aerobic environments resulting in the formation of mobile and bioavailable metal- SO_4^{2-}- compounds.

The pH level directly influences the composition of surface functional groups which dictate sorption affinity. This is attributed to the completion of the co-ordination shell with OH- groups in the presence of H_2O. Solution pH facilitates H^+ release or binding from OH^- groups (decreasing (acidification) or increasing (alkalinization) the pH level, respectively (Stollenwerk 2003)), developing a surface charge, and controlling further ion exchange processes. This change in metal species, induced by redox reactions may liberate/mobilize metals complexed with OM, increasing their bioavailability and consequently, their toxicity. The substrate of wetland systems, containing carbonated minerals, has been demonstrated to increase the pH level by metal precipitation processes (García et al. 2010). The combination of pH and E_h mediate various processes such as the formation of S^{2-}, further facilitating the formation of soluble metal species within wetland systems. Salinity has been demonstrated to mediate the retention and release of metals into bioavailable metal species. Salinity may increase contaminant bioavailability and mobility via the complexation of contaminants with salt-derived anions, such as SO_4^{2-} (Acosta et al. 2011), resulting in the release of contaminants previously bound with soil particles. Therefore, a decrease in pH and an increase in substrate salinity will increase the bioavailability and mobility of heavy metals in wetland systems. Various studies have demonstrated the ability of natural and constructed wetlands to increase the pH level (thereby decreasing the bioavailability and mobility, and thus the toxicity of harmful contaminants in solution) through hydro-geochemical (e.g., presence of acid-consuming minerals such as carbonates) and biological (bioremediation and phytoremediation) processes (Figure 4).

2.2. Biological Remediation

Biological remediation, i.e., use of micro-organisms (bacteria- and fungi-bioremediation) and plants (phytoremediation), have been demonstrated to effectively remediate a range of contaminants (ITRC 2009). The use of such biological systems are typically cheaper, more efficient to maintain, no chemical additions are required, promotes biodiversity, and enhances other ecosystem services (Kivaisi 2001). It must be noted that numerous studies have concluded that microbial bioremediation plays a greater role in the biological remediation of contaminants compared with phytoremediation. However, this does not negate the vital role of plants in contaminant remediation processes (e.g., phytostabilization and phytoextraction), as well as providing a microenvironment conducive to the effective bioremediation of contaminants by a range of microbial communities hosted by the plant roots. These bioremediation (i.e., biotechnologies) and phytoremediation (i.e., phytotechnologies) mechanisms, involved in contaminant remediation, will be discussed (Figure 4).

2.2.1. Microbial Bioremediation

AMD Bioremediation

Micro-organisms, including bacteria and fungi (such as arbuscular mycorrhizal fungi – AMF), are the major biological remediation component contributing to the remediation of a range of contaminants in wetland systems (Khan et al. 2009). Numerous environmentally important species of bacteria have been demonstrated to be involved in the remediation of contaminants within AMD as well as in the generation of AMD (Figure 2). *Pseudomonas* spp. have been demonstrated to reduce the harmful contaminant Cr^{6+} to a less toxic Cr^{3+} species, whereas *P. fluorescens*, has been demonstrated to biotransform Hg to a less harmful, volatile species. *Pseudomonas fluorescens* utilizes organomercurial lyase, an enzyme, which cleaves the C-Hg bond, resulting in the release of Hg^{2+}. Subsequently, another enzyme, mercuric reductase, transforms the Hg^{2+} to Hg^0, a volatile Hg species which is removed from the system via evapo(transpi)ration processes (Figure 4) (Schiering et al. 1991; Bhattacharya et al. 2006). The biological and hydro-geochemical features of wetlands are conducive to the methylation of metals [i.e., the addition of a methyl group to metals] such as Hg, tin (Sn) and tellurium (Te), and metalloids, including As and selenium (Se). This biomethylation results in the formation of volatile metal derivatives which, under certain hydro-geochemical processes include their removal from wetlands via evapo(transpi)ration. Although the methylation of metals and/ or metalloids may enhance the removal of toxic contaminants from wetland systems, the methylation of some metals, e.g., transformation of Hg to methylmercury (CH_3Hg^+/ $MeHg^+$), may increase the metal's bioavailability for plant and animal uptake, as well as its accumulation in biological tissue (due to the lipophilic nature of CH_3Hg^+), increasing its degree of toxicity in aquatic ecosystems. Sulfate-reducing bacteria have been demonstrated to be major contributors to Hg methylation in aquatic ecosystems (Bright et al. 1994). Moreover, mineral-degrading, acidophilic (acid-tolerant) microbial species, found in mine wastewater, are involved in the oxidation of Fe (including *Leptospirillum ferrooxidans* and *L. ferriphilum*), oxidation of S and Fe (including *Acidithiobacillus thiooxidan*), as well as the reduction of Fe [*Sulfobacillus* spp. (oxidation and reduction) and *Acidiphilium* spp.] (Hallberg and Johnson 2005).

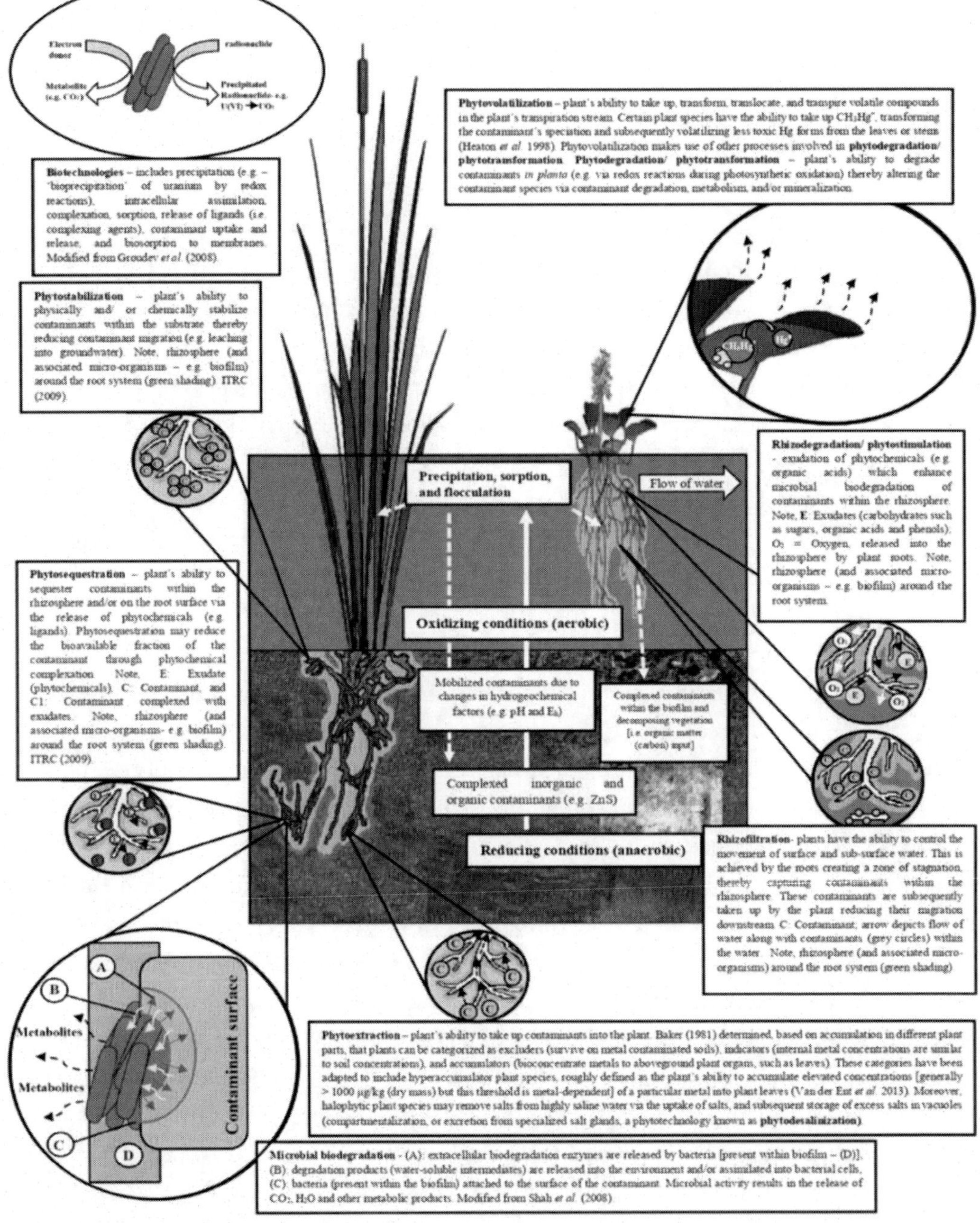

Figure 4. Potential biological remediation, including microbial bioremediation (bacteria- and fungi-based) and phytoremediation (plant-based), mechanisms involved in the remediation of inorganic and organic contaminants in a wetland system. The remediation potential of wetlands depends on hydro-geochemical and biological factors, as well as the interaction thereof. The number/ type of bio- and phytotechnology implemented depends on the plant species, element composition, and environmental factors. Some images have been modified from Favas et al. (2014). References: ITRC (2009); Kramer et al. (2007; 2010); Prasad (2013); Van der Ent et al. (2013).

These microbial species differ in their presence and density in wetland systems based on the wetland's hydro-geochemical (including the presence of anaerobic or aerobic substrate conditions), biological (plant species composition), and other environmental (e.g., temperature and humidity) factors. Singhakant et al. (2009[b]) demonstrated that organic compounds, released by the decomposing roots of *Vetiveria zizanioides* (vetiver grass) along with microbial interactions, were involved in the co-precipitation of As and Fe under reducing conditions within CWs. Desulfurization by microbial communities are mediated by the type of substrate present and O_2 availability as microbial-induced precipitation of Cd-, Cu-, and Zn-sulfide compounds has been demonstrated to occur under anaerobic conditions whereas the precipitation of Al and co-precipitation of As (Batty et al. 2002; Lizama et al. 2011) take place under aerobic conditions.

Biofilms - the assemblage of microbial communities embedded in an extracellular polymeric substance (EPS) matrix made up of biomolecules (including polysaccharides and proteins) - have been demonstrated to enhance bioremediation (and phytoremediation) processes (Das and Kazy 2014). Microbial bioremediation mechanisms have been used to target various contaminants including excess nitrogen levels (denitrification and ammonification processes), sulfur levels (desulfurization), and heavy metals including Fe and Mn (via redox reactions) (Kosolapov et al. 2004; Johnson and Hallberg 2005; Groudev et al. 2008; Koschorreck 2008). Biotechnologies, i.e., microbial bioremediation mechanisms, include biosorption (active and passive processes), metal precipitation by redox reactions (e.g., SO_4^{2-} reduction), as well as the methylation of metals (Figure 4) (Kosolapov et al. 2004). Manobala et al. (2018) concluded that U, as uranyl ions (UO^{2+}), were effectively removed by the radioactive-tolerant bacteria species, *Deinococcus radiodurans*, in which sorption processes played a major role in U removal. More specifically, this bioadsorption process may have involved the sorption of U to negatively charged exopolysaccharides. The hydro-geochemical conditions of wetlands (such as aerobic zones) are conducive to the abiotic oxidation, and subsequent precipitation (and/or retention), of Fe, Mn, and Al in such systems. Microbial action furthers such oxidation processes, resulting in the hydroxylation of these metals. This change in metal species increases their binding affinity for other heavy metals, such as Co, Cr, Cu, and Ni, thereby rendering the heavy metal(s) biological inert within wetland systems.

Heavy metal (im)mobilization by microbial activity occurs via metal chelation by siderophores and metabolites, redox reactions, (de)methylation, and/or autotrophic/heterotrophic leaching, by micro-organisms (Lloyd and Lovley 2001; Gadd 2004). This change in metal species (i.e., im/mobilization of metals) directly influences their transport in aquatic ecosystems (Kosolapov et al. 2004). It must be noted that certain bacteria species increase the bioavailability, and therefore the degree of toxicity of certain heavy metals such as Hg. This highlights the importance of identifying particular hydro-geochemical and biological (e.g., microbial communities) factors present within wetland systems. Certain biological processes involved in the remediation of AMD increase alkalinity and render toxic contaminants biologically inert, thereby reducing the negative impacts associated with AMD generation (Johnson and Hallberg 2005). The following equations describe these processes;

$$6HCO_3^-(aq) + 6H_2O \rightarrow C_6H_{12}O_6 + 6O_2 + 6OH^- \quad (1)$$

A weak base, bicarbonate – HCO_3^-, is utilized by photosynthetic micro-organisms which produce a stronger base, namely OH^-, increasing the pH level. This reaction produces glucose ($C_6H_{12}O_6$), a carbon-based compound providing an energy source for further microbial activity (e.g., metabolism).

$$Fe(OH)_3 + 3H^+ + e^- \rightarrow Fe^{2+} + 3H_2O \quad (2)$$

This equation represents an electron donor reaction in which the electron donor (e^-), supplied by organic substrate, increases the pH level by the reduction of Fe^{3+} in solid phase.

$$SO_4^{2-} + 2CH_2O + 2H^+ \rightarrow H_2S + 2H_2CO_3 \quad (3)$$

The dissimilatory pathway of SO_4^{2-} reduction to S^{-2}, by SRB, alkalinizes the pH level of AMD-affected water by transforming and reducing the type of acid in solution. This reaction immobilizes heavy metals with regards to the formation of metal (e.g., Zn)-S^{2-} precipitates as described by the following reaction:

$$Zn^{2+} + H_2S \rightarrow ZnS + 2H^+ \quad (4)$$

Sulfate is a major contaminant present within AMD where the remediation of AMD must consider the removal of SO_4^{2-}. Micro-organisms play a vital role in the sulfur cycle, viz. dissimilatory SO_4^{2-} reduction (reduction of SO_4^{2-} to S^{2-}) [e.g., $4H^+ + SO_4^{2-} + H+ \rightarrow HS^- + 4H_2O$], assimilatory SO_4^{2-} (assimilation of reduced S^{2-} in biomass and biomolecules), hydrogen sulfide (H_2S) release along with the mineralization of organic acids, and the oxidation of S^{2-} as electron acceptors (Sánchez-Andrea et al. 2014). This highlights the importance of the role of bacteria in the speciation of S in AMD wastewater. However, it must be noted that the presence of SRB may increase the toxicity of certain contaminants, such as Hg. For example, SRB methylate Hg forming CH_3Hg^+, a Hg-compound with an elevated degree of toxicity. King et al. (2000) demonstrated that SRB [requiring SO_4^{2-} as an electron acceptor to mediate S^{2-} formation (as a result of respiration)] did not methylate Hg in the absence of SO_4^{2-} under controlled experimental conditions. The presence of a submerged macrophyte reduced soluble Hg in solution. This was attributed to the increase in the suspended particulate matter as well as creating microenvironments conducive to enhancing demethylation processes by microbial communities (King et al. 2002). This demonstrates the importance of considering various biogeochemical factors, present in wetland systems, especially AMD-contaminated water where elevated concentrations of SO_4^{2-} are present. Moreover, metals in lower layers of TSFs may become more bioavailable for organism uptake as such tailings have been exposed to a higher degree of S-oxidizing *Thiobacillus* bacteria (Tutu et al. 2008). The optimum pH range for SRB activity is around neutral; however, obligate anaerobic bacterial species growing in acidic environments have the ability to alkalinize their environment via bicarbonate production, precipitating metals out of solution (García et al. 2010). Thus, bioremediation processes reduce SO_4^{2-} concentration, thereby lowering the salinity, the mobility of bioavailable heavy metals, as well as the generation of AMD.

Plastic Bioremediation

The efficacy of plastic biodegradation is mediated by the physicochemical properties of the polymer, viz. - polymer characteristics (e.g., mobility, type of functional groups, and molecular weight) and the type of micro-organism(s) required for biodegradation (Xu et al. 2005; Artham and Doble 2008). Plastic biodegradation involves the enzymatic hydrolysis of plastics, forming functional groups on the plastic's surface (Figure 4). The presence of these functional groups decreases the polymer's hydrophobicity due to the degradation of the main polymer chains thereby lowering the molecular weight of the degraded polymer. This decrease in molecular weight (and increase in hydrophilicity) enables microbial assimilation and/or the further biodegradation of the polymer into monomers (i.e., depolymerization) or even into organic intermediates (Arutchelvi et al. 2008; Shah et al. 2008). The biodegradation of the monomers into these organic intermediaries, such as alcohols and acids, increases their bioavailability for assimilation in microbial cells and incorporation into metabolic functioning. Plastic has also been demonstrated to be biodegraded by numerous microbial species including *Rhodococcus ruber* and *Psychrobacter* spp., as well as microbial species, associated with mangrove environments (i.e., highly saline environments with varying frequency and duration of saturation) including *Pseudomonas* spp., *Micrococcuss* spp., *Staphylococcus* spp., and *Streptococcus* spp. (Kathiresan 2003).

Polyethylene biodegradation involves changes in the physicochemical properties of the plastic, namely the molecular weight, surface functional groups, crystalline properties, and the topography (i.e., changes in the surface structure) (Restrepo-Flórez et al. 2014). Numerous studies have demonstrated the biodegradation of PE, PVC, PUR, PS, and PP. Highly chlorinated compounds are generally more resistant to aerobic degradation processes, compared with anaerobic processes. Moreover, the amphiphilic structure of surface-active compounds, known as biosurfactants (forming the biofilm), increase the surface area for hydrophobic (H_2O-insoluble) substances as well as alter the structure of microbial cells in which biosurfactants facilitate the attachment process to the polymer's hydrophobic surface (Tribedi and Sil 2013). The process involved in PE biodegradation involves the (i) reduction of the polymer's molecular weight (increases enzymatic action for degradation and transport of molecules for transport across bacteria cell membranes), and (ii) molecule oxidation (Yoon et al. 2012). Once the polymer is reduced, the oxidation transforms hydrocarbons into carboxylic acids (Figure 5). Subsequently, these carboxylic acids are incorporated into and metabolized by ß-oxidation and the citric acid cycle (CAC – also known as the Kreb's cycle) – the final pathway for carbohydrate, amino acid, and lipid oxidation (see Akram 2014).

It must be noted that PE biodegradation is a result of synergistic interactions between abiotic (e.g., photo-oxidation) and biotic (microbial biodegradation) factors (Figure 5). This highlights the importance of various hydro-geochemical and biological variables, as well as the interaction thereof, required for effective PE degradation - variables present in wetland systems fluctuating through space and time.

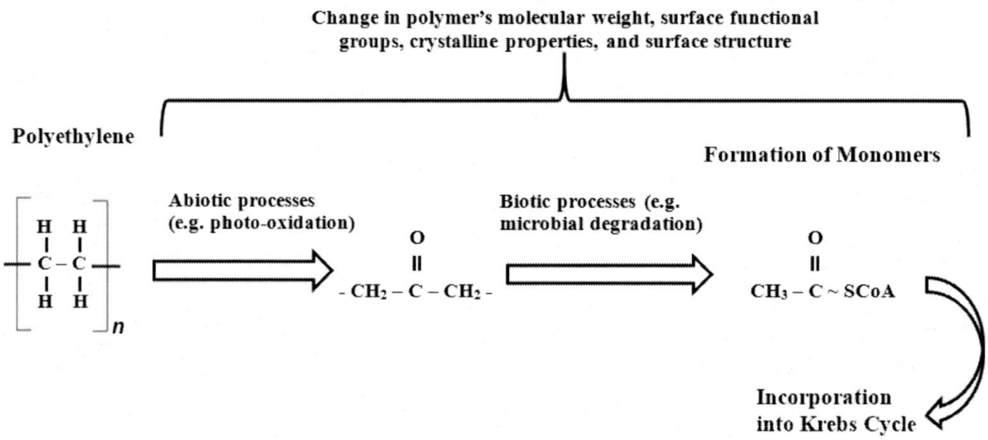

Figure 5. Proposed biodegradation pathway of polyethylene (PE). A carbonyl group (C=O) has been added to the polymer chain via the abiotic process of photo-oxidation. It must be noted that various abiotic (photo-oxidation, surface abrasion, O_2 and H_2O diffusion in the PE structure) and biotic [enzyme-induced changes in PE molecular structure, production of surface-active compounds (biosurfactants) for biofilm development, and the assimilation of monomers in bacterial cells] processes are involved in the biodegradation of PE. Adapted from Albertsson et al. (1987).

Potin et al. (2004) demonstrated that various fungi species, present in constructed wetland systems, including *Absidia cylindrospora*, *Cladosporium sphaerospermum*, and *Ulocladium chartarum* had the ability to degrade PAH (via sorption processes) to varying degrees. Polycyclic aromatic hydrocarbons have an affinity for substrate binding rendering such organic compounds resistant to microbial biodegradation (Bojes and Pope 2007). Phthalate esters, namely dimethyl phthalate (DMP), di-(2-ethylhexyl) phthalate (DEHP), DBP and diisobutyl phthalate were found in the sediment of various wetlands dominated by different plant species (e.g., *Phragmites australis*, *Deyuexia angustifolia*, and *Carex lasiocarpa*) (Wang et al. 2017). Thus, it may be inferred that differences in the redox state (i.e., O_2 availability) between sample sites may have contributed to the retention of these phthalate esters in sediments by the interplay between hydro-geochemical (e.g., redox reactions resulting in contaminant-sediment sorption) and biological (microbial biodegradation conditions such as the presence and activity of certain microbial communities) processes. These processes have been demonstrated by the addition of carbon sources, such as acetate, which has been involved in enhancing the microbial bioremediation of organic contaminants, previously recalcitrant to such degradation processes. Plant root exudates, containing acetate, have also been demonstrated to promote the phytoremediation of these organic compounds (Martin et al. 2014).

Contaminated soil and water are generally inhabited by micro-organisms, such as bacteria, fungi, and/or algae, which have the ability to tolerate, metabolize, and/or degrade contaminant(s) present at that particular contaminated site. Moreover, the change in the composition and concentrations of contaminants (limiting factor) over space (i.e., within different sections of the wetland system) and time, as well as the hydrological regime, influences the number of type of microbial communities, as well as plant species, present within the wetland system. For example, a study conducted by Balasooriya et al. (2008) showed that plant and microbial community structures changed along a hydrological gradient within the same wetland system. This may be attributed to the ability of certain organisms to tolerate, metabolize, and/or degrade contaminants (thereby remediating such contaminants) up to a particular threshold. However, if the concentration of contaminants exceeds this tolerance threshold, the health and survival of the organism may be compromised. Species of *Pseudomonas* and *Rhodococcus* have been demonstrated to degrade the PAH naphthalene, via oxidation reactions using dioxygenase enzymes. Naphthalene degradation products, such as naphthalene dihydrodiols have an elevated bioavailability and therefore degree of toxicity. However, PAHs have various biodegradation pathways (Figure 6). Hydro-geochemical factors influencing PAH bioremediation include, (i) pH [in which sub-optimal pH levels compromise the biodegradation efficacy of certain acidic-intolerant microbial species], (ii) change in temperature which fluctuates seasonally [where an increase in temperature increases the solubility of PAH], (iii) O_2 availability (i.e., aerobic vs. anaerobic conditions) [in which O_2 is a vital constituent of enzymes (e.g., dioxygenase) involved in PAH biodegradation processes (Figure 6)], (iv) nutrient availability [including the presence of nitrogen (N), P, and potassium (K)] enhancing microbial metabolism and therefore PAH biodegradation, and (v) PAH bioavailability [in which the rate and degree of bioremediation are mediated by physicochemical and microbial factors] (Breedveld and Sparrevik 2000; Bamforth and Singleton 2005).

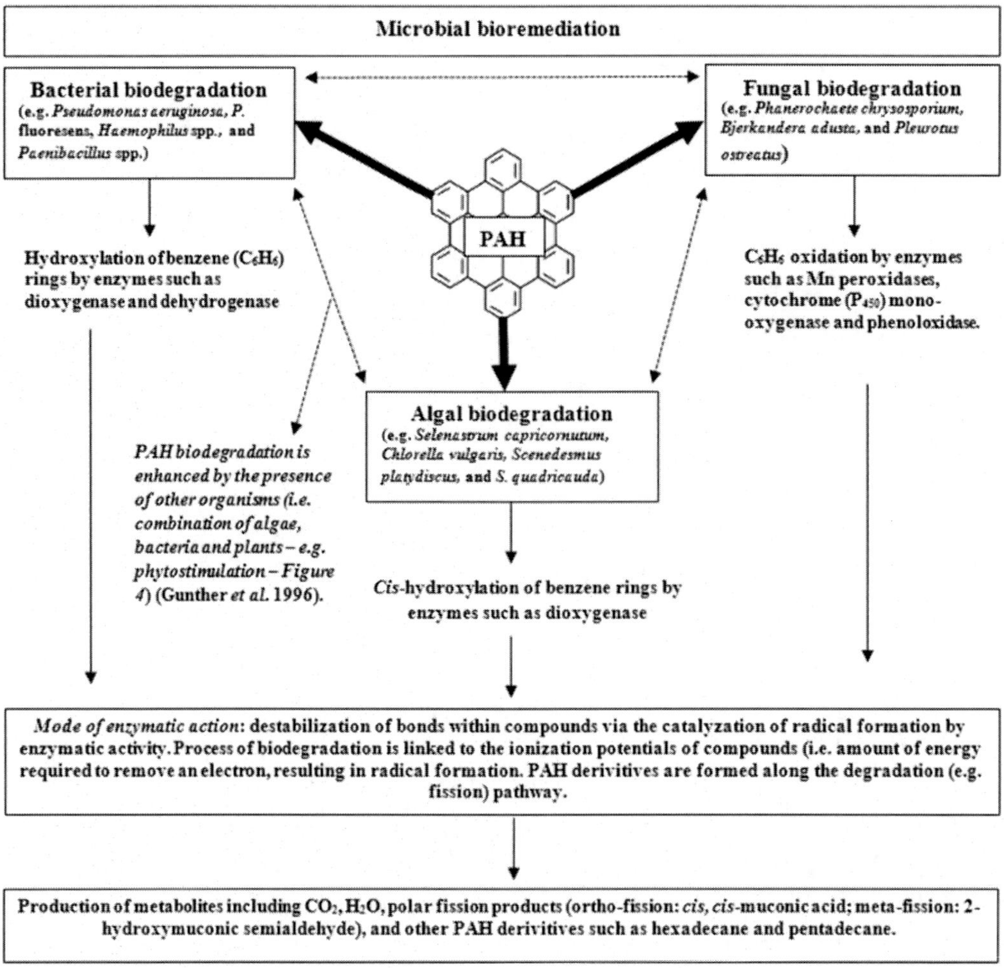

Figure 6. Microbial bioremediation of polycyclic aromatic hydrocarbons (PAH). Aerobic bacterial bioremediation involves the use of dioxygenase enzymes in order to oxidize the initial benzene ring, forming *cis*-dihydrodiols (which are subsequently dehydrogenated to form dihydroxylated intermediates). These intermediaries are metabolized via catechols, releasing CO_2 and H_2O as metabolic by-products. Note, dotted lines represent the combination of micro-organisms which has been demonstrated to enhance bioremediation. Wetland systems are comprised of a combination of micro-organisms and thus, a particular contaminant may be exposed to a combination of micro-organisms at any single point in space and time. Adapted from Cerniglia (1993), and Haritash and Kaushik (2009).

Microbial reductive dechlorination of polychlorinated dibenzo-p-dioxins and dibenzofurans, such as TCDD, has been demonstrated to reduce the degree of toxicity of these relatively recalcitrant compounds, inert to biodegradation processes (Ahn et al. 2008). This is attributed to the removal of chlorine (Cl) from such compounds. Treatment process, combining hydro-geochemical (e.g., use of palladium nanoparticles used for initial reductive dechlorination under anoxic conditions) and biological remediation (oxidative biomineralization using *Sphingomonas wittichii* under aerobic conditions) mechanisms, enhanced the dechlorination of an isomer of TCDD thereby reducing its degree of toxicity in the environment (Bokare et al. 2012). This highlights the elevated remediation potential of combining hydro-geochemical and biological remediation mechanisms thereby enhancing the biodegradation of PAHs and chlorinated dioxins, such as TCDD. These studies exhibit the ability to utilize wetland systems, comprising of various aforementioned remediation mechanisms, thereby creating an environment conducive to the remediation of a range of contaminants by such systems.

Wetland plants, also known as helophytes, play a vital role in the remediation of contaminants (phytoremediation) in the wetland system, as well as facilitating the bioremediation of contaminants by micro-organisms. Within a wetland system, bacteria and fungi populations are influenced by plant composition and more specifically, various plant processes. These processes include, but are not limited to, the release of exudates (e.g., carbon-compounds – a source of energy for microbial communities), signaling molecules, and the aeration of root systems creating aerobic conditions conducive to enhanced microbial bioremediation (Figure 4) (Saggai et al. 2017). It must also be noted that wetland substrate provides surfaces for direct contaminant sorption as well as supporting media for the establishment, growth, and survival of plant and microbial communities (Sheoran and Sheoran 2006).

Phytoremediation

Phytoremediation is defined as the use of the plant species' ability/ies to render toxic metal ions biologically inert via extracting, accumulating, volatilizing, and/or sequestering contaminants *in planta* and/or in the rhizosphere, as well as stabilizing substrate surfaces and controlling the vertical migration of contaminants within surface- and/or groundwater (USPA 2000; Keeling et al. 2003; Arnold et al. 2007; ITRC 2009; Manousaki and Kalogerakis 2011). Phytotechnologies utilized in wetland systems include phytostabilization, phytoextraction, phytovolatilization, and rhizofiltration (Figure 4). Various phytotechnologies utilize the plant's ability to remove contaminants by sorption, cation-exchange processes, changes in rhizosphere pH and E_h (induced by the release of organic compounds into the rhizosphere), selective and/or general uptake of contaminants contributing to the rhizosphere acidity/alkalinity, as well as the diffusion of O_2 into the rhizosphere by plant roots (Figure 4). Organisms are exposed to numerous abiotic and/or biotic stresses at any point in time and space. In light of this, organisms used for the remediation of numerous contaminants must be able to cross-tolerate such stresses. For example, halophytes, plant species with an elevated ability to tolerate highly saline ($\geq$ 200mM NaCl) environments have been utilized for the phytoremediation of metal- and highly saline-contaminated sites due to their ability to cross-tolerate (i.e., their ability to tolerate osmotic and ionic stresses induced by heavy metal and salinity exposure) the combination of these stresses (e.g., Van Oosten and Maggio 2015). Therefore, halophytic plant species must be considered for the remediation of AMD (highly saline and heavy metal-contaminated water).

The plant's root system provides a suitable surface area (for the adhesion of microbial communities), and an environment conducive to microbial activity via the diffusion of O_2 into the rhizosphere, creating aerobic and anaerobic microenvironments. These processes influence the hydro-geochemical factors within the rhizosphere, as the change in pH, E_h, and OM content affects contaminant speciation and mobility in wetland systems and thus, microbial bioremediation (Kosolapov et al. 2004). The change in pH may mobilize certain contaminants from the wetland substrate thereby increasing the contaminant's bioavailability for plant uptake. Moreover, the presence of plants may alter the redox state in the rhizosphere of wetland substrate (transforming the contaminant sink to a source of contamination) through the liberation of contaminants from soil particles (present mainly in anoxic and reduced conditions) by the oxidation of sediments by plants. The release of O_2 by plant roots may also enhance the precipitation of metals from solution due to increased microbial activity. The precipitation of Fe (oxy)(hydr)oxides in the rhizosphere may result in the formation of Fe plaques, leading to the immobilization of contaminants, such as As and Zn, by sorption processes as some contaminants have an elevated adsorption affinity for these Fe-plaques (Otte et al. 1989; Otte et al. 1995). Moreover, symbiotic relationships between plant roots and root-colonizing bacteria may enhance phytoremediation efficacy. Biofilms, present on the root surface, form a symbiotic relationship with the plant where the plant releases carbon-based compounds, as well as aerates the rhizosphere. In return, the plant receives resistance to pathogenic attack and bioavailable nutrients (through microbial transformations) necessary for plant uptake, growth, and survival (Figure 4) (Saggai et al. 2017). The aforementioned plant-induced processes have been demonstrated to significantly increase the assimilation of contaminants within wetland substrate (Doyle and Otte 1997), compared with non-vegetated wetlands, thereby reducing the mobility (and toxicity to fauna and flora) within aquatic and terrestrial ecosystems.

In order for metal ions to be taken up by plants, the metal ion must be bioavailable. Metal ion mobilization, modulated by biogeochemical factors such as rhizosphere acidification/alkalinization and lixiviant exudation, increases the mobility and available fraction of metals for plant uptake. Complex and intrinsic genetic, molecular, cellular, and physiological factors, as well as cross-talk between these levels, mediate the degree of contaminant uptake and tolerance. At a physiological level, plants overcome heavy metal toxicity by excluding or including (avoidance and tolerance) toxic metal ions. The compartmentalization, complexation, precipitation, and binding of toxic metal ions stabilize the heavy metal, rendering the metal biologically inert. This, in turn, reduces the direct and indirect negative impacts of toxic metal ions on plant functioning and survival (ITRC 2009; Marschner 2011). The degree of resistance to toxic metal species and concentrations is attributed to the number and efficacy of metal ion exclusion and/or inclusion strategies possessed by the particular plant species and/or genotype. Tolerance mechanisms may also include the plant's ability to effectively and efficiently regulate the binding of metal ions to organic and/or inorganic compounds during metal uptake and translocation thereby maintaining the plant's metal homeostasis network [plant's ability to control the transportation of a certain concentration of metal ions (e.g., Fe) required by metal-requiring molecules at a specific point in time whilst avoiding the accumulation and transport of metals in/across metal-sensitive areas of plant cells)] (Kramer et al. 2007).

Numerous wetland plant species have been studied in order to determine their remediation potential. Plant species such as *Typha latifolia*, *Scirpus validus*, *Phragmites australis,* and *Oryza sativa* have been demonstrated to form plaques in their root epidermis, via the production of metal (hydr)oxide precipitates which reduce the translocation of metals between plant parts and tissues. *Phragmites australis*, a cosmopolitan plant species, is (hyper)tolerant to various abiotic stresses, such as elevated contaminant levels. Furthermore, *P. australis* has demonstrated its ability to effectively remediate Ni from wastewater whilst cross-tolerating elevated concentrations of Cr, Cu, Cd, Pb, and Zn. The extraction of certain heavy metals by *P. australis* may also create conditions conducive to the survival of other plant species intolerant of certain contaminants. Cattails (*Typhus angustifolia*) and sedges (*Carex* spp.) have been successfully used in CWs (Somodi and Botta-Dukát 2004). Due to shallow water bodies (< 30 cm), aerobic wetlands allow for sufficient time for the oxidation and precipitation of metals, in which plant species, such as *Typha* spp. and *Phragmites* spp. regulate the flow of water thereby stabilizing the contaminants which are subsequently accumulated to varying degrees as metal precipitates in wetland systems.

3. The Case for Constructed Wetlands

Constructed wetland systems, typically, have an elevated remediation potential when compared with natural wetlands, due to the ability to control certain variables of the wetland system [*viz.* - the hydraulic regime (influencing contaminant-contact time and aerobic conditions) and the use of certain plant and microbial communities (i.e., ability to target certain contaminants)] optimal for the remediation of environmentally important contaminants (Horne and Fleming-Singer 2005). It must be noted that results obtained from various studies differ based on their location (e.g., cold vs. warm climatic conditions), type of bioremediation (different mechanisms of bioremediation utilized relative to composition of organisms and geochemical features), CW design (including controlled variables such as flow rate), as well as contaminants focused on (i.e., competitive ion adsorption). Although the remediation efficacy of natural wetland systems has been studied relative to certain contaminants, the main mechanism contributing to this efficacy may not be accurately distinguished due to the influence of and interaction between other unquantified/unidentified hydro-geochemical and biological factors (as well as the interplay between these factors). These factors, such as the presence of particular microbial communities, may enhance the efficacy of the studied mechanism (e.g., phytoremediation potential of a particular plant species) while decreasing, and overlooking, the efficacy of other unquantified factors contributing to the overall remediation potential of the wetland system.

Numerous studies have been conducted to determine the heavy metal remediation potential of natural and constructed wetland systems. Groudev et al. (2008) demonstrated the ability of natural wetlands and CW to increase the pH level (2.15 - 4.15 increased to 6.51 – 7.92), while effectively reducing U, radium (Ra), Cu, Zn, Cd, Pb, Ni, Co, Fe, Mn, and As (e.g., U was reduced from 0.28 – 4.82 to < 0.10 mg/L; Fe was reduced from 73-1072 mg/L to < 1-14 mg/L) to permissible levels for agricultural and industrial use. The studied plant species, *viz. P. australis*, *T. latifolia*, and *S. lacustris* accumulated U, Ra, Cu, Cs, Ni, Co, Pb, Mn, and As, in which elevated element concentrations were accumulated in roots compared with shoots (i.e., low metal root-to-shoot translocation factor). Manganese (Mn^{2+}) and Fe^{2+} were precipitated via oxidation by heterotrophic (e.g., *Leptothrix* spp.) and acidophilic Fe-oxidizing bacteria (e.g., *Acidithiobacillus ferrooxidans*), respectively. Natural wetlands reduced the levels of certain contaminants, present within the AMD, within permissible levels. The remediation efficacy of the wetland systems was attributed to the microbial dissimilatory SO_4^{2-} reduction and sorption (by OM) processes, in which metals were precipitated as metal-S^{2-}. Uranium was precipitated out of solution via reduction processes (Figure 4 – "*Biotechnologies*") whereas Ra was removed by sorption processes. Results from this study were in accordance with Sheoran (2005) in which a natural wetland increased the pH level, whilst removing Cu (37.41 and 75.76%), Fe (47.86 and 78.02%), Co (60.68 and 64.27%), Zn (66.41 and 88.37%), Pb (50.14 and 60.86%), Ni (48.68 and 64.39%), and Mn (64.81 and 85.21%) at two sampled locations, respectively, over three seasons (pre-monsoon, post-monsoon, and winter). The removal of heavy metals from the wetland systems, at both locations, was lowest during winter highlighting the effect of uncontrolled abiotic factors (i.e., temperature) on the biological remediation of multiple contaminants. Cutright et al. (2012) demonstrated that metal uptake differed between plant species (i.e., *P. australis*: Fe >> Al > Mg > Mn; *T. latifolia*: Fe >> Mg = Mn > Al; and *Solidago* spp.: Mg > Mn > Fe > Al) in which *P. australis* hyperaccumulated Al. Although the aforementioned studies demonstrated the remediation efficacy of natural wetlands, a study conducted by Nyquist and Gregor (2008) concluded that temperature and contaminant contact time are limiting factors negatively impacting the remediation potential of wetland systems. This highlights the inability to accurately compare studies differing in environmental conditions.

Various *in situ* hydro-geochemical factors may reduce the efficacy of biological remediation. The biodegradation of plastics occurs at varying rates depending on the type of substrate (Shah et al. 2008). Fluctuating (a)biotic factors (such as various types of substrates which differ in their physicochemical properties), as well as the interplay between these factors, which (i) differ between studies (i.e., differences in substrate biogeochemistry), and/or (ii) were not quantified/controlled for at the time of these studies, limit the comparability and repeatability between these studies conducted on natural wetland systems (Golan-Goldhirsh et al. 2004). Furthermore, remediation by wetland systems involves a complex series of processes in which distinguishing a specific pathway for the removal (i.e., fate) of a contaminant of interest may be impeded due to other interacting processes (Sheoran and Sheoran 2006). The influence of these unquantified variables compromises the results obtained from these studies. The removal of a contaminant(s) of interest may also be compromised due to fluctuating environmental variables not controlled for and/or not considered as important variables, resulting in a gap in knowledge. However, in order to overcome these challenges, experiments can be conducted under controlled conditions, such as the use of CWs. Numerous, published studies have aimed to determine the fate of a particular constituent involved in the remediation process under controlled conditions - e.g., Li et al. (2016) determined the phytoextraction, phytotransformation, and rhizodegradation of ibuprofen (a pharmaceutical pain killer) by *Typha angustifolia* in a constructed (horizontal subsurface flow) wetland under controlled conditions (e.g., temperature and water flow).

Constructed wetlands are advantageous, compared with natural systems, due to the control of various hydro-geochemical and biological factors, namely their hydrological regime (flow direction and pathway (horizontal (HF) or vertical (VF), free surface water or subsurface flow), and the selection of effective substrate, vegetation (emergent, submerged, or floating macrophytes), and micro-organisms optimal for targeted contaminant remediation (Figure 7) (Stefanakis 2018). Due to the role of substrate in contaminant retention in natural systems (sorption properties in precipitation and ion-exchange processes), various types of natural (e.g., minerals) and/or synthetic (e.g., activated carbon) substrates are used. Moreover, other variables influencing effective contaminant biological remediation may be controlled, namely O_2 supply (in which artificial aeration has been demonstrated to enhance contaminant removal processes – Nivala et al. 2012), types of plant species (e.g., *P. australis* may have an elevated ability to remediate a targeted contaminant, such as Ni, whilst simultaneously tolerating a range of other contaminants, compared with other commonly occurring plant species including *Typha* spp. and *Scirex* spp.), and the rate, frequency, and duration of water flow. De Matos and Zhang (2011) studied the remediation potential of two different CW systems (HF and VF). Horizontal flow CW design increased the pH level from 4 to 6.8 as well as removed 90% and 30% Fe and Mn, respectively, whereas the VF system resulted in the removal of 71%, 80% and 60% of Fe, Al, and Mn, respectively. Moreover, 87% of all metals of interest were removed from synthetic AMD by an HF CW (Gikas et al. 2013). This highlights the efficacy of CWs by controlling various hydro-geochemical and biological factors, optimizing the remediation efficacy of CWs relative to targeted contaminants.

Figure 7. Example of a constructed wetland (CW), namely the Vertical Up-Flow Modular Constructed Wetland Wastewater Treatment System (VUF-MCW-WTS) utilizing the plant species *Cyperus* spp., for wastewater remediation. The VUF-MCW-WTS system is advantageous over other traditional CW systems (e.g., horizontal flow CWs) as (i) the design of the cells and system (e.g., no substrate is used) increases contaminant-root contact time (a limiting factor for all CWs utilizing substrtae) thereby increasing the system's remediation potential, (ii) a 48-cell system, which occupies less than $2m^2$, is capable of treating 2 500 L of water/hr (Turton 2015[c]), (iii) a low maintanance, scalable system, (iv) this is a novel and patented (Patent No. 2017/03538) CW system (based on latest scientific literature), (v) numerous abiotic and biotic factors can be controlled, such as the hydrological regime, more easily compared with traditional CW systems, and (vi) the VUF-MCW-WTS can be easily moved/ installed and does not require the clearance of large areas of vegetation and/or the excavation of land (activities which may require environmental and/or water use authorizations as per the applicable environmental legislation (such as the National Environmental Management Act (NEMA, Act No. 107 of 1998) and National Water Act (NWA, Act No. 36 of 1998)).

The hydraulic (HLR), along with the pollutant (PLR), loading rate and hydraulic retention time (HRT) are major contributors to the remediation efficacy of CWs. This is mainly related to the optimum contaminant-contact type with the substrate, microbial communities, and the root system that they are associated with. Shearon and Shearon (2006) demonstrated that the uptake of Fe, Cu, Zn, Pb, Co, Ni, and Mn by *T. angustifolia*, *Desmostachya bipinnata*, and *Sacharum bengalense* increased as the contaminant exposure period (i.e., contaminant-root contact time) increased where the maximum metal removal occurred within the first 24 hrs. Moreover, the HLR and HRT were also influenced by the plant species and substrate utilized in the system. In larger CW systems, such as reed beds, plants (generally phreatophytes) are utilized to control the movement of sub-surface water, as well as the contaminants within the water, by phytohydraulics (a phytotechnology incorporating rhizofiltration). The rhizosphere reverses the hydraulic gradient forming an area of stagnation, trapping contaminants within this area thereby reducing their migration downstream (LeDuc and Terry 2005; ITRC 2009).

The nature of plastics, i.e., recalcitrant nature along with a high molecular weight and complex structure, impedes plastic's ability to be biodegraded (Hadad et al. 2005). However, the hydro-geochemical and biological factors incorporated in CWs may increase the ability to biodegrade such compounds due to physical, chemical, and other biological processes, which can be manipulated to optimize such processes, and thus optimize certain biodegradation pathway(s). Studies have demonstrated the ability of *Clostridium sphenoides* and *Desulfosporosinus* spp. to remove significant concentrations of U under anaerobic conditions (Francis et al. 2004; Suzuki and Banfield 2004; Suzuki and Suko 2006). However, these studies concluded that maintaining anaerobic conditions at an industrial scale is challenging due to fluctuating conditions, such as conditions present in natural wetlands. This presents an opportunity to utilize CWs, such as the VUF-MCW-WTS (Figure 7), with anaerobic chambers, for the removal of U and other heavy metals present in AMD, on larger scales. At a smaller scale, plants generate oxidizing or reducing microenvironments around their rhizospheres via the release of organic compounds or sorption of metals (Cohen 2006). This may liberate immobilized contaminants, increasing their bioavailability for plant uptake (phytoextraction) and other phytoremediation processes (Figure 7).

In conclusion, the complexity in distinguishing a particular process of contaminant removal within a natural wetland system may be impeded by various hydro-geochemical (e.g., pH level, Eh, hydrological regime, and type of substrate) and biological (e.g., plant and microbial species composition) factors, which fluctuate through space and time. Moreover, for the effective remediation of particular contaminants, such as POPs, a combination of hydro-geochemical and biological factors optimal for the remediation of the targeted contaminant(s) is required. The main factors limiting the microbial bioremediation of plastics, and their sorbed constituents, are the plastic's physicochemical properties, namely the high molecular weight and hydrophobicity (reducing the colonization and subsequent biodegradation). This highlights the need for the implementation of controlled, interacting hydro-geochemical (e.g., photo-oxidation) and biological factors, which enhance the colonization and biodegradation of plastics, and their sorbed constituents, by micro-organisms in wetland systems. Wetland systems, and more specifically CWs which can optimize such mechanisms, possess a range of hydro-geochemical and biological mechanisms that can be optimized for the remediation of contaminants of interest. Therefore, future studies, aimed at determining the remediation efficacy of wetland systems, under controlled conditions, relative to contaminants present in AMD and plastics (such as POPs) will benefit the scientific, and non-scientific, community by creating holistic solutions to solve complex challenges associated with the contamination of water by AMD, plastics, and their sorbed constituents.

REFERENCES

Acosta, J. A., Jansen, B., Kalbitz, K., Faz, A. & Martínez-Martínez, S. (2011). "Salinity increases mobility of heavy metals in soils." *Chemosphere*, *85*, 1318-1324.

Ahn, Y. B., Liu, F., Fennell, D. E. & Häggblom, M. M. (2008). "Biostimulation and bioaugmentation to enhance dechlorination of polychlorinated dibenzo-p-dioxins in contaminated sediments." *FEMS Microbiology Ecology*, *66*, 271-281.

Akcil, A. & Koldas, S. (2006). "Acid Mine Drainage (AMD): causes, treatment and case studies." *Journal of Cleaner Production.*, *14*, 1139-1145.

Akram, M. (2014). "Citric acid cycle and role of its intermediates in metabolism." *Cell Biochemistry and Biophysics*, *68*, 475-478.

Albertsson, A. C., Andersson, S. O. & Karlsson, S. (1987). "The mechanism of biodegradation of polyethylene." *Polymer Degradation and Stability*, *18*, 73-87.

Andrady, A. L. & Neal, M. A. (2009). "Applications and societal benefits of plastics." *Philosophical Transactions of the Royal Society B: Biological Sciences*, *364*, 1977-1984.

Andrady, A. L. (2011). "Microplastics in the marine environment." *Marine Pollution Bulletin*, *62*, 1596-1605.

Antunes, J. C., Frias, J. G. L., Micaelo, A. C. & Sobral, P. (2013). "Resin pellets from beaches of the Portuguese coast and adsorbed persistent organic pollutants." *Estuarine, Coastal and Shelf Science*, *130*, 62-69.

Arnold, C. W., Parfitt, D. G. & Kaltreider, M. (2007). "Phytovolatilization of oxygenated gasoline-impacted groundwater at an underground storage tank site via conifers." *International Journal of Phytoremediation*, *9*, 53-69.

Artham, T. & Doble, M. (2008). Biodegradation of aliphatic and aromatic polycarbonates. *Macromolecular Bioscience*, *8*(1), pp. 14-24.

Arutchelvi, J., Sudhakar, M., Arkatkar, A., Doble, M., Bhaduri, S. & Uppara, P. V. (2008). Biodegradation of polyethylene and polypropylene. *Indian Journal of Biotechnology*, *97*, pp. 9-22.

Baker, A. J. (1981). "Accumulators and excluders -strategies in the response of plants to heavy metals." *Journal of Plant Nutrition*, *3*, 643–654.

Balasooriya, W.K., Denef, K., Peters, J., Verhoest, N. E. C. & Boeckx, P. (2008). Vegetation composition and soil microbial community structural changes along a wetland hydrological gradient. *Hydrology and Earth System Sciences Discussions, European Geosciences Union*, *12*, 277-291.

Bamforth, S. M. & Singleton, I. (2005). "Bioremediation of polycyclic aromatic hydrocarbons: current knowledge and future directions." *Journal of Chemical Technology & Biotechnology: International Research in Process, Environmental & Clean Technology*, *80*, 723-736.

Batty, L. C., Baker, A. J. & Wheeler, B. D. (2002). "Aluminium and phosphate uptake by *Phragmites australis*: the role of Fe, Mn and Al root plaques." *Annals of Botany, 89*, 443-449.

Bhattacharya, T., Banerjee, D. K. & Gopal, B. (2006). "Heavy metal uptake by Scirpus littoralis schrad. from fly ash dosed and metal spiked soils." *Environmental Monitoring and Assessment, 121*, 363-380.

Bojes, H. K. & Pope, P. G. (2007). "Characterization of EPA's 16 priority pollutant polycyclic aromatic hydrocarbons (PAHs) in tank bottom solids and associated contaminated soils at oil exploration and production sites in Texas." *Regulatory Toxicology and Pharmacology, 47*, 288-295.

Bokare, V., Murugesan, K., Kim, J. H., Kim, E. J. & Chang, Y. S. (2012). "Integrated hybrid treatment for the remediation of 2, 3, 7, 8-tetrachlorodibenzo-p-dioxin." *Science of the Total Environment, 435*, 563-566.

Breedveld, G. D. & Sparrevik, M. (2000). "Nutrient-limited biodegradation of PAH in various soil strata at a creosote contaminated site." *Biodegradation, 11*, 391-399.

Bright, D. A., Brock, S., Reimer, K. J., Cullen, W. R., Hewitt, G. M. & Jafaar, J. (1994). "Methylation of arsenic by anaerobic microbial consortia isolated from lake sediment." *Applied Organometallic Chemistry, 8*, 415-422.

Buddhawong, S., Kuschk, P., Mattusch, J., Wiessner, A. & Stottmeister, U. (2005). "Removal of arsenic and zinc using different laboratory model wetland systems." *Engineering in Life Sciences, 5*, 247-252.

Cerniglia, C. E. (1993). "Biodegradation of polycyclic aromatic hydrocarbons." *Current Opinion in Biotechnology, 4*, 331-338.

Cohen, R. R. (2006). "Use of microbes for cost reduction of metal removal from metals and mining industry waste streams." *Journal of Cleaner Production, 14*, 1146-1157.

Cronk, J. K. & Fennessy, M. S. (2016). *Wetland Plants: Biology and Ecology*. CRC press.

Cutright, T. J., Senko, J., Sivaram, S. & York, M. (2012). "Evaluation of the phytoextraction potential at an acid-mine-drainage-impacted site." *Soil and Sediment Contamination: An International Journal, 21*, 970-984.

Das, R. & Kazy, S. K. (2014). "Microbial diversity, community composition and metabolic potential in hydrocarbon contaminated oily sludge: prospects for *in situ* bioremediation." *Environmental Science and Pollution Research, 21*, 7369-7389.

De Matos, M. C. C. F. & Zhang, Z. (2011). "Mining acid rock passive treatment for closure of a lignite mine in Spain: Achieving necessary water quality for discharge into lakes and drinking water reservoirs. *Lakes & Reservoirs: Research & Management, 16*, 195–204.

Do Sul, J. A. I. & Costa, M. F. (2014). "The present and future of microplastic pollution in the marine environment." *Environmental Pollution, 185*, 352-364.

Doyle, M. O. & Otte, M. L. (1997). "Organism-induced accumulation of iron, zinc and arsenic in wetland soils." *Environmental Pollution, 96*, 1-11.

Farrell, L. A., Hamann, R. & Mackres, E. (2012). "A clash of cultures (and lawyers): Anglo Platinum and mine-affected communities in Limpopo Province, South Africa." *Resources Policy*, *37*, 194-204.

Farrell, P. & Nelson, K. (2013). "Trophic level transfer of microplastic: *Mytilus edulis* (L.) to *Carcinus maenas* (L.)." *Environmental Pollution*, *177*, 1-3.

Favas, P. J., Pratas, J., Varun, M., D'Souza, R. & Paul, M. S. (2014). "Phytoremediation of soils contaminated with metals and metalloids at mining areas: potential of native flora." *Environmental Risk Assessment of Soil Contamination*, *3*, 485-516.

Francis, A. J., Gillow, J. B., Dodge, C. J., Harris, R., Beveridge, T. J. & Papenguth, H. W. (2004). "Uranium association with halophilic and non-halophilic bacteria and archaea." *Radiochimica Acta*, *92*, 481-488.

Gadd, G. M. (2004). "Microbial influence on metal mobility and application for bioremediation." *Geoderma*, *122*, 109-119.

García, J., Rousseau, D. P., Morato, J., Lesage, E. L. S., Matamoros, V. & Bayona, J. M. (2010). "Contaminant removal processes in subsurface-flow constructed wetlands: a review." *Critical Reviews in Environmental Science and Technology*, *40*, 561-661.

Geyer, R., Jambeck, J. R. & Law, K. L. (2017). "Production, use, and fate of all plastics ever made." *Science Advances*, *3*, 1-5.

Gikas, P., Ranieri, E. & Tchobanoglous, G. (2013). "Removal of iron, chromium and lead from waste water by horizontal subsurface flow constructed wetlands." *Journal of Chemical Technology & Biotechnology*, *88*, 1906-1912.

Glaister, B. J. & Mudd, G. M. (2010). "The environmental costs of platinum – PGM mining and sustainability: Is the glass half-full or half-empty?" *Minerals Engineering*, *23*, 438–450.

Golan-Goldhirsh, A., Barazani, O., Nepovim, A., Soudek, P., Smrcek, S., Dufkova, L., Krenkova, S., Yrjala, K., Schröder, P. & Vanek, T. (2004). "Plant response to heavy metals and organic pollutants in cell culture and at whole plant level." *Journal of Soils and Sediments*, *4*, 133-140.

Groudev, S., Georgiev, P., Spasova, I. & Nicolova, M. (2008). "Bioremediation of acid mine drainage in a uranium deposit." *Hydrometallurgy*, *94*, 93-99.

Gruau, G., Dia, A., Olivie-Lauquet, G. & Serrat, E. (2000). "The effects of organic matter and seasonal redox dynamics on chemical weathering: Constraints from natural wetland studies." *Journal of Conference Abstracts*, *5*, 463.

Gunther, T., Dornberger, U. & Fritsche, W. (1996). "Effects of ryegrass on biodegradation of hydrocarbons in soil" *Chemosphere*, *33*, 203–215.

Hadad, D., Geresh, S. & Sivan, A. (2005). "Biodegradation of polyethylene by the thermophilic bacterium *Brevibacillus borstelensis*." *Journal of Applied Microbiology*, *98*, 1093-1100.

Hallberg, K. B. & Johnson, D. B. (2005). "Microbiology of a wetland ecosystem constructed to remediate mine drainage from a heavy metal mine." *Science of the Total Environment, 338*, 53-66.

Haritash, A. K. & Kaushik, C. P. (2009). "Biodegradation aspects of polycyclic aromatic hydrocarbons (PAHs): a review." *Journal of Hazardous Materials, 169*, 1-15.

Heaton, A. C., Rugh, C. L., Wang, N. J. & Meagher, R. B. (1998). "Phytoremediation of mercury-and methylmercury-polluted soils using genetically engineered plants." *Journal of Soil Contamination, 7*, 497-509.

Henken, K. R. & Hutchison, A. (2009). "Arsenic chemistry." In: Henken, K.R. (Ed.), *Arsenic Environmental Chemistry, Health Threats and Waste Treatment.* John Wiley & Sons Ltd., Chichester, pp. 9–6

Horne, A. J. & Fleming-Singer, M. (2005). *Phytoremediation using constructed treatment wetlands: an overview*, pp. 329-377. Science Publishers, Enfield, NH.

Interstate Technology Regulatory Council, (ITRC). (2009). "*Phytotechnology technical and regulatory guidance and decision trees*" Available at: http://www.itrcweb.org/ Guidance/GetDocument? documentID=64.

Johnson, D. B. & Hallberg, K. B. (2005). "Acid mine drainage remediation options: a review." *Science of the Total Environment, 338*, 3-14.

Kathiresan, K. (2003). "Polythene and plastics-degrading microbes from the mangrove soil." *Revista de Biologia Tropical, 51*, 629-633.

Keeling, S. M., Stewart, R. B., Anderson, C. W. N. & Robinson, B. H. (2003). "Nickel and cobalt phytoextraction by the hyperaccumulator *Berkheya coddii*: implications for polymetallic phytomining and phytoremediation." *International Journal of Phytoremediation, 5*, 235-244.

Khan, S., Ahmad, I., Shah, M. T., Rehman, S. & Khaliq, A. (2009). "Use of constructed wetland for the removal of heavy metals from industrial wastewater." *Journal of Environmental Management, 90*, 3451-3457.

King, J. K., Harmon, S. M., Fu, T. T. & Gladden, J. B. (2002). "Mercury removal, methylmercury formation, and sulfate-reducing bacteria profiles in wetland mesocosms." *Chemosphere, 46*, 859-870.

King, J. K., Kostka, J. E., Frischer, M. E. & Saunders, F. M. (2000). Sulfate-reducing bacteria methylate mercury at variable rates in pure culture and in marine sediments. *Applied Environmental Microbiology, 66*, 2430-2437.

Kivaisi, A. K. (2001). "The potential for constructed wetlands for wastewater treatment and reuse in developing countries: a review." *Ecological Engineering, 16*, 545-560.

Koschorreck, M. (2008). "Microbial sulphate reduction at a low pH." *FEMS Microbiology Ecology, 64*, 329–342.

Kosolapov, D. B., Kuschk, P., Vainshtein, M. B., Vatsourina, A. V., Wiessner, A., Kästner, M. & Müller, R. A. (2004). "Microbial processes of heavy metal removal from carbon-deficient effluents in constructed wetlands." *Engineering in Life Sciences* 4:403-411.

Krämer, U. (2010). "Metal hyperaccumulation in plants." *Annual Review of Plant Biology*, *61*, 517–34.

Krämer, U., Talke, I. N. & Hanikenne, M. (2007). Transition metal transport. *FEBS Letters*, *581*, 2263–2272.

LeDuc, D. L. & Terry, N. (2005). "Phytoremediation of toxic trace elements in soil and water." *Journal of Industrial Microbiology and Biotechnology*, *32*, 514-520.

Li, Y., Zhang, J., Zhu, G., Liu, Y., Wu, B., Ng, W. J., Appan, A. & Tan, S. K. (2016). "Phytoextraction, phytotransformation and rhizodegradation of ibuprofen associated with *Typha angustifolia* in a horizontal subsurface flow constructed wetland." *Water Research*, *102*, 294-304.

Lizama, K., Fletcher, T. D. & Sun, G. (2011). "Removal processes for arsenic in constructed wetlands." *Chemosphere*, *84*, 1032-1043.

Lloyd, J. R. & Lovley, D. R. (2001). "Microbial detoxification of metals and radionuclides." *Current Opinion in Biotechnology*, *12*, 248-253.

Manobala, T., Sudhir K. Shukla, Subba Rao, T. & Dharmendira Kumar, M. (2019). "A new uranium bioremediation approach using radio-tolerant *Deinococcus radiodurans* biofilm." *Journal of Biosciences*, *44*, 122.

Manousaki, E. & Kalogerakis, N. (2011). "Halophytes—An emerging trend in phytoremediation." *International Journal of Phytoremediation*, *13*, 959–969.

Marschner, H. (2011). *Marschner's mineral nutrition of higher plants*, Academic Press.

Martin, B. C., George, S. J., Price, C. A., Ryan, M. H. & Tibbett, M. (2014). "The role of root exuded low molecular weight organic anions in facilitating petroleum hydrocarbon degradation: current knowledge and future directions." *Science of the Total Environment*, *472*, 642-653.

Matsumoto, T., Nagashima, Y., Kusuhara, H., Ishizaki, S., Shimakura, K. & Shiomi, K. (2008). "Evaluation of hepatic uptake clearance of tetrodotoxin in the puffer fish *Takifugu rubripes*." *Toxicon*, *52*, 369-374.

Maurice, P. A., Cabaniss, S. E., Drummond, J. & Ito, E. (2002). "Hydrogeochemical controls on the variations in chemical characteristics of natural organic matter at a small freshwater wetland." *Chemical Geology*, *187*, 59-77.

McAdams, C. L. & Aquino, J. T. (1994). "Dioxin: impact on solid waste industry uncertain." *Waste Age*, *25*, 11.

Naicker, K., Cukrowska, E. & McCarthy, T. S. (2003). "Acid mine drainage arising from gold mining activity in Johannesburg, South Africa and environs." *Environmental Pollution*, *122*, 29-40.

Net, S., Sempere, R., Delmont, A., Paluselli, A. & Ouddane, B. (2015). "Occurrence, fate, behavior and ecotoxicological state of phthalates in different environmental matrices." *Environmental Science & Technology*, *49*, 4019-4035.

Niering, W. A. & Lowe, C. H. (1985). "Vegetation of the Santa Catalina Mountains: community types and dynamics" In *Plant community ecology: Papers in honor of Robert H. Whittaker*, (pp. 159-184). Springer, Dordrecht.

Nivala, J., Knowles, P., Dotro, G., García, J. & Wallace, S. (2012). "Clogging in subsurface-flow treatment wetlands: measurement, modeling and management." *Water Research*, *46*, 1625-1640.

Nyquist, J. & Gregor, M. (2008). "A field study of constructed wetlands for preventing and treating acid mine drainage." *Ecological Engineering*, *35*, 630–642.

Ollis, D. J., Snaddon, C. D., Job, N. M. & Mbona, N. (2013). *Classification System for Wetlands and other Aquatic Ecosystems in South Africa. User Manual: Inland Systems.* SANBI Biodiversity Series 22. South African National Biodiversity Institute, Pretoria.

Otte, M. L., Kearns, C. C. & Doyle, M. O. (1995). "Accumulation of arsenic and zinc in the rhizosphere of wetland plants." *Bulletin of Environmental Contamination and Toxicology*, *55*, 154-161.

Otte, M. L., Rozema, J., Koster, L., Haarsma, M. S. & Broekman, R. A. (1989). "Iron plaque on roots of *Aster tripolium* L.: interaction with zinc uptake." *New Phytologist*, *111*, 309-317.

Potin, O., Rafin, C. & Veignie, E. (2004). "Bioremediation of an aged polycyclic aromatic hydrocarbons (PAHs)-contaminated soil by filamentous fungi isolated from the soil." *International Biodeterioration & Biodegradation*, *54*, 45–52.

Prasad, M. N. V. (2013). *Heavy metal stress in plants: from biomolecules to ecosystems*, Springer Science & Business Media.

Restrepo-Flórez, J. M., Bassi, A. & Thompson, M. R. (2014). "Microbial degradation and deterioration of polyethylene–A review." *International Biodeterioration & Biodegradation*, *88*, 83-90.

Saggaï, M. M., Aïnouche, A., Nelson, M., Cattin, F. & El Amrani, A. (2017). "Long-term investigation of constructed wetland wastewater treatment and reuse: Selection of adapted plant species for metaremediation." *Journal of Environmental Management*, *201*, 120-128.

Sánchez-Andrea, I., Sanz, J. L., Bijmans, M. F. & Stams, A. J. (2014). "Sulfate reduction at low pH to remediate acid mine drainage." *Journal of Hazardous Materials*, *269*, 98-109.

Schiering, N., Kabsch, W., Moore, M. J., Distefano, M. D., Walsh, C. T. & Pai, E. F. (1991). Structure of the detoxification catalyst mercuric ion reductase from *Bacillus sp.* strain RC607. *Nature*, *352*(6331), p. 168.

Shah, A. A., Hasan, F., Hameed, A. & Ahmed, S. (2008). "Biological degradation of plastics: a comprehensive review." *Biotechnology Advances*, *26*, 246-265.

Sheoran, A. S. (2005). "Performance of a natural wetland treating acid mine drainage in arid conditions." *Mine Water and the Environment*, *24*, 150-154.

Sheoran. A. S. & Sheoran, V. (2006). "Heavy metal removal mechanism of acid mine drainage in wetlands: A critical review." *Minerals Engineering*, *19*, 105–116.

Sheridan, C., Akcil, A., Kappelmeyer, U. & Moodley, I. (2018). "A Review on the Use of Constructed Wetlands for the Treatment of Acid Mine Drainage." Pp. 249-262 in: Stefanakis, A.I. ed. 2018. *Constructed Wetlands for industrial wastewater treatment.* John Wiley & Sons.

Singhakant, C., Koottatep, T. & Satayavivad, J. (2009a). "Fractional analysis of arsenic in subsurface-flow constructed wetlands with different length to depth ratios." *Water Science and Technology*, *60*, 1771-1778.

Singhakant, C., Koottatep, T. & Satayavivad, J. (2009b). "Enhanced arsenic removals through plant interactions in subsurface-flow constructed wetlands." *Journal of Environmental Science and Health Part A*, *44*, 163-169.

Somodi, I. & Botta-Dukát, Z. (2004). "Determinants of floating island vegetation and succession in a recently flooded shallow lake, Kis-Balaton (Hungary)." *Aquatic Botany*, *79*, 357-366.

Stefanakis, A. I. ed. (2018). *Constructed Wetlands for industrial wastewater treatment.* John Wiley & Sons.

Stollenwerk, K. G. (2003). "Geochemical processes controlling transport of arsenic in groundwater: a review of adsorption." In *Arsenic in ground water*, (pp. 67-100). Springer, Boston, MA.

Singer, P. C. & Stumm, W. (1970). "Acid mine drainage: rate-determining step." *Science*, *167*, 1121–1123.

Stumm, W. & Morgan, J. J. (1996). "*Aquatic Chemistry, Chemical Equilibria in Natural Waters*." Third Edition, Wiley, New York, USA.

Suzuki, Y. & Banfield, J. F. (2004). "Resistance to, and accumulation of, uranium by bacteria from a uranium-contaminated site." *Geomicrobiology Journal*, *21*, 113-121.

Suzuki, Y. & Suko, T. (2006). "Geomicrobiological factors that control uranium mobility in the environment: Update on recent advances in the bioremediation of uranium-contaminated sites." *Journal of Mineralogical and Petrological Sciences*, *101*, 299-307.

Teuten, E. L., Rowland, S. J., Galloway, T. S. & Thompson, R. C. (2007). "Potential for plastics to transport hydrophobic contaminants." *Environmental Science & Technology*, *41*, 7759-7764.

Thompson, R. C., Moore, C. J., Vom Saal, F. S. & Swan, S. H. (2009). "Plastics, the environment and human health: current consensus and future trends." *Philosophical Transactions of the Royal Society B: Biological Sciences*, *364*, 2153-2166.

Tiner, R. W. (2016). *Wetland indicators: A guide to wetland formation, identification, delineation, classification, and mapping*. CRC press.

Tribedi, P. & Sil, A. K. (2013). "Low-density polyethylene degradation by *Pseudomonas sp*. AKS2 biofilm." *Environmental Science and Pollution Research*, *20*, 4146-4153.

Turner, A. & Holmes, L. A. (2015). "Adsorption of trace metals by microplastic pellets in fresh water." *Environmental Chemistry*, *12*, 600-610.

Turton, A. R. (2013). "Debunking persistent myths about AMD in the quest for a sustainable solution." *SAWEF Paradigm Shifter No.1. Johannesburg: South African Water, Energy and Food Forum (SAWEF).*

Turton, A. R. (2015a). "When gold mining ends – an environmental catastrophe for Johannesburg?" In: *New South Africa Review* (5). Johannesburg: Wits University Press.

Turton, A. R. (2015b). "Untying the Gordian Knot: unintended consequences of environmental and water policy for the gold mining industry in South Africa." *Water International*, *41*, 330-350.

Turton, A. R. (2015c). "*Concept Document for the Vertical Up- Flow Modular Constructed Wetland Water Treatment System (VUF-MCW-WTS) for Grey Water*" as defined in South African Patent Application No 2015/03442.

Turton, A. R. (2016). "Managing the unintended consequences of mining: acid mine drainage in Johannesburg." in: Eslamian, S. 2016. *Urban Water Reuse Handbook.* CRC Press, pp. 593-604.

Tutu, H., McCarthy, T. S. & Cukrowska, E. (2008). "The chemical characteristics of acid mine drainage with particular reference to sources, distribution and remediation: The Witwatersrand Basin, South Africa as a case study." *Applied Geochemistry*, *23*, 3666-3684.

United States Protection Agency (USPA). (2000). *Introduction to Phytoremediation.* U.S. Environmental Protection Agency, (February), pp. 1–72.

Van der Ent, A., Baker, A. J., Reeves, R. D., Pollard, A. J. & Schat, H. (2013). "Hyperaccumulators of metal and metalloid trace elements: facts and fiction." *Plant and Soil*, *362*, 319-334.

Van Oosten, M. J. & Maggio, A. (2015). "Functional biology of halophytes in the phytoremediation of heavy metal contaminated soils." *Environmental and Experimental Botany*, *111*, 135-146.

Verma, H. R., Singh, K. K. & Mankhand, T. R. (2016). "Dissolution and separation of brominated epoxy resin of waste printed circuit boards by using di-methyl formamide." *Journal of Cleaner Production*, *139*, 586-596.

Vollenweider, R. A. (1975). "Input-output models with special reference to the phosphorus loading concept in limnology." *Schweizerische Zeitschrift für Hydrologie Hydrobiologie*, *37*, 53-84.

Wang, H., Liang, H. & Gao, D. (2017). "Occurrence and distribution of phthalate esters (PAEs) in wetland sediments." *Journal of Forestry Research*, *28*, 1241-1248.

Wang, S. & Mulligan, C. N. (2006). "Natural attenuation processes for remediation of arsenic contaminated soils and groundwater." *Journal of Hazardous Materials*, *138*, 459-470.

Winde, F. (2006). "Inventory of sources and pathways in the catchment." In: Coetzee, H., Winde, F., and Wade, P.W. *An assessment of sources, pathways, mechanisms and risks of current and potential future pollution of water and sediments in gold-mining areas of the Wonderfonteinspruit catchment*. Water Research Commission, WRC Report No. 1214/1/06, pp 35-53.

Xu, X. R., Li, H. B. & Gu, J. D. (2005). "Biodegradation of an endocrine-disrupting chemical di-n-butyl phthalate ester by *Pseudomonas fluorescens* B-1." *International Biodeterioration & Biodegradation*, *55*, 9-15.

Yoon, M., Jeon, H. & Kim, M. (2012). "Biodegradation of polyethylene by a soil bacterium and AlkB cloned recombinant cell." *Journal of Bioremediation and Biodegradation*, *3*, 1-8.

Zedler, J. B. & Kercher, S. (2005). "Wetland resources: status, trends, ecosystem services, and restorability." *Annual Review of Environment and Resources*, *30*, 39-74.

In: Advances in Hydrogeochemistry Research
Editors: S. Saeid Eslamian and Faezeh Eslamian
ISBN: 978-1-53617-829-6

Chapter 3

IMPACT OF GROUNDWATER POLLUTION-QUALITY ON REGIONAL ECONOMICS: A CASE STUDY IN THE ARAB REGION

Souha El Khanji[1,*] and Saeid Eslamian[2,3]
[1]Lebanese International University, Lebanese University, Middle East Enlight Research, Beirut, Lebanon
[2]Department of Water Engineering, Collage of Agriculture, Isfahan University of Technology, Isfahan, Iran
[3]Center of Excellence in Risk Management and Natural Hazards, Isfahan University of Technology, Isfahan, Iran

ABSTRACT

Groundwater is a vital source of water in the Arab region which accounts for almost 80% of water resource in some Arab countries. In the Arab countries, water resources are the most vulnerable; its availability and quality are vital to serve the socio-economic development of the region. Population growth, industrial and agricultural activities, in addition to the arid climate and the pollution issue in groundwater aquifers caused overexploitation of groundwater. Once groundwater is contaminated, it is very difficult and costly to treat the issue and to regenerate safe groundwater. Therefore, groundwater pollution from non-point sources is a serious environmental issue that affects the economy due to irreversibility of the damage. This study explores the effect of groundwater exploitation and quality on the economic growth for 22 Arab countries. We use Nitrous oxides conceptually as an indicator of water quality; which is a result of denitrification of Nitrates, derived from the fact that Nitrates in groundwater cause health problems and add high costs in treating drinking water. Regression analysis illustrates the significant impact of groundwater withdrawal and the Nitrous Oxide, together with the

* Corresponding Author's Email: souha.el-khanji@bath.edu; souha.khanji@liu.edu.lb.

interaction factor of the two variables on economic growth of the Arab countries. In fact, these call for better future scenarios and policies for groundwater sustainability in the Arab region.

Keywords: groundwater quality, groundwater quantity, aquifers, nitrate contamination, nitrous oxide, per capita income, economic growth

1. INTRODUCTION

Humanity is facing important challenges in the 21st century; sustainable economic development, reserving natural resources, and sustaining the environment. The scarcity of water resources around the world is becoming a more prominent issue due to several factors including population growth, global warming, climate change, industrialization and urbanization that exert a further problem by adding additional stress on surface and groundwater (GW) resources. In addition to this stress is the diminishing quality of the water resources, which adds a further burden on this renewable resource, where the cost of providing safe water for drinking and for different economic sectors is increasing, and that adds another burden on the economic growth and development. El Khanji and Hudson (2016) conducted a panel data analysis for 175 countries and found that although water quantity affects the economic growth, water quality appeared to have a higher impact on the economic growth and development on the short and the long run.

Water, as a renewable resource, is subjected to pollution and it is a part of the environment, where studies find that environmental degradation due to industrial pollution can cost the economy nearly 5% of the GDP (Pearce and Warford, 1993). Furthermore, Yonggua *et al.*, (2001) explored this effect and concluded that industrial pollution affects negatively on different economic sectors (56% is the cost on the GDP from the agricultural sector alone) and costs the GDP for their area of study (Chongqing) about 1.2%. Another study by Reddy and Behara (2016) who assessed the economic costs of industrial water pollution on agricultural production, human health and livestock, estimated those costs finding that the impact of industrial pollution on the rural communities is quite substantial in monetary terms.

As a result, water pollution for all water resources poses a serious international challenge due to its impact on a large number of economic activities and developments, where water withdrawal for one sector has a spillover effect and enhances water withdrawal for another economic sector (El Khanji, 2017). Thus, provision for safe water and sanitation became a major concern and a challenge for governments and for international societies around the world. Nowadays, the most stressing challenge is determining the balance between water demand and supply. Water quality (either for surface or for GW), contributes highly in the deterioration of resources, which may subject the human survival for a hazard. It is one of the most common serious problems in arid areas.

Any country is considered as water stressed when annual water supply drops below 1,700 m3/ person (WWAP, 2012). The MENA region embodies 22 Arab countries - in addition to Turkey and Iran - suffers from a low level of precipitations (around 600mm/year), and it is considered as an arid area and as one of the driest regions in the world. Water resources in this region are considered as scarce and are highly affected by the climate change (IPCC, 2007). Figure 1 illustrates renewable water resources per capita for different regions in the world, and this can show that the renewable annual freshwater resources per capita in the Arab countries is about 290.78 m3/ year/ inhabitant, which is the lowest among other regions. The level of water stress for the Arab countries, illustrated in Figure 2, is estimated to be 9219.52 percentage points, which is very high compared to other regions. Khater (2003) projected the availability of almost 1.4% of the world's renewable freshwater for 6% of the world's population living in the Arab region. Khatar also highlighted the effect of water scarcity that plays a role in population density and distribution.

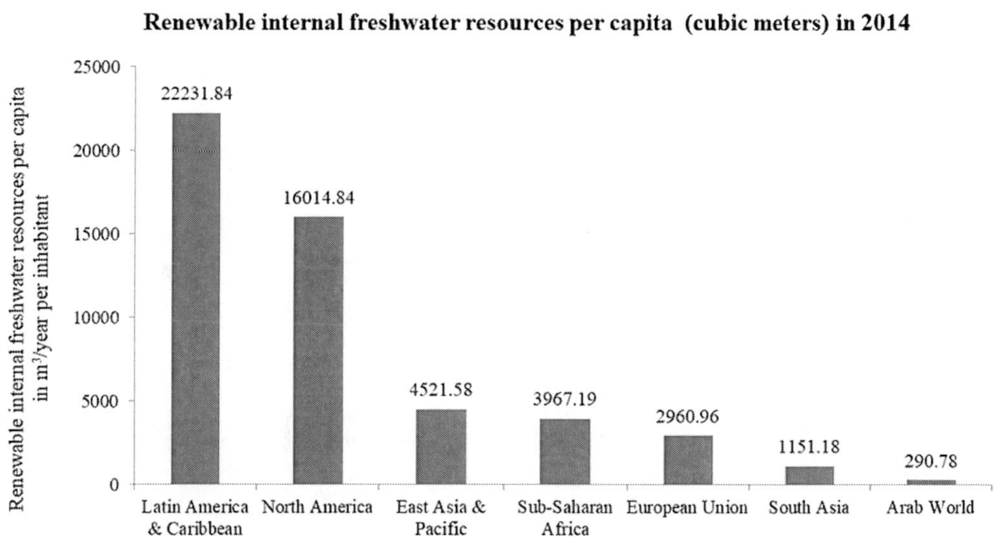

Source: World Bank Development Indicators.[1]

Figure 1. Renewable water resources per capita in cubic meters per region.

[1] Renewable internal freshwater resources is defined as internal river flows and groundwater from rainfall in the country. Renewable internal freshwater resources per capita are calculated using the World Bank's population estimates. The unit of calculation is m^3/year per inhabitant. (Source: World Bank Development Indicators, last accessed in June 2019).

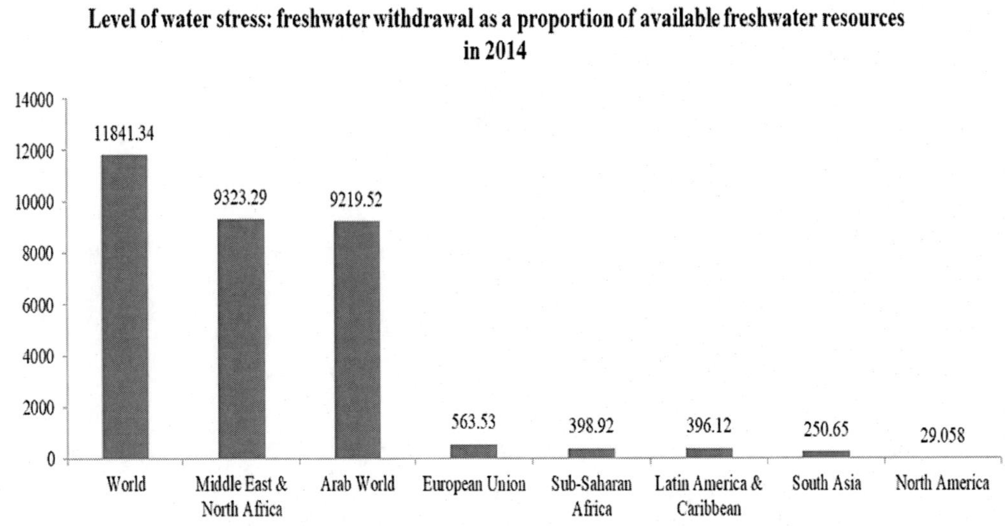

Source: World Bank Development Indicators.

Figure 2. Level of water stress in different regions[2] in 2014.

The main issue here is how to give the GW quality management the same importance as the surface water. So far, most of the countries manage them separately. In addition, it is about how to maintain a sustainable development in this region while it is burdened with the population growth accompanied with the projected economic growth. In GCC[3] countries, particularly, population growth is accompanied with accelerated economic growth demanding high quantity of water, which adds more stress on water resources. Water resources in the Arab countries are not just scarce but are also expensive to extract and supply. Resources are managed in a centralized mode where the management has entirely been focusing on the supply side (Brooks, 1999). There is a vital need for a database for GW that would help in monitoring the quality and the quantity of GW in the Arab region.

We, in this study, aim to explore the effect of groundwater exploitation and the quality on the economic growth for the 22 Arab countries using regression analysis.

2. Literature Review

Water security is important to get the required economic development in the region. The Sustainable Development Goals, SDG 6 moved beyond calling for safe access to water and sanitation, SDG 6 calls for several elements that strengthen water security by managing

[2] The indicator is computed by dividing total water withdrawal by total actual renewable water resources minus environmental requirements and expressed in percentage points (World Bank, last accessed June 2019).

[3] Gulf Cooperation Council.

sources of water pollution, managing water quality, reusing of treated water and integrated water resource management.

GW embodies 30% of fresh water in the world. Generally, the resources of groundwater are snow and rain and it is mostly allocated in the underground aquifers. GW is the main fresh water supply for most of the European countries (Water Economy Prospects for 1990 and 2000, 1982). It is also a main source for municipal water supply in the United States (Survey of operating and financial characteristics, 1977). The importance of the amount of fresh water and the quality of water, currently, is directly connected with the human welfare and human survival (Gleick, 2012).

Overexploitation of underground water resources is the result of a random drilling of wells without planning. GW needs further social sustainable management to avoid any conflict for the end users of GW resources, in addition to the environmental management to avoid the overexploitation and the depletion of the nonrenewable water aquifers (Jagannathan *et al.*, 2009). Environmental management does not depend only on quantity management or quota allocation of the resources, it depends on managing the quality of the resources as well, where quality of the freshwater resources proved to be the highest impact on the economy (El Khanji and Hudson, 2016). Although GW is less subjected to a direct source of pollution than the surface water, there is a difficulty in monitoring and eliminating the pollution. In the near future, the most dominant issue for the international society is the quality and the quantity of water resources. In fact, accessing GW resources to manage and monitor quality is not an easy task; that is, many geographic and natural obstacles prevent societies from carrying out close monitoring for this natural resource.

GW is an important source of water in Arab region, which accounts for substantial amount (80%) of water resource in some Arab countries where water resources are the most vulnerable. Water scarcity in this region, population growth, agricultural activities, the arid climate, the exploitation of non-renewable GW resources in addition to the pollution issue in GW aquifers trigger the need for better sustainable groundwater management. Due to the high evaporation and the fluctuations or the infrequent rainfall, the volume of the renewable aquifers and groundwater resources is very limited in this region. In general, both the renewable and the non-renewable groundwater resources are heavily subjected to exploitation (FAO and AQUASTAT, 2012).

2.1. Nature and Sources of Groundwater in the Arab World

The Arab countries considered as arid and semi-arid areas, scarce water resources exert a pressure on sustainable development. Khater (2003) has explained the nature of the aquifers in the Arab world. In this region, there are two kinds of aquifers; sandy or calcareous facies. Most of the aquifers are nonrenewable and the renewable ones rely on the amount of precipitation in the area; most of the aquifers are shared between two or more countries in the region. There are two types of GW in Jordan, the renewable and the nonrenewable fossil groundwater aquifers that receive a very small amount of recharge; generally, fossil aquifers are non-renewable (El-Naqa and Al-Shayeb, 2009). Groundwater resources in Jordan are scarce and found mostly in the Yarmouk, Amman-Zarqa and the Dead Sea basins.

Djibouti is characterized by two kinds of aquifers, the volcanic rock aquifers and the unconsolidated sedimentary aquifers (Jalludin and Razack, 2004). Groundwater is usually recharged directly from the leakage of infrequent rainfall or indirectly from passing river flows in wadis. GW in Djibouti is affected by dry seasons during which GW levels decline.

The major sources of GW in Lebanon are:

> "Kesrouan Limestone Formation aquifer (Jurassic age), Sannine-Maameltein Limestone Formation aquifer (two formations forming a single aquifer of Cenomanian and Turonian age respectively), the Eocene limestone aquifer (Eocene), the Neogene and Quaternary deposits of the Bekaa reserves, in addition to the following formations that are of importance only in the Mediterranean Province which are the Jebel Terbol Limestone Formation aquifer (Miocene age), the Abeih and Mdeirej (Aptian and Albian age) groundwater reserves, the Chouf Sandstone groundwater reserves, and the recent deposits groundwater reserves." (from MoE/LEDO/ECODIT report, 2001, p. 113)

There are two kinds of groundwater resources in Libya: the shallow aquifers that recharge from rainfall and surface runoff and are located in the Northern underground basins such as Jiffarah Plain system, Al Jabal Al Akhdar system, and Al Hamada basin. The second type is the non-renewable deep aquifers (fossil water), which are located in most of the Southern half of Libya such as Murzuq basin, Kufra basin, and Sarir basin (General Water Authority, 2006).

Egypt, on the other hand, is characterized by three types of aquifers: fractured-bedrock, weathered-mantle and fluvial aquifers, and the highest yielding aquifers are the fractured bedrock. The major groundwater aquifer systems in Egypt are many. First, the Nile aquifer that is recharged by infiltration of excess irrigation water from the Nile river, and it provides about 85 percent of the total groundwater (AfDB, 2015). Second, Nubian sandstone aquifer is a fossil groundwater shared with Libya, Chad and Sudan. Third, the Fissured carbonate aquifer that is spread over more than half of Egypt. Fourth, the Moghra aquifer is recharged both by rainfall and inflows from the Nile. Fifth, the Coastal aquifer: on northern and western coasts, which is recharged by rainfall. Finally, the Hardrock aquifer which is located in the Eastern deserts and the Southern Sinai (FAO, 2016).

Mauritania lies in Sahara Desert with limited rainfall and water resources, most of which are GW. Somalia has variable kinds of rocks that store the GW. The rocks can be crystalline basement rocks that store small quantities of water or the alluvial sediments with different depths and different volumes of water. GW yield is less than 0.5 $l/s/km^2$, and that makes it enormously poor in groundwater reserves. Aquifers that are estimated to serve more development and groundwater utilization are: Major Togga'a alluviums (inter-granular aquifer of Quaternary age), Jurassic lime-stones (karstic aquifer), Auradu lime-stones (karstic aquifer of Lower Eocene age), and Karkar lime-stones (karstic aquifer of Upper Eocene) from (FAO-SWALIM, 2012).

2.2. Quantity Deterioration of Groundwater in the Arab Region

Most of the population and municipalities in the MENA region depend on aquifers and groundwater for drinking water and for irrigation. In the Arab region, especially, the shortages of water resources are a major concern, and it is a big obstacle for economic growth and socio-economic development, given the fact that water resources in this region are considered as the lowest available resources per inhabitant in the world (Osman-Elasha, 2010). This issue intensified due to the rapid population growth in this area, the per capita amount of freshwater supply would decline by 85%, from 4,462 m3 in 1950 to 682 m3 in 2025 (Khater, 2003). In rural areas, the supply of fresh water depends on pumping groundwater to meet the demand and the agricultural purposes. Moreover, technological advancements facilitated accessing deep groundwater, and that led to excessive exploitation of groundwater resources. Figure 3 illustrates the very low value of renewable groundwater per capita for most of the Arab countries. Yemen is one of the countries, where the rate of pumping of groundwater exceeds the natural recharge of underground aquifers (World Bank, 2007). In meeting the EU standards and to participate in international farms supply, farmers in Egypt ditch the highly polluted surface water that led to the depletion of most of the groundwater (Jagannathan *et al.*, 2009). The quantity and the quality of GW in the UAE deteriorated due to over pumping and scarcity in precipitation. Agricultural activities in the UAE extract more than 70% of groundwater resources (Murad, 2010). In Iraq, the country of Tigris and Euphrates rivers, groundwater usage for agricultural is increasing and the number of wells has increased (Abdullah and Umrikar, 2014).

Saudi Arabia has very limited water resources with an average rainfall of less than 100 mm/year. This country depends mostly on groundwater (Abdel-Sater *et al.,* 2017; Hereher *et al.*, 2012) and surface water resources and of course on the desalinated sea water in addition to the treated waste water (Abdel-Sater *et al.*, 2017). About 80% of the extracted usable water is obtained from the groundwater (Aly *et al.*, 2013). Groundwater resources of Saudi Arabia are depleting very fast. Most of the withdrawn water comes from deep fossil type aquifers. These resources, according to estimates, may not last more than 25 years. In Bahrain, continuous development caused depletion of its groundwater resources, and excessive pumping deteriorated GW resources and caused a lowering in aquifer's potentiometric levels during the period from 1925 to 1991 (Zubari *et al.*, 1993).

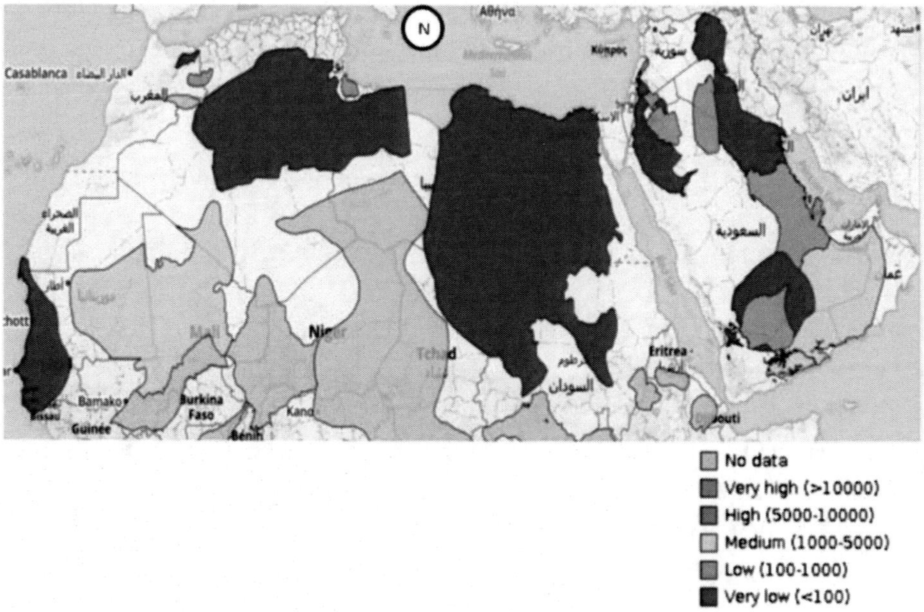

Source: Global Groundwater Information System (GGIS) from International Groundwater Resources Assessment Centre scale:1:32M.

Figure 3. Renewable groundwater resources per capita (m3/yr/capita).

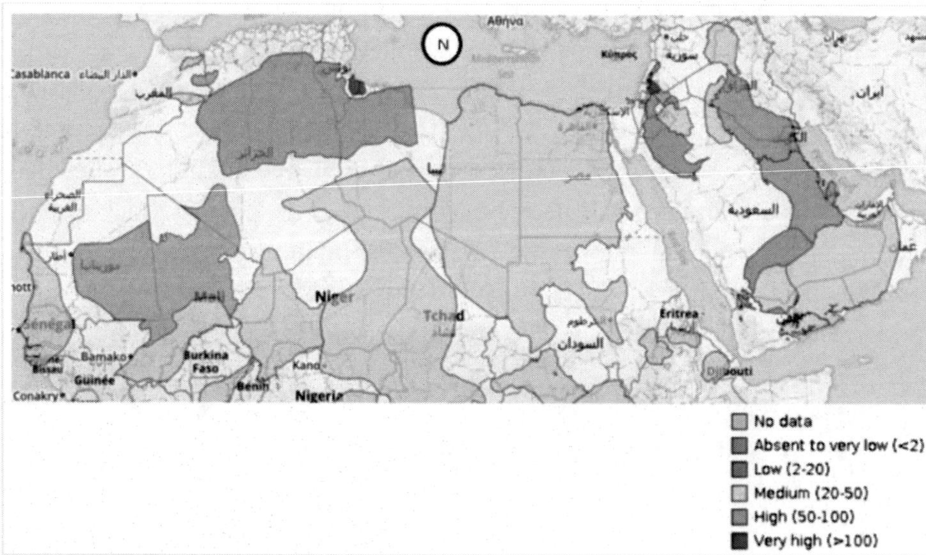

Source: Global Groundwater Information System (GGIS) from International Groundwater Resources Assessment Centre, scale:1:32M.

Figure 4. Groundwater depletion.

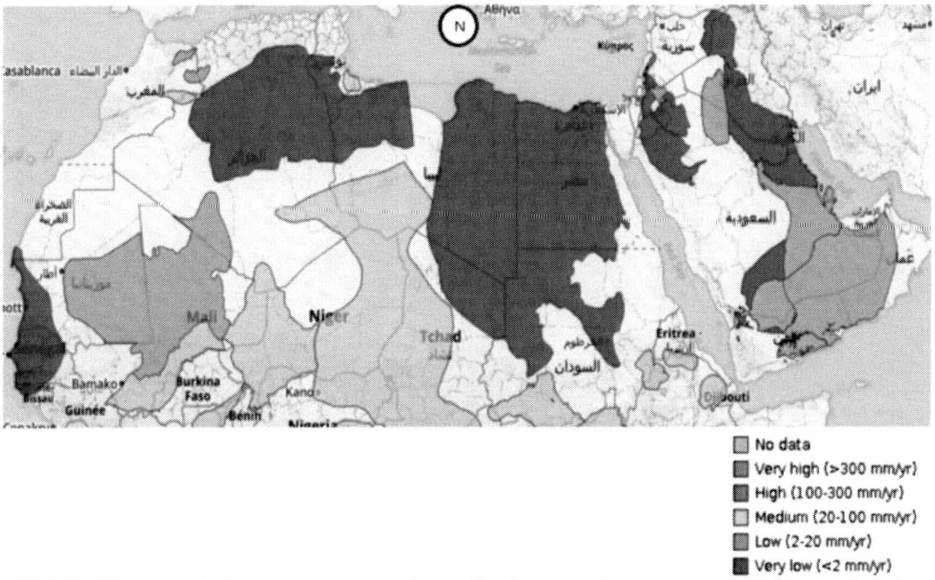

Source: Global Groundwater Information System (GGIS) from International Groundwater Resources Assessment Centre, scale:1:32M.

Figure 5. Groundwater recharge depth (mm/yr).

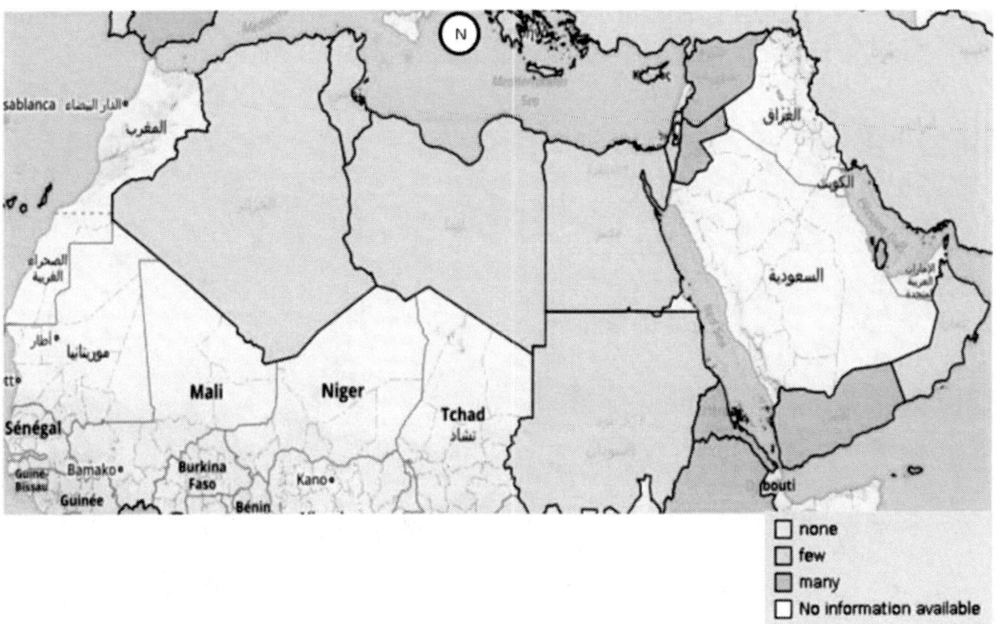

Source: Global Groundwater Information System (GGIS) from International Groundwater Resources Assessment Centre, scale:1:32M.

Figure 6. Reported zones with groundwater depletion.

GW is the main source of water in Jordan, which is considered as one of the tenth poorest countries in water resources in the world (Salameh, 1996). Lebanon, a country that is well endowed with GW and surface water resources, due to the lack of water management and to meet the agricultural and the industrial demand, the extraction rate of GW exceeded the natural recharge rate creating serious salt water intrusion problems requiring urgent actions to prevent the over-pumping in the coastal areas (Akkr, 1992). In Libya, about 98% of water consumption is obtained from GW (Shahin, 2003), and in Djibouti people depend heavily on GW for drinking water and for irrigation purposes (Awaleh *et al.,* 2017). Figures 4, 5 and 6 explain more about the depletion, the recharge depth and the extent of depletion in the Arab world.

The international society and research centers in addition to the NASA have done a lot in developing measures for groundwater globally, and the reason behind these assessments is to get a full figure of how to meet the increasing water demand (Shiklomanov and Penkova, 2003; Alley, 2006). Although most of the recent assessments revealed a huge depletion of groundwater resources (Voss *et al.*, 2013; Castle *et al.*, 2014), groundwater stress worsens between 1990 and 2010. According to an assessment undertaken by NASA's Gravity Recovery and Climate Experiment (GRACE), satellites discovered the loss of 117 million-acre feet, which is equivalent to 144 km^3 in the Tigris and Euphrates river basins. This assessment covers a period from January 2003 through December 2009, so researchers concluded that about 60% of the reason of this loss is due to uncontrolled groundwater pumping (Voss *et al.*, 2013). Figure 7 illustrates the level of deterioration for underground water in Tigris and Euphrates river basins. The red color stands for drier conditions.

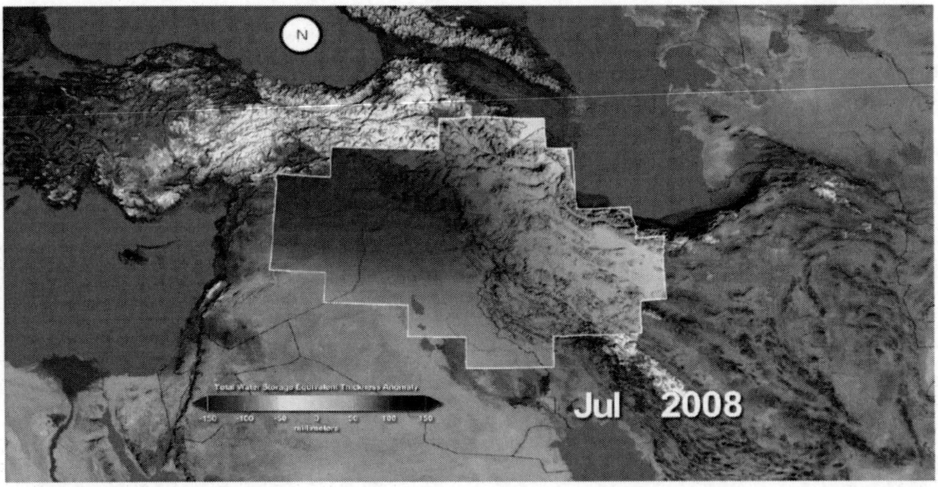

Source: NASA/UC Irvine/NCAR.

Figure 7. Variations in total water storage from normal, in millimeters, in the Tigris and Euphrates river basin[4].

[4] Retrieved from: https://www.nasa.gov/mission_pages/Grace/news/grace20130212.html.

Adding to these issues, urbanization, centralization and population density, the unorganized agricultural activity and the expansion of agriculture to meet the demand, have all exerted a pressure on water resources in the region. The high annual population growth adds more stress on water demand, and most importantly the economic development, which adds more stress on the limited groundwater resources. Generally, the overexploitation of GW is projected to increase substantially, given the fact of weather fluctuations and the need to combat the increasing food insecurity.

2.3. Quality Deterioration of Groundwater in the Arab Region

Health risks emerge by contaminated fresh water resources. In the region, many countries have no strict policies with the use of fertilizers and pesticides that contaminate groundwater resources. In addition to contamination resulted from untreated municipal wastewater and the dumping of industrial effluents into natural rivers and lakes before treating the wastewater, all these activities affect human health and consequently the governmental public spending on health sector, which add a burden on the economy. Some Arab countries were aware of this impact and there is a wide literature discussing the quality of groundwater resources (see Table 1). Table 1 contains a literature review explaining the quality of GW in the Arab countries. In Saudi Arabia and Libya, groundwater plays a big role in supplying water especially in rural areas. In Saudi Arabia, the groundwater is considered as an important source of drinking water (Alabdula'aly *et al.*, 2010). A study by Al-Naim (2014) explored the effect of over pumping on groundwater salinity in Ha'il region in Saudi Arabia, where quality deterioration is resulted from excess pumping. Previous studies illustrated how groundwater salinity increases with increased exploitation or overuse (O'Hara, 1997).

Groundwater quality depends on the aquifers' boundary conditions and on a geological formation (Aljawad *et al.*, 2018). Groundwater is subjected to contamination and pollution due to different chemicals, oils, gasoline, agricultural chemicals and other man-made products that can leak from the soil to the underground and consequently to the aquifers, where the groundwater is located. The quality of groundwater depends highly on soluble chemical compositions. Therefore, the quality of GW depends on the contamination from the atmosphere. El Alfy *et al.* (2015) conducted a study for assessing groundwater pollution in Saudi Arabia and described how the pollution risk for this resource is high; it also illustrated how urbanization impacts negatively on groundwater quality due to contamination from both wastewater and the un-rehabilitated sewage system.

Table 1. Literature review for research on the quality of groundwater in the Arab countries

Country	Total renewable water resources	Total renewable groundwater	Quality of groundwater in Literature
Algeria	292 m^3/inhab/yr (AQUASTAT, 2014)	1.517 km^3/yr	• A study carried by Hamlat and Guidoum (2018) studying groundwater quality of 12 aquifers situated in Northwestern Algeria using water quality index and multivariate statistics methods to find poor and medium quality of GW for the majority of the studied aquifers. Water scarcity conditions affect the water of some aquifers that has degraded and will continue to degrade with the planned development in the study area. In addition, salt water affected some freshwater aquifers, which need further attention in terms of monitoring and for sustainable management. • Rouabhia *et al.* (2008) found salinity and Nitrate in GW of the aquifers located near the city of Tebessa in eastern Algeria apparently due to the use of nitrate fertilizers in the farms. Also, some pollution is caused by the effluent of the industrial activities.
Bahrain	160 m^3/inhab/yr (AQUASTAT, 2005)	0.112 km^3/yr	• Zubari *et al.* (1994) demonstrated the decline in GW quality beginning from 1960 and registered an intrusion occurred by the seawater into the aquifers. • An assessment carried by (Mahmood, 1993) found GW to be saline and unsuitable for human consumption and for irrigation; deeper aquifers consist of high values of the hydrogen sulphide because the deeper geological formations contain oil rich in sulphur.
Comoros	1519 m^3/inhab/yr (AQUASTAT, 2004)	1km^3/yr	• Comte *et al.* (2016) investigated the management of costal groundwater in Comoros Archipelago and in other coastal areas; they found high salinity in water wells in the La Grille and Karthala massifs (generally above 2 g L^{-1}). • Quality of water in Comoros is very poor. In Grande Comore, groundwater is polluted from infiltration of land-based sewage and leachate, and salinized from invasion and leakage of salt-water. Generally, groundwater resources are polluted and suffer from salinity; rivers in Anjouan and Mohéli are also subject to pollution and sedimentation (UNEP, 2010, project document).
Djibouti	421 m^3/inhab/yr (AQUASTAT, 2004)	0.015 km^3/yr	• Salinity of GW is very high in Djibouti and water quality is a major concern, where the limited resources are mostly obtained from the available GW resources, aquifers are affected by seawater intrusion or fossil saltwater contamination (Boschetti *et al.*, 2018). • Rapid urbanization and the lack of adequate sanitation cause degradation of GW quality in urban areas. Drinking water contains toxic elements (As and Se) with a high amount in aquifer affected by seawater intrusion. (Ahmed *et al.*, 2017).

Country	Total renewable water resources	Total renewable groundwater	Quality of groundwater in Literature
			• Another study found Nitrate concentrations are high due to natural factors or due to contamination by anthropological activities (Awaleh *et al.*, 2017).
Egypt	699 m^3/inhab/yr (AQUASTAT, 2014)	1.5 km^3/yr	• GW is the second source of water in Egypt after the Nile River. Masoud (2014) found high salinity in 83% of GW samples which is not suitable for irrigation, very high nitrate concentrations that exceeds the allowed value (50 mg/l) due to hydrologic discharges and related to wastewater leakage. • GW in West of Sohag is the second source for irrigation. 44% suffer from high salinity, and most of the underground samples are characterized by high hardness that makes it not suitable for domestic and industrial use (Ismail & El Rawy, 2018). While in Assiut, the TDS for GW are less than 3000 mg/l; hence, it is suitable for irrigation (El Tahlawi *et al.*, 2014).
Iraq	2625 m3/inhab/yr (AQUASTAT, 2005)	3.28 km^3/yr	• Al-Mussawi (2014) evaluated the groundwater quality in Umm Er Radhuma aquifer (located in Iraqi Western Desert area between the Euphrates River and the Saudi Arabian borders) using irrigation water quality index (IWQI) to find a high restriction water quality cover 46.8% (about 28136 km^2) of total study area. • Dawood (2018) used GIS spatial distribution approach for all the parameters to analyze GW quality in Basrah finding retardation in quality due to environmental pollution.
Jordan	161 m3/inhab/yr (AQUASTAT, 2005)	0.54 km^3/yr	• Ta'any *et al.* (2007) evaluated quality of groundwater resources of the Yarmouk Basin where it is found good for domestic, industrial and agricultural purposes. • Al-Khashman and Jaradat (2013) assessed quality of GW in Ma'an area to find it is not entirely fit for direct drinking but can be used for irrigation.
Kuwait	7.4 m^3/inhab/yr (AQUASTAT, 2005)	0.02 km^3/yr	• There are no laws that limit the quantity, so farmers can pump from groundwater due to political obstacles (AlAli, 2008) that led to deterioration of both quality and quantity of GW. • Nearly 8% of the domestic water supply and most of irrigation needs ere dependent on GW. GW is subject to overexploitation and to pollution due to oil refineries and oil-related plants (Al-Senafy *et al.*, 2016).
Lebanon	1259 m^3/inhab/yr (AQUASTAT, 2005)	3.2 km^3/yr	• GW is contaminated from pesticides and nitrates in addition to contamination from leachate leakage due to industrial activities. Extraction rate of GW exceeded the natural recharge rate, creating serious salt water intrusion problems (Akkr, 1992). • El Moujabber *et al.* (2006) studied Seawater intrusion in the Choueifat-Rmeyle region to find that the regions of Choueifat, Jiye and Rmeyle are subject to seawater intrusion that increases the salinity problems. This problem worsens with excessive pumping. The salinity level is inadequate for the cultivation of many crops and consequently reduces the agro diversity in the region.

Table 1. (Continued)

Country	Total renewable water resources	Total renewable groundwater	Quality of groundwater in Literature
Libya	112 m^3/inhab/yr (AQUASTAT, 2015)	0.6 km^3/yr	• Abd El-Aziz (2017) assessed the quality of GW in North- Western area of Libya by examining its physio-chemical characteristics finding that most GW samples were not suitable for drinking and household uses due to their high levels of most cations and anions, total hardness, EC and TDS. Moreover, most of the GW is unsuitable for irrigation due to its high salt content.
Mauritania	3826 m^3/inhab/yr (AQUASTAT, 2004)	0.3 km3/yr,	• Sewage systems frequently contaminate groundwater supply, where GW is found to be affected by pollution and salt intrusion, many deep-water samples, finding that although water quality was acceptable in many samples, several samples had very high nitrate values (80–200 mg/L) and salinity is as up as 1 g/L (Ebih *et al.*, 2011). • Mapping of GW salinity by Friedel and Finn (2008) revealed that most of Mauritania consists of fresh water and intrusion of salt water occurs along the Atlantic Coast.
Morocco	879 m3/inhab/yr (AQUASTAT, 2013)	10 km3/yr	• Kherrati *et al.* (2015) run physical, chemical, metal and bacteriological analysis for quality of GW used as drinking water in the city of Kenitra and the results were good. However, high concentrations of nitrates (over 210.8 mg/L) and other metals (185 μg/l for lead; 58.98 for nickel; 187.3 μg/L for iron; 2204 μg/L for zinc) in some samples are worrying and can cause serious health issues. • Most of the wells located in the Essaouira Basin particularly in Kourimat perimeter is contaminated with nitrates, concentrations over 400mgl^{-1}. GW is affected badly by agricultural fertilizers and by human and livestock wastes (Laftouhi *et al.*, 2003).
Oman	545 m3/inhab/yr (AQUASTA, 2005)	1.3 km3/yr	• Al-Kalbani *et al.* (2017) assessed quality of 11 drinking water resources in Jabal Al Akhdar to find the water suitable for drinking but also found high nitrate concentrations in some groundwater wells. • fertilizers and pesticides are considered to be a serious hazard to GW quality where, as in most of the Sultanate, groundwater is free of charge (AQUASTAT-Water Report 34, 2009) and that causes nitrates pollution for GW.
Qatar	71 m^3/inhab/yr (AQUASTAT, 2005)	0.058 km^3/yr	• Shamrukh *et al.* (2012) evaluated Aruma aquifer South-West of Qatar and found groundwater is brackish in quality with total dissolved solids ranges from 4000 to 6000 mg/l. In general, GW quality makes drinking water supply unsuitable without treatment and potentially suitable for agricultural activities.

Country	Total renewable water resources	Total renewable groundwater	Quality of groundwater in Literature
			• Kuiper *et al.* (2015) registered high levels of molybdenum in GW and it was brackish to saline, which needs precautions when used for irrigation and may affect soil structure and crop yield.
Saudi Arabia	98 m^3/inhab/yr (AQUASTAT, 2005)	2.2 km^3/yr	• Although in Saudi Arabia the quality of water is not a big concern, quality of both ground, surface water still exists where microbial, and chemical pollution creating problems in some areas of the kingdom (Al-Turki, 2010(chemical contamination can be due to usage of fertilizers, pesticides, and animal waste. • Alabdula'aly *et al.* (2010) conducted a study evaluating the concentration of nitrates in groundwater in Saudi Arabia to find that the nitrates level exceeded the acceptable amount and they recommend a close monitoring for agricultural activities.
Somalia	1377m^3/inhab/yr (AQASTAT, 2004)	3.3 km^3/yr	• GW quality is not suitable as drinking water. Based on the chemical analyses of 511 samples hardness, calcium, magnesium, sodium and potassium do not meet WHO-standards. In addition, salt exceeds 1 g/l, which is not suitable for human consumption (FAO-SWALIM, 2012), also GW pollution is a threat for humans and livestock in Somaliland and Puntland.
Sudan	996 m^3/inhab/yr (AQUASTAT, 2013)	3 km3/yr	• The main supply for water in Khartoum city is wells recharged by aquifers in the Nubian Sandstone, where the boreholes lie in residential areas and are subject to contamination by sewage disposal into the water table in the upper alluvial deposits. Pollution is evident by the presence of E. coli bacteria. (Abd Alraheem, 2000). • Mohammad *et al.* (2015) also tested the GW that supply the Khartoum city and found that almost 50% of drinking water contaminated with total coliform, thermos-tolerant coliform, and the E. coli.
Syria	882 m3/inhab/yr (AQUASTAT, 2005)	6.174km3/yr	• Abou Zakhem and Hafez (2015) assessed a sample of twenty-seven GW samples from the municipal drinking water supply in Damascus Oasis finding high value of Heavy Metal index (HPI) due to industrial, agricultural and urban activities. • Another study that uses different approaches to identify the sources of GW nitrate pollution finding almost 51.8% of the wells contaminated with high level of 50 mg/L for drinking water (Abou Zakhem and Hafez, 2015).
Tunisia	419.7 m^3/inhab/yr (AQUASTAT, 2013)	1.595 km^3/yr	• In rural areas in Tunisia, GW is always a source of drinking water, nitrates contamination became a major issue and concern recently. A study by Zghibi *et al.* (2013) identified nitrates as high as about 50 mg/l in the central region of Korba (North- East of Tunisia) due to the use of nitrogen fertilizers, domestic and industrial effluents.

Table 1. (Continued)

Country	Total renewable water resources	Total renewable groundwater	Quality of groundwater in Literature
			• Uncontrolled dumping sites (Henchir El Yahoudia landfill) leachate analysis shows high contamination with nitrogen, heavy metals and organic matter coincides with clay-richest layers. The leakage of these pollutants reaches groundwater between 0.5 and 5 m caused pollution with high rates of organic matter (BOD, COD and nitrogen) and heavy metals (Marzougui and Ben Mammou, 2006).
UAE	48.29 m^3/inhab/yr (AQUASTAT, 2005)	0.12 km^3/yr	• Murad *et al.* (2011) explored the GW composition and salinization of Al-Ain area and its surrounding desert fringes detecting intrusion of seawater, also affected by recharge water from the infiltration of rainfall in inter dune areas and sewage effluent water from the residential areas.
West Bank and Gaza	51 m^3/inhab/yr (AQUASTAT, 2005)	0.75 km^3/yr	• Agricultural activities and the excess use of fertilizers cause GW contamination. Nitrate level in GW is used as an indicator to pollution and in many areas has exceeded the permissible nitrate level (45 mg/l) as identified by the WHO. In addition, untreated wastewater causes contamination of GW as well (Isaac *et al.,* 1995). • Scarcity of water causes extensive pumping for groundwater from the coastal natural aquifer causing seawater intrusion. Only 4 percent of the 180 m3 abstracted per year meets drinking water quality standards. Most of the wells are high in chloride (World Bank, 2018).
Yemen	100 m^3/inhab/yr (AQUASTAT, 2005).	1.5 km^3/yr	• Al-Mikhlafi (2010) assessed the suitability of GW in volcanic aquifers in rural Yemen by examining water quality in one hundred thirty boreholes, finding that these boreholes contain fluoride, levels of nitrate, some heavy metals, total dissolved solids, and sulfates that could pose a health risk for users. Concentrations of NO^{-3} and Cl^- indicate sewage and/or agricultural runoff. High level of chemical constituents in GW indicated degradation.

Total renewable water resources and groundwater resources are collected from AQUASTAT website, country profile database, last accessed June 2019.

Most of the GCC countries suffer from deterioration in both the quantity and the quality of groundwater (Figures 3, 4, 5, and 6) particularly from salinity. Adding to this exploitation, the non-point source of pollution that occurs due to human activities. Consequently, the contamination and the pollution of groundwater, which is a vital source of drinking water in the Arab world, exert serious social, health and environmental problems, in addition to the pressing issue on the economy of the region. Sustainable management of this resourec is a neccessity, and this can occur by creating economic incentives, setting up the appropriate tariffs and increase taxes on the polluter by enforcing polluter pay principle.

3. Groundwater Pollution and Economic Growth

A vast literature and gravity of models found that environmental degradation is inevitable when economic development is taking place in any country (Munasinghe,1999 ; Shafik, 1994; Guo and Ma, 2008; Neumayer, 1998; Lieb, 2004). Environmental Kuznet curve explained the link between the economic growth and the environmental degradation. Most frequently, aquifers are found to be contaminated with Nitrates which contained in many fertilizers used for agricultural activities, and they cause high salinization in South Africa (Du Plessis, 2017). In arid and semi-arid regions, excess Nitrates in groundwater can be due to anthropic pollution in the urban areas or to the use of fertilizers in the agricultural areas.

3.1. Nitrates (NO_3) and Nitrous Oxide (N_2O) Contamination of Groundwater

Nitrates are considered to be the most common pollutants of groundwater (Eslamian and Lavaei, 2009). Yet, Nitrate contamination is not a regional problem, but it is an international one. This contamination occurs due to agricultural activities, industrial activities, sewage systems and untreated wastewater. Policies are applied in different regions of the world to control the nitrate level in water. In Europe, the Nitrates Directive (91/676/EEC) is the widely applied current policy.

In addition to the leakage from wastewater and sewages that affect groundwater negatively, as we previously mentioned, the over-use of Nitrogen fertilizers in agriculture to meet the demand of the increasing population proved their polluting effect on soil, water and air, and consequently, it led to the hazard effect on human health and the environment (Lord and Anthony, 2002; Schroeder *et al.*, 2004; Eslamian and Lavaei, 2009). Nitrate is mostly easily transferable through both water and soil since it is highly soluble in water; therefore, any excess nitrate from sewages, agricultural fertilizers can be transferred into underground aquifers and surface waters bodies (Wick *et al.*, 2012). Additionally, Crusberg and Eslamian (2017) explain how drought also facilitates wildfires and the resulted ash which is rich in phosphates and nitrates dissolved in waterways with rain that consequently contaminates GW. The levels of nitrate in water can exceed the maximum contamination level or MCL of 10 mg nitrate/L set by the U.S. Environmental Protection Agency; this nitrate contaminates groundwater in the same area.

Generally, Nitrates (NO_3) are considered to be one of the major pollutants of drinking water resources. Water contaminated with nitrates is the most common cause of blue baby syndrome[5], in addition to another health problems including Thyroid disease, type I childhood diabetes, and central nervous system. Some birth defects are associated with nitrate contaminated water ingestion, methemoglobinemia in infants, stomach cancer in adults (Knobeloch *et al.*, 2000; Hall *et al.*, 2001; Yang *et al.*, 1998; Ward *et al.*, 2018; Johnson and Kross, 1990). This nitrate contamination problem is common in societies that use wells for drinking water, where this groundwater resource is contaminated due to the use of fertilizers.

Wakida and Lerner (2005) debated that although Nitrate is correlated to agricultural sources of pollutant of GW, and it is expected to be at a high concentration in rural areas, explorations proved the barely difference in Nitrates concentration between rural and urban areas. Further exploration proved that Nitrates are also contaminating GW in urban aquifers due to "*wastewater disposal (on-site systems and leaky sewers), solid waste disposal (landfills and waste tips).*"

Excess nitrates in aquifers cause the emission of N_2O (Weymann *et al.*, 2008). It is a continuous cycle in the nature, which is initiated by human activities, nitrogen from chemical fertilizers causes organic changes to agricultural soils and consequently facilitates the N_2O emission through microbial nitrification and denitrification processes (Minamikawa *et al.*, 2013). The emission of nitrous oxide (N_2O) contributes not only to environmental deterioration and global increase in GHG, but it also decreases the quality of the GW (Hakeem et al., 2016).

3.2. Costs of Groundwater Pollution: How Groundwater Pollution Affects the Environment and Consequently the Economy

It is always an advantage to express the environmental pollution in economic terms and always effective to determine the effect of environmental deterioration on the income of the economy. Generally, it is a repetitive procedure, to increase production of both agricultural and industrial, so the pollution (Nitrogen, nitrates) increases, then soil and GW become contaminated, and nitrous oxide emission evolved. Therefore, GW contamination exerts a threat to the agricultural production, human health and to the amount of supply for the needed water for different economic sectors. In the USA, the standard amount of nitrate in water is 10 mg/L according to the federal health National Primary Drinking Water Standard. Groundwater quality also affects surface water quality that gains substantial volume of water from groundwater (Nathan and Evans, 2011).

In the past, no treatment or barely a slight treatment was required for groundwater around the globe. Industrialization and urbanization in addition to different human

[5] It is caused by lack of oxygen in infants, this syndrome is more common to occur in rural areas, when water is contaminated with nitrates, in the body the nitrates changes into nitrites that couples with hemoglobin and prevent the hemoglobin from carrying oxygen to the brain.

behaviors have affected this trend negatively; some areas may become polluted to the extent that treating them is very expensive. Several areas around the world have installed treatments at their groundwater sources, which influence the operating costs and the charges of water will be raised for the final consumer.

The restoration of GW quality after the deterioration of this quality is expensive with low success rates (Al-Senafy *et al.*, 2016). Groundwater is affecting the economy in different ways, and aquifers usually determine the economic values of the extracted groundwater. The characteristic of the aquifer and the quality of the water depend on the type of the aquifer. The cost here is not only the cost of contamination and treatment; it is the cost of the side effects on the human health, on the irrigated lands and on the ecosystem, as well as on the deterioration of the environment as a whole. Furthermore, the cost of prevention of pollution of groundwater is high as well. What is meant by water quality is to what extent the water is suitable for use. The cost of contamination by pesticides and industrial organics is high since it can spread the contamination (WHO/UNEP, 1991).

4. Analysis

We discussed in previous sections how overexploitation of groundwater is accompanied with and related to the deterioration of its quality (Zubari *et al.*, 1994). In the analysis, the aim is exploring whether we have an impact and to what extent and whether this impact is significant or not? This question motivates this empirical work.

The study uses panel data (cross-sectional time-series data) covering the period from 1980–2016 for 22 countries of the Arab World namely: Algeria, Bahrain, Comoros, Djibouti, Egypt, Iraq, Jordan, Kuwait, Lebanon, Libya, Mauritania, Morocco, Oman, Qatar, Saudi Arabia, Somalia, Sudan, Syria, Tunisia, UAE, West bank and Gaza and Yemen. The analysis and the empirical model explore whether there is an impact on growth or not, so a simple linear regression model is used that depicts the aim of the topic:

$$Log\ GDP_{it} = GW\ withdrawal_{it} + \beta_1 N2O\ emission_{it} + \beta_2 Food\ production\ index_{it} + \beta_3 Fertilizer\ consum_{it} + \beta_4 Arable\ land_{it} + \beta_5 Government\ effect_{it} + \varepsilon_{it}$$

Where i is for country, t is for time (year) and ε stands for the error term.

The economic growth is represented by GDP per capita; GW withdrawal is added to the model to test its impact on growth and to see its effect on the economy. To represent quality, N_2O emission is added as well as a conceptual indicator of the spillover effect of pollution and environmental degradation on groundwater, several reasons stand behind this choice:

First, previous research proved that the declining in groundwater levels is directly affecting and deteriorating the quality of water (AlAli, 2008). Also, the focus is on nitrate contamination in this chapter for several reasons, one of which is the enormous amount of studies that discussed the Nitrates pollution of groundwater and the cost of treatment for nitrate from groundwater.

Second, fertilized soils lead to denitrification and N_2O emission[6] which is facilitated by soil moisture (Smart and Bloom, 2001; Zhao *et al.,* 2019). Some studies found that nitrogen loss when nitrogen fertilizers are used is very high resulting in higher nitrous oxide emission and nitrate leaching; consequently, nitrate drained into water bodies to cause the decrease of water resources quality. Nagatake *et al.* (2018) found that the N_2O emission and the nitrate nitrogen (NO_3^-–N) leaching during the chemical fertilizers for grass renovation are higher than renovation without fertilizers. In this concern, scientists had rarely investigated the N_2O emission of nitrate- contaminated groundwater on the economy. Although all the recent studies found a decrease of the nitrous oxide emission factor from aquifers and agricultural drainage water (EF5-g) from 0.015 to 0.0025 (IPCC, 2006), the argument is that nitrates contamination is also resulted from industrial and human wastes. As mentioned before, excess Nitrates in GW are found in both the urban areas (anthropic pollution) with almost the same concentration as in the rural areas.

Third, concerning the lack of GW quality data, GW quality data for the Arab world is not available which calls for further action from governments to set up databases. Setting up databases would facilitate the monitoring of groundwater quality especially the fact that the water quality and the deterioration of freshwater resources in the region is a serious issue. Most of the groundwater quality studies collect their individual data and some researches calculate their water quality index, and this is at a micro level. For macro studies of groundwater resources quality, the lack of this kind of data is a restriction that needs to be resolved. Pollution is not a micro issue, it is rather a macro issue, so no borders can prevent pollution from migrating underground from one country to another or from one sea to another. Generally, the Nitrogen presence in the agro ecosystem has increased due to human activities[7]. Finally, the most important reason for using N_2O data is the availability of the data from a credible source, the World Bank Development Indicators.

Fertilizer consumption is added to the model for its direct connection with GW quality, throughout the previous sections, the effect of fertilizers on GW contamination is described extensively, for that adding this variable to the model is a necessity to see its effect on growth within the model of GW quality. Several facts are behind this choice, and fertilizers are used for improvement of agricultural yield to meet the increase of population. Moreover, McArthur & McCord (2017) documented a clear evidence that agronomic inputs, fertilizers, water and cereal yield are strongly linked and connected. Their analysis and results show an increase of half ton in primary yields leads to 14 -19

[6] Nitrous oxide (N_2O) is a key atmospheric greenhouse gas that contributes to global climatic change.
[7] Reducing fertilization can significantly reduce the nitrogen losses.

percent increase in GDP per capita and a 4.6 to 5.6 percentage point lower labor share in agriculture. Their findings suggest a structural change in the agricultural sector and its input in GDP and economic development, where fertilizers are found to play a main role in productivity. Therefore, in this study, fertilizer consumption is a conceptual indicator for both agricultural sector productivity and for the increase in yield to meet the increase demand due to the increase in population.

Groundwater withdrawal data is collected by the AQUASTAT. Food production index is added to this model because it is directly connected to water, and it is a reason for sustaining human survivor, food production and economic development, so it is a conceptual indicator for population growth, agricultural productivity and agro-labor productivity.

Table 2. Summary of data

Variable	Definition and source
Government effectiveness	Government Effectiveness captures perceptions of the quality of public services, the quality of the civil service and the degree of its independence from political pressures, the quality of policy formulation and implementation, and the credibility of the government's commitment to such policies. Score from -2.5 to 2.5. Source: Worldwide Governance Indicators
Food production index (2004-2006 = 100)	Food production index covers food crops that are considered edible and that contain nutrients. Coffee and tea are excluded because, although edible, they have no nutritive value. Source: World Bank Development Indicators
GDP per capita (constant 2010 USD	Source: World Bank Development Indicators
Fresh groundwater withdrawal	Annual gross amount of water extracted from aquifers. 10^9 m^3/yr. Source: AQUASTAT database
Arable land (% of land area)	Arable land includes land defined by the FAO as land under temporary crops. Source: World Bank Development Indicators.
Fertilizer consumption (% of fertilizer production)	Fertilizer consumption measures the quantity of plant nutrients used per unit of arable land. Fertilizer products cover nitrogenous, potash, and phosphate fertilizers (including ground rock phosphate). Source: World Bank Development Indicators.
Nitrous oxide emissions (thousand metric tons of CO2 equivalent)	N_2O emissions are emissions from agricultural biomass burning, industrial activities, and livestock management. Source: World Bank Development Indicators.
Fresh groundwater withdrawal	Primary and secondary in 10^9 m^3/year. Source: AQUASTAT database

Arable land area is added for its direct connection with the quantity of GW withdrawal. The World Bank (2004c) warns that the extensive irrigation due to expansion in agricultural production in GCC countries caused an exploitation and a non-reversible damage for the nonrenewable groundwater aquifers, adding this variable is also an indicator for agricultural input in the economy and the withdrawal of GW.

Table 3. Descriptive statistics

Variable	Obs	Mean	Std. Dev.	Min	Max
Gdp pc (con. 2010 USD)	618	12284.38	18648.29	707.7524	113682
Government effectiveness	396	-0.49315	0.82182	-2.44588	1.509872
Nitrous oxide emissions	713	5201.994	11484.07	39.26646	122754.7
Food production index	795	85.72597	31.74324	7.05	203.34
Fertilizer consumption	207	209.7356	846.182	0	6996.209
Arable land	777	7.650643	9.882964	0.043141	34.92746
Groundwater withdrawal	88	3.098768	4.583304	0.018	21.54

Government effectiveness represents the quality of the public sector provision by the governments. It also represents the level that governments can affect environmental and natural resource reservations, for that it is added to the model. The data of all variables of interest, definition of data and sources are summarized in Table 2, while the descriptive statistics for the variables are included in Table 3.

5. Results and Discussions

The goal here is exploring the impact of GW quality on the economic growth not the effect of the economic growth on quality; the model is a repetition for static setup to exogenous regressors in analyzing the impact of water quality on the economic growth. Table 4 illustrates the results of OLS regression analysis. Column 1 presents the impact of both GW quantity and GW quality on GDP per capita within the model. In column 2, the interaction term is introduced to the model, which is calculated by using both the GW quantity and the N_2O emission. Column 3 is to explore the impact of water withdrawal in quadratic form on the economy. Finally, Column 4 contains both the GW quality and quantity in quadratic form with the interaction term. The results indicate the following:

Government effectiveness is significant at 1% significant level in the four models, which is not surprising, and good governance facilitates and adopt the development of policies that enhance the economic growth and performance. From the results, the 1% increase in government effectiveness can be generated from 6.7 to 16% increase in the economic growth.

Groundwater withdrawal is positive and highly significant in the four models; a 1 unit increase in GW withdrawal increases the GDP between 2.2% and 6.7%. From the results of the linear and the quadratic forms in Columns 3 and 4, there is an evidence of the presence of an inverted U curve in the relationship between GW withdrawal and the GDP per capita, and this is explained as growth increases with GW withdrawal until the withdrawal reaches the maximum, where the deterioration of the resources is very high, then the economy cannot grow anymore; therefore, it starts to decrease.

Nitrous oxide is significant and negative, but once we add it in a quadratic form, it is significant in its quadratic form. From Column 1, a 1 unit increase in GW pollution affects the GDP negatively by 7% and by 12% (Column 3), and this is explained by increase in the treatment costs that affect negatively and add costs to the production and the usage of the GW. Costs were previously explained as costs of contamination and treatment, in addition to the costs generated by the side effects on the human health and the deterioration of the ecosystem.

Results show that the *interaction term* is negative and significant. When it is added in the model, the N_2O turns positive, which explains a lot here, and it is an indication to what was discussed before that GW exploitation leads to GW deterioration in quality, which is illustrated with the model that includes the interaction term. From columns 2 and 4, it is clear that the interaction term is highly significant at 1% significant level. The negative sign means that GW withdrawal and N_2O emission are interrelated in their effect on growth, which is agreeable with most of the literature and the research findings indicating the negative influence of the groundwater withdrawal with the consequence of pollution, the nitrous oxide, on the GDP per capita. The sign of the interaction means the more GW extracted, the more N_2O emission evolved. Moreover, the effect of the N_2O on growth differ with the amount of groundwater withdrawn. Consequently, the effect of Nitrous oxide on growth is not limited on the coefficient of N_2O in the model but also depends on the values of the coefficient of the interaction term and the amount of groundwater withdrawn, which is an evidence of the effect of groundwater quality on the economic growth.

Pollution increases as long as there is economic growth and it is moving upward in its quadratic form, once deterioration is at the highest level (due to the irreversibility of the damage and the deterioration of the natural resource). Figure 8 illustrates the plot of the prediction line of the linear regression of Log GDP per capita on the interaction term, which is downward sloping; the figure supports the impact of GW quality on GDP per capita. Therefore, the effect of N_2O on growth depends on the level of GW withdrawal. Interaction is negative that means the higher the GW withdrawal, the higher is the pollution and the less is the amount of renewable GW, which is feasible and agreeable with the fact. This exactly illustrates the deterioration of GW. The highest the pollution is and the less the quality is, the less is the amount available for extraction.

Fertilizer consumption is not significant in the four models, which indicates that Nitrous oxide emission is not just a result of using fertilizers, but there are other factors that increase Nitrogen level in water and are difficult to assess at macro level, which calls for further accurate investigations for the amount of Nitrogen in water.

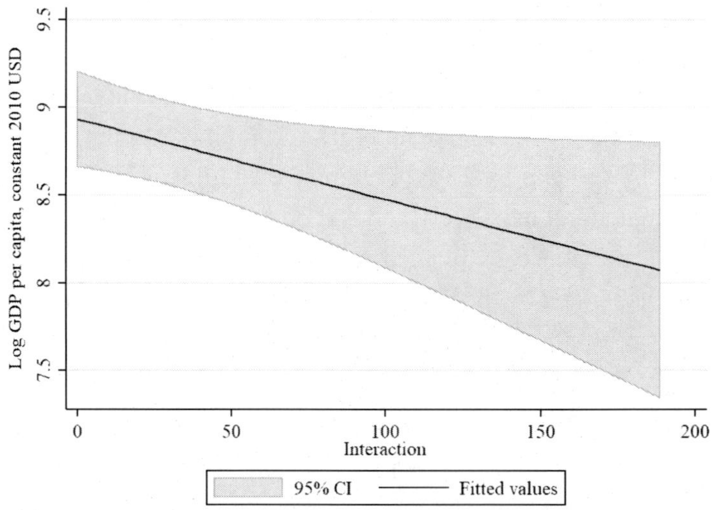

Figure 8. The prediction for Log GDP per capita from a linear regression of lGDP per capita on interaction and plots the resulting line, along with a confidence interval.

Food Production Index is negative and significant at 10% significant level (in Columns 2 and 4), which is not surprising. Although Food Production Index increases from 82.6 in 2000 to 118.8 in 2013, several Arab countries have limited production in agriculture owing to the limited economic resources. Also, low levels of technology, inadequate crop patterns, rapid population growth, limited water resources, the lack of adequate management of natural resources, arid environment and political conflict made food insecurity a big issue for many Arab countries (SDGs, Goal 2: Zero hunger)[8].

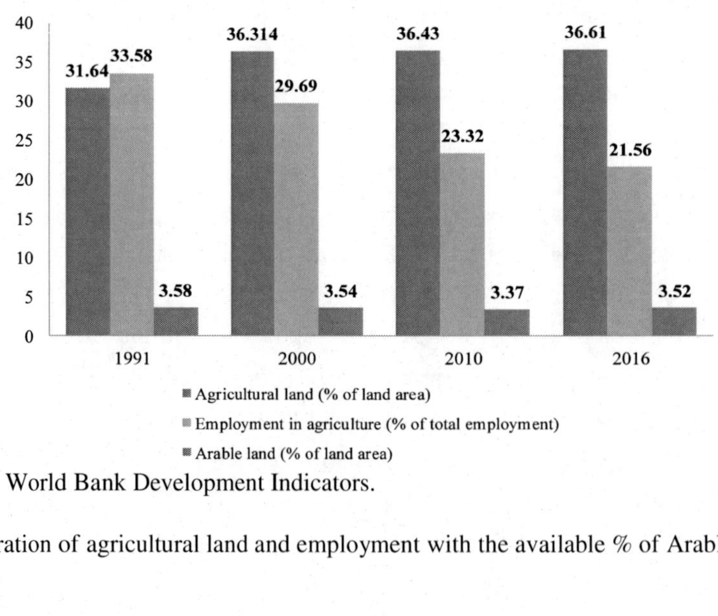

Source of data: World Bank Development Indicators.

Figure 9. Illustration of agricultural land and employment with the available % of Arable land area in Arab region.

[8] http://www.arabstates.undp.org/content/rbas/en/home/sustainable-development-goals/goal-2-zero-hunger.html.

Table 4. Regression analysis for GW quality and quantity on Log GDP per capita

	Column 1	Column 2	Column 3	Column 4
	Log GDP pc	Log GDP pc	Log GDP pc	Log GDP pc
Constant	10.15***	10.13***	10.03***	10.05***
	(15.88)	(16.89)	(10.66)	(10.94)
Government effectiveness	0.0652***	0.113***	0.134***	0.154***
	3.73	5.84	8.36	9.43
Groundwater withdrawal	0.0651***	0.0213***	0.0521***	0.0553***
	5.76	5.95	5.36	5.79
GW withdrawal squared			-0.0194***	-0.0117**
			(5.32)	(2.84)
N_2O emissions	-0.0687***	0.141**	-0.114***	0.0345
	(9.93)	3.01	(9.15)	0.85
N_2O emissions squared				0.0513*
				(2.25)
Interaction term[9]		-0.0287***		-0.0377***
		(4.68)		(4.55)
Food production index	0.00779	-0.0110*	-0.0121	-0.0134*
	(1.48)	(2.29)	(1.80)	(2.13)
Fertilizer consumption	-0.0127	-0.0887	-0.108	-0.145
	(-0.12)	(0.83)	(0.93)	(1.22)
Arable land area	-0.0741***	-0.110***	-0.0785***	-0.100***
	(6.38)	(8.95)	(6.96)	(8.94)
N	84	84	67	67
R-sq	0.578	0.639	0.672	0.697
adj. R-sq	0.545	0.605	0.633	0.649
rmse	0.691	0.643	0.651	0.637
F-test	89.92***	66.73***	73.13***	88.05***

Post estimations diagnosis: Breusch-Pagan/Cook-Weisberg test (H0: the error variances are all equal). Chi2 (1) = 6.80***, Chi squared indicate that the heteroskedasticity is present, it is corrected by using robust variance estimates. The VIF for the first model is 1.33 which is acceptable. F- test is significant, so we reject the null hypothesis. Corrolation test between lgdp per caita and N_2O emissions is (-0.3270***) and between lgp per capita and interaction term is (-0.1966***).

Arable land area is highly significant at 1% significant level, and it is negatively significant. A 1-unit increase in Arable land decreases the economic output from 7.7% to 11.6%. Worldwide, agriculture shows a decrease in its input in GDP. Employment in agricultural sector has declined as well compared to other sectors in the economy (services and industry). There is a tradeoff between agricultural expansion and the limited GW resources because most of the agricultural expansion in Arab countries is ecologically unsustainable. Some countries in the Arab region are suffering from desertification; GCC countries are investing in crop production abroad (agricultural and arable land) while the employment in agriculture declines as seen in Figure 9. Therefore, several factors make agricultural production in the Arab countries limited or restricted in addition to different

[9] Interaction term is calculated: GW withdrawal x N_2O emission.

economic factors (increase in the cost of treatment of polluted GW) and the ecosystem deterioration (deterioration of GW resources).

6. Conclusion

Literature is rich with the effect of water resources on economic growth and lacks the studies that deal with the effect of groundwater quality on this growth on the short and the long run. So far, this impact was ignored; however, based on the findings of this study, we can argue that the groundwater quality and the extracted groundwater resources have a high impact on the economy.

The Arab region has an arid climate, and the management of groundwater reserves must be developed efficiently with strict policies which so far have been missed. Since groundwater is linked with surface water, farmers can replace the groundwater with surface water by maintaining international standards for surface water quality and by applying strict environmental regulations. Industries and municipalities can initiate an action by taking precautions in dealing with their wastes and effluents.

Results prove the high pollution costs of groundwater on the economy and the environment of the Arab countries; results also prove that groundwater exploitation is always accompanied with groundwater contamination and quality deterioration. Most of the Arab countries should start recognizing the vital need to set up new policies and scenarios in managing their resources. Groundwater management reforms, strict legislations, and sustainable policies are highly needed to avoid the exploitation of this socio-economic resource. Investing in desalination and international treatment procedures can reduce the cost of pollution and the deterioration of the ecosystem. Results can be seen on the long run, and that would enhance the required sustainable development.

There is an essential need for monitoring groundwater contamination levels, and this would help in providing information for direct action and for risk management in treating the contaminated sites effectively and would reduce the cost of action when the problem is diagnosed at the beginning. In addition, groundwater allocation data is a tool that can alleviate the influence of aquifers deterioration. Therefore, gaining good quality data for groundwater quality and quantity for this region proved to be a high priority.

7. References

Abd Alraheem M. A. A. 2000. Pollution in the water supply wells of Khartoum, Sudan. *Bull. Eng. Geol. Env.*, 58, 257-264.

Abd El-Aziz, S. H. (2017). Evaluation of groundwater quality for drinking and irrigation purposes in the north-western area of Libya (Aligeelat). *Environ Earth Sci*, 76-147. https://doi.org/10.1007/s12665-017-6421-3.

Abdel-Sater, A. M., Al-khabbas, M. H., Alahmad, N. R., Yousef, W. M., Alsomadi, R. H., Iqbal, T. (2017). Quality assessment of groundwater and agricultural soil in Hail region, Saudi Arabia. *Egyptian J. Aquat. Res.* 43, 55-64.

Abdullah, K. and Umrikar, B. (2014). Groundwater Quality and its Suitability for Domestic and Agricultural Use in Khanaqin Town, Iraq. *IOSR Journal of Applied Geology and Geophysics,* 2, 57-66. doi: 10.9790/0990-02225766.

Abou Zakhem, B. and Hafez, R. (2015). Hydrochemical, isotopic and statistical characteristics of groundwater nitrate pollution in Damascus Oasis (Syria). *Environ Earth Sci,* 74, Issue 4, 2781–2797. https://doi.org/10.1007/s12665-015-4258-1.

Abou Zakhem, B. and Hafez, R. (2015). Heavy metal pollution index for groundwater quality assessment in Damascus Oasis, Syria. *Environ Earth Sci,* 73:6591–6600. doi: 10.1007/s12665-014-3882-5.

Abu Naser, A. A., Ghbn, N. and Khoudary, R. (2007). Relation of nitrate contamination of groundwater with methaemoglobin level among infants in Gaza. *EMHJ - Eastern Mediterranean Health Journal,* 13 (5), 994-1004. http://www.who.int/iris/handle/10665/117340.

AfDB. (2015). *Environmental and social management framework* (*ESMF*). National Drainage Programme Egypt. African Development Bank.

Ahmed, A. H., Rayaleh, W. E., Zghibi, A. and Ouaddane, B. (2017). Assessment of chemical quality of groundwater in coastal volcano-sedimentary aquifer of Djibouti, Horn of Africa. *Journal of African Earth Sciences* 131, 284-300. doi: 10.1016/j.jafrearsci.2017.04.010.

Akkr, N. (1992). The seepage of seawater into the aquifers along the Lebaneseborders. p. 14. In *Proc. of the Conf. about Water Resources in Lebanon*, Beirut, Lebanon. 27ñ28 Nov. 1992. National Council for Scientific Research, Beirut, Lebanon.

Alabdula'aly, A., Al-Rehaili, A., Al-Zarah, A., Khan, M. (2010). Assessment of nitrate concentration in groundwater in Saudi Arabia. *Environ Monit Assess,* 161(1–4),1–9.

Al-Ahmadi, M. E. (2013). Groundwater quality assessment in Wadi Fayd, Western Saudi Arabia. *Arab. J. Geosci.* 6, 247–258.

AlAli, E. H. (2008). Groundwater history and trends in Kuwait. *WIT Transactions on Ecology and the Environment,* 112, 153-164.

Aljawad, S., Al-Ansari, N., Adamo, N., Sissakian, V. K., Laue, J., Knutsson, S. (2018). Groundwater Quality and Their Uses in Iraq. *Journal of Earth Sciences and Geotechnical Engineering,* 8 (3), 123–144.

Al-Kalbani, M. S., Price, M. F., Ahmed, M. et al. (2017). Environmental quality assessment of groundwater resources in Al Jabal Al Akhdar, Sultanate of Oman *Appl Water Sci,* 7, 3539. https://doi.org/10.1007/s13201-017-0621-6.

Alley, W. M. (2006), Tracking U.S. groundwater reserves for the future?, *Environment,* 48(3), 10–2.

Al-Mikhlafi, A. S. (2010). Groundwater quality of Yemen volcanic terrain and their geological and geochemical controls. *Arab J Geosci,* 3: 193. https://doi.org/10.1007/s12517-009-0068-7.

AL-Mussawi, W. H. (2014). Assessment of Groundwater Quality in UMM ER Radhuma Aquifer (Iraqi Western Desert) by Integration Between Irrigation Water Quality Index and GIS. *Journal of Babylon University Engineering Sciences,* 1 (22), 201-217.

Al-Naim, A. (2014). Effect of Excess Pumping on Groundwater Salinity and Water Level in Hail Region of Saudi Arabia. *Research Journal of Environmental Toxicology,* 8: 124-135. doi: 10.3923/rjet.2014.124.135.

Al-Senafy, M., Fadlelmawla, A., Mukhopadhyay, A., Al-Khalid, A., Al-Fahad, K. (2016). Groundwater Protection in Kuwait: Design of monitoring network and recommended path. *Nineteenth International Water Technology Conference*, IWTC19. Sharm ElSheikh, 21-23 April 2016.

Al-Turki, A. L. (2010). Assessment of well Drinking water Quality in Hail. Region of North central Saudi Arabia. *J. Agric. Vet. Sci* 2 (2), 101–110.

Awaleh, M. O., Baudron, P., Soubaneh, Y. D., Boschetti, T., Hoch, F. B., Egueha, N. M., Mohamed, J., Dabar, O. A., Masse-Dufresne, J. and Gassani, J. (2017). Recharge, groundwater flow pattern and contamination processes in an arid volcanic area: Insights from isotopic and geochemical tracers (Bara aquifer system, Republic of Djibouti). *Journal of Geochemical Exploration* 175, 82-98. doi: 10.1016/j.gexplo.2017.01.005.

Bakir, H. (1999). Water Demand Management and Pollution Control: Strategic but neglected elements in Integrated Water Resources Management Processes. In *Proceedings 4th Gulf Water Conference on CD-ROM,* Bahrain, Feb. 1999. Bahrain: Water Science and Technology Association.

Bakir, H. (2002). Water Demand Management and Pollution Control: Key to Securing and Safeguarding the Water Supplies of MENA in the 21st Century," *Proceedings of Water Demand Management Workshop*, CEHA, Amman, Jordan. http://www.emro.who.int/Publications/CEHA-ESCWA.pdf.

Biswas, K. D. (2003). "*Water Quality Management-A National Concern,*" A Report from Central Pollution Control Board Delhi. India.

Boschetti, T., Awaleh, M. O., Barbieri, M. (2018). Waters from the Djiboutian Afar: A Review of Strontium Isotopic Composition and a Comparison with Ethiopian Waters and Red Sea Brines. *Water,* 10, 1700.

Brooks, D. B. (1999). "*Setting the Scene,*" Presentation to Donors' Meeting, Water Demand Management Research Network, IDRC- Cairo, 22nd November, 1999.

Castle, S. L., Thomas, B. F., Reager, J. T., Rodell, M., Swenson, S. C. and Famiglietti, J. S. (2014), Groundwater depletion during drought threat-ens future water security of the Colorado River Basin, *Geophys. Res. Lett.,* 41, 5904–5911, doi: 10.1002/2014GL061.

Comte, J. C., Cassidy, R., Obando, J., Robins, N., Ibrahim, K., Melchioly, S., Mjemah, I., Shauri, H., Bourhane, A., Mohamed, I., Noe, C. (2016). Challenges in groundwater resource management in coastal aquifers of East Africa: investigations and lessons learnt in the Comoros Islands, Kenya and Tanzania. *Journal of Hydrology: Regional Studies* 5, 179–199.

Crusberg, T. C., Eslamian, S., 2017, Drought and Water Quality, Ch.11 in *Handbook of Drought and Water Scarcity, Vol. 2: Environmental Impacts and Analysis of Drought and Water Scarcity*, Ed. by Eslamian S. and Eslamian F., Francis and Taylor, CRC Press, USA, 205-218.

Dawood, A. (2018). GIS Approach for Spatial Distribution Analysis of Groundwater Quality at South-West Part of Basrah. *Journal of Engineering,* 24(8), 81-95. doi: 10.31026/j.eng.2018.08.07.

Du Plessis, A. (2017). Global Water Quality Challenges. In *Freshwater Challenges of South Africa and its Upper Vaal River* (pp. 13-44). Springer International Publishing. https://doi.org/10.1007/978-3-319-49502-6_2.

Ebih, O. M. E. H. S., Gaye, A. O., Haiba, B. Y. S., Zouari, K., & Ito, M. (2011). *Studies on Groundwater in Mauritania* (INIS-XA--11E0047). International Atomic Energy Agency (IAEA): IAEA.

El Alfy, M., Lashin, A., Al-Arifi, A. & Al-Bassam, A. (2015) Groundwater characteristics and pollution assessment using integrated hydrochemical investigations GIS and multivariate geostatistical techniques in arid areas. *Water Resour. Manag.* 29 (15), 5593–5612. https://doi.org/10.1007/s11269-015-1136-2.

El Khanji, S. (2017). An exploration of the interaction between socio-economic productivity and water withdrawal *Environ Dev Sustain.* 19: 653. https://doi.org/10.1007/s10668-016-9757-z.

El Khanji, S., & Hudson, J. (2016). Water utilization and water quality in endogenous economic growth. *Environment and Development Economics,* 21(5), 626-648. doi: 10.1017/S1355770X16000061.

El Moujabber, M., Bou Samra, B., & Darwish, T., Atallah, T. (2006). Comparison of Different Indicators for Groundwater Contamination by Seawater Intrusion on the Lebanese Coast. *Water Resources Management.* 20. 161-180. doi: 10.1007/s11269-006-7376-4.

El Tahlawi, M. R., Mohamed, M. A., Boghdadi, G. Y., Rabeiy, R. E., Saleem, H. A. (2014). Groundwater quality assessment to estimate its suitability for different uses in Assiut Governorate, Egypt. *Int J Recent Technol Eng* 3, 2277–3878.

El-Naqa, A. & Al-Shayeb, A. (2009). Groundwater Protection and Management Strategy in Jordan. *Water Resources Management.* 23. 2379-2394. 10.1007/s11269-008-9386-x.

Eslamian, S. S. and N. Lavaei, 2009, Modelling Nitrate Pollution of Groundwater using Artificial Neural Network and Genetic Algorithm in an Arid Zone, *International*

Journal of Water, Special Issue on Groundwater and Surface Water Interaction (GSWI), Vol. 5, No. 2, 194-203.

FAO, AQUASTAT (2012) *Food and Agriculture Organization of the United Nations*. http://www.fao.org/nr/water/aquastat/main/index.stm. Website accessed on June 2019.

FAO. (2016). AQUASTAT website. *Food and Agriculture Organization of the United Nations* (FAO). Website accessed on June 2019.

FAO-SWALIM. (2012).*Hydrogeological Survey and Assessment of Seletced Areas in Somaliland and Puntland*. Technical Report No. W-20, FAO-SWALIM(GCP/SOM/049/EC) Project, Nairobi, Kenya.

Friedel, M. J. and Finn, C. (2008). *Hydrogeology of the Islamic Republic of Mauritania: U.S. Geological Survey*, Open-File Report 2008-1136. 32 p.

General Water Authority. *State of Water Report*; General Water Authority: Tripoli, Libya, 2006.

Gleick, P. (2012). *The World's Water: the biennial report on freshwater resources*, vol. 7. Island Press, Washington. doi: 10.1007/978-1-59726-228-6. Available online at: http://link.springer.com.library.smu.ca:2048/book/10.1007/978-1-59726-228-6/page/1#section=1000042&page=3&locus=64.

Global Groundwater Information System (GGIS) from International Groundwater Resources Assessment Centre. https://www.un-igrac.org/global-groundwater-information-system-ggis. Last accessed June 2019.

Guo, L. and Ma, H. (2008). Conflict between Developing Economic and Protecting Environment, *Journal of Sustainable Development,* 1(3), 91-97.

Hakeem, K. R., Sabir, M., Ozturk, M., Akhtar M. S., Ibrahim F. H., Ashraf, M., Ahmad M. S. A. (2016). Nitrate and Nitrogen Oxides: Sources, Health Effects and Their Remediation. Reviews of Environmental Contamination and Toxicology. *Reviews of Environmental Contamination and Toxicology,* 242:183-217. https://doi.org/10.1007/398_2016_11.

Hall, M. D., Shaffer, M. J., Waskom, R. M., and Delgado, J. A. (2001). Regional nitrate leaching variability: what makes a difference in Northeastern Colorado, *J. Am. Water Resour. Assoc.,* 37, 139–150.

Hamlat, A. and Guidoum, A. (2018). Assessment of groundwater quality in a semiarid region of Northwestern Algeria using water quality index (WQI). *Appl Water Sci,* 8, 220. https://doi.org/10.1007/s13201-018-0863-y.

ICARDA and AusAID. (2011). *Water and agriculture in Egypt*. International Center for Agricultural Research in the Dry Areas and Australian Government.

International Panel on Climate Change: *2006 IPCC guidelines for national greenhouse gas inventories*, Prepared by the National Greenhouse Gas Inventories Programme, edited by: Egglestone H. S., Buendia L., Miwa, K., Ngara T., and Tanabe, K., IGES, Japan, 2006.

Isaac J., Qumsieh, V., Owewi, M. (1995). *Assessing the Pollution of the West Bank Water Resources*. Applied Research Institute –Jerusalem (ARIJ).

Ismail, E. and El-Rawy, M. (2018). Assessment of groundwater quality in West Sohag, Egypt. *Desalination and water treatment.* 123. 101–108.

Jagannathan, N. V., Mohamed, A. S., Kremer, A. (2009). Water in the Arab World. Middle East and North Africa Region and The World Bank. *Management Perspectives and Innovations.* World Bank, Washington, DC.

Jalludin, M. and Razack, M. (2004). Assessment of hydraulic properties of sedimentary and volcanic aquifer systems under arid conditions in the Republic of Djibouti (Horn of Africa). *Hydrogeology Journal,* 12, 159-170. doi 10.1007/s10040-003-0312-2.

Johnson, C. J. and Kross, B. C. (1990). Continuing importance of nitratecontamination of groundwater and wells in rural areas. *Am J Ind Med;* 18: 449-56.

Khater, A. R. (2003) Intensive groundwater use in the Middle East and North Africa. In: Llamas MR and Custodio E (eds.) *Intensive Use of Groundwater: Challenges and Opportunities*. A. A. Balkema, Lisse/Abingdon, Exton, Tokyo, 355–386.

Kherrati, I., Alemad, A., Sibbari, M., Ettayea, H., Ezziani, K., Saidi, Y., Benchikh, M., Alzwi, S., Chiguer, H., Zgourdah, Z., Bourass, A., Daifi, H., Elrhouat, O., Elkharrim, K., Belghyti, D. (2018). Health Risk of Maâmora's Groundwater Pollution in Morocco. *Natural Resources,* 2015, 6, 290-305 http://dx.doi.org/10.4236/nr.2015.64026.

Knobeloch, L., Salna, B., Hogan, A., Postle, J., Anderson, H. (2000). Blue babies and nitrate-contaminated well water. *Environ Health Perspect;* 108, 675–8.

Kuiper, N., Rowell, C., Shomar, B. (2015). High levels of molybdenum in Qatar's groundwater and potential impacts. *J. Geochem. Explor* 150, 16e24. http://dx.doi.org/ 10.1016/j.gexplo.2014.12.009.

Kuroda, K. and Fukushi, T. (2008). Groundwater Contamination in Urban Areas. From book *Groundwater Management in Asian Cities: Technology and Policy for Sustainability,* 125-149. doi: 10.1007/978-4-431-78399-2_7.

Laftouhi, Nour-Eddine., Vanclooster, M., Jalal, M., Witam, O., Aboufirassi, M., Bahir, M., Persoons, E´. (2003). Groundwaternitrate pollution in the Essaouira Basin (Morocco). *Comptes Rendus Geosciences* 335, 307–317.

Lieb, C. M. (2004). The Environmental Kuznets Curve and Flow versus Stock Pollution: The Neglect of Future Damages. *Environ Resource Econ*, 29: 483. https://doi.org/ 10.1007/s10640-004-1046-x.

Lord I. and Anthony S. G. (2002). Agricultural nitrogen balance and water quality in the UK. *Soil Use and Management.* 18(4), 363–369.

Mahmood, H. R. (1993). *Groundwater chemistry and water table variations in Bahrain.* A Doctoral Thesis. https://dspace.lboro.ac.uk/2134/11707.

Marzougui, A. and Ben Mammou, A. (2006). Impacts of the dumping site on the environment: Case of the Henchir El Yahoudia Site, Tunis, Tunisia. C. R. *Geoscience* 338 (16), 1176–1183. doi:10.1016/j.crte.2006.09.020.

Masoud A. A. (2014) Groundwater quality assessment of the shallow aquifers west of the Nile Delta (Egypt) using multivariate statistical and geostatistical techniques. *J Afr Earth Sci* 95:123–137. doi:10.1016/j.jafrearsci.2014.03.006.

McArthur, J. W., and McCord, G. C. (2017). Fertilizing growth: Agricultural inputs and their effects in economic development. *Journal of development economics,* 127, 133–152. doi: 10.1016/j.jdeveco.2017.02.007.

Millar, N., Doll, J. E., Robertson, G. P. (2014). *Management of nitrogen fertilizer to reduce nitrous oxide (N_2O) emissions from field crops.* Lansing, Michigan: MSU Extension Bulletin.

Minamikawa, K., Eguchi, S., Nishimura, S., Ihara, H., Maeda, M., Yagi, K., Komada, M. (2013). Groundwater induced emissions of nitrous oxide through the soil surface and from subsurface drainage in an Andosol upland field: A monolith lysimeter study. *Soil Science and Plant Nutrition,* 59, (1), 87-95. doi: 10.1080/00380768.2012.740606.

Ministry of Water and Irrigation (MWI). (2001). *Groundwater Management action plan for the Amman-Zarqa Basin highlands.* Internal Report of the MWI, 64 pp.

MoE/LEDO/ECODIT (2001). *Lebanon state of the environment report.* Prepared by ECODIT.

Mohammed, B. H., Elhassan. A. M., Qafisheh, N., L Gaz, H. (2015). Vulnerability of Khartoum State (Sudan) Groundwater to Contamination. *Applied Journal of Environmental Engineering Science*, [S.l.], v. 1, n. 1, p. 36-45. ISSN 2509-2065.

Munasinghe, M. (1999). Is environmental degradation an inevitable consequence of economic growth: tunneling through the environmental Kuznets curve. *Ecological Economics,* 29, 89–109.

Murad, A. A. (2010). An Overview of conventional and non-conventional water resources in Arid region: Assessment and constrains of the United Arab Emirates (UAE). *J. Water Res. Prot.,* 2, 181–190.

Murad, A. A. Gerish, M. H. Mahgoub, F. M. Hussein, S. (2011). Physiochemical processes affecting the geochemistry of carbonate aquifer of Southeastern Al-Ain Area, United Arab Emirates (UAE), *Water Air Soil Pollut.* 214, 653–665. doi: 10.1007/s11270-010-0453-6.

Nagatake A., Mukumbuta I., Yasuda K., Shimizu M., Kawai M., Hatano R. (2018). Temporal Dynamics of Nitrous Oxide Emission and Nitrate Leaching in Renovated Grassland with Repeated Application of Manure and/or Chemical Fertilizer. *Atmosphere,* 9 (12):485.

Nathan, R. and Evans, R. (2011). Groundwater and surface water connectivity. In R. Grafton & K. Hussey (Eds.), *Water Resources Planning and Management* (pp. 46-67). Cambridge: Cambridge University Press. doi:10.1017/CBO9780511974304.006.

Neumayer, E. (1998). Is Economic Growth the Environment's Best Friend?, *Zeitschrift für Umweltpolitik und Umweltrecht*, 21(2), 161–176.

O'Hara, S. L. (1997). Irrigation and land degradation: Implications for agriculture in Turkmenistan, central Asia. *J. Arid Environ.*, 37: 165-179.

Osman-Elasha, B. (2010) *Mapping of climate change threats and human development impacts in the Arab region*. Research Papers Series 03/2010; UNDP, Arab Human Development Report. http://www.arab-hdr.org/publications/other/ahdrps/paper02-en.pdf. Accessed June 2019.

Ouda, O. K. M., Shawesh, A., Al-Olabi, T. et al. (2013). Review of domestic water conservation practices in Saudi Arabia. *Appl Water Sci* (2013) 3: 689. https://doi.org/10.1007/s13201-013-0106-1.

Pearce, D. W., Warford, J. J. 1993. *World Without End: Economics Environment and Sustainable Development*. Oxford University Press (published for the World Bank).

Reddy, V. R., Behera, B. (2006). Impact of water pollution on rural communities: aneconomic analysis. *Ecological Economics* 58 (3), 520–537

Rinaudo J. D., Blanchin R., Arnal C., Elsass P., Meilhac A., and Loubier S. (2004). Assessing the cost of groundwater pollution: the case of diffuse agricultural pollution in the upper Rhine valley aquifer. *Water Science and Technology* 52 (9) - pp. 153-162.

Rouabhia A., Fehdi Ch., Baali F., Djabri L., Rouabhi R. (2008). Impact of human activities on quality and geochemistry of groundwater in the Merdja area, Tebessa, Algeria. *Environ Geol.* doi: 10.1007/s00254-008-1225-0.

Salameh, E. (1996). *Water quality degradation in Jordan.* Royal society for the conservation of nature, Amman.

Schroeder J. J., Scholefield D., Cabral F., Hofman G. (2004). The effect of nutrient losses from agriculture on ground and surface water quality: the position of science in developing indicators for regulation. *Environmental Science & Policy*. 2004;7:15–23.

Shafik, N. (1994). Economic Development and Environmental Quality: An Econometric Analysis. *Oxford Economic Papers*, 46, new series, 757-773. Retrieved from http://www.jstor.org/stable/2663498.

Shahin, M. (2003). *Hydrology and Water Resources of Africa*, Groundwater Resources of Africa; Kluwer Academic Publishers: New York, NY, USA; p. 529.

Shamrukh, M., Al-Muraikhi, A. A. & Al-Hamar, I. Y. (2012). *Exploring of Deep Groundwater in the Southwest Aquifer of Qatar.* doi: 10.13140/2.1.3191.5521. The 10th Gulf Water Conference.

Shiklomanov, I. A., and Penkova, N. V. (2003), Methods for assessing and forecasting global water use and water availability, in *World Water Resources at the Beginning of the 21st Century 2*, edited by I. A. Shiklomanov and J. C. Rodda, 452 pp., Cambridge Univ. Press, Cambridge, U. K.

Smart, D. R., and Bloom, A. J. (2001). Wheat leaves emit nitrous oxide duringnitrate assimilation. *Proc. Nat. Acad. Sci. USA*. 98:7875–7878.

Stephan R. M. (2007). Legal Framework of Groundwater Management in the Middle East (Israel, Jordan, Lebanon, Syria and the Palestinian Territories). In: Shuval H., Dweik H. (eds.) *Water Resources in the Middle East.* Springer, Berlin, Heidelberg.

Taany R. A., Batayneh, A. & Jaradat, R. (2007). Evaluation of groundwater quality in the Yarmouk Basin, North Jordan. *Journal of Environmental Hydrology*. 15.

UNEP. (2010). *Adapting water resource management in the Comoros to expected climate change*. Project Document. Accessed online: https://www.google.ca/url?sa=t&rct=j&q=&esrc=s&source=web&cd=3&ved=2ahUKEwisnL_R_PLiAhVNAGMBHYddCAcQFjACegQIAhAC&url=https%3A%2F%2Finfo.undp.org%2Fdocs%2Fpdc%2FDocuments%2FCOM%2FUNEP_Comoros_2%2520June_2010%2520(submission).1.doc&usg=AOvVaw3OwxCkjprssnwTfBuLIZxG.

Voss, K. A., Famiglietti, J. S., Lo, M., De Linage, C., Rodell, M. & Swenson, S. C. (2013). Groundwater depletion in the Middle East from GRACE with implications for transboundary water management in the Tigris-Euphrates-Western Iran region, *Water Resour. Res.*, 49, 904–914, doi:10.1002/wrcr.20078.

Wakida, F. T., Lerner, D. N. (2005). Non-agricultural sources ofgroundwater nitrate: a review and case study. *Water Res.* 39(1), 3–16.

Ward, M. H., Jones, R. R., Brender, J. D., de Kok, T. M., Weyer, P. J., Nolan, B. T., Villanueva, C. M. and van Breda, S. G. (2018). Drinking Water Nitrate and Human Health: An Updated Review. *International journal of environmental research and public health,* 15(7), 1557. doi: 10.3390/ijerph15071557.

Water UN. (2012) *Global Annual Assessment of Sanitation and drinking-water (GlAAS) 2012 report: the challenge of extending and sustaining services*. UN, New York.

Weymann, D., Well, R., Flessa, H., von der Heide, C., Deurer, M., Meyer, K., Konrad, C., and Walther, W. (2008). Groundwater N_2Oemission factors of nitrate-contaminated aquifers as derived fromdenitrification progress and N_2O accumulation, *Biogeosciences,* 5, 1215–1226, doi:10.5194/bg-5-1215-2008, 2008.

Wick K., Heumesser C., Schmid E. (2012). Groundwater nitrate contamination: factors and indicators. *J Environ Manage*. 111(3), 178-86.

World Bank (2007). *Making the Most of Scarcity: Accountability for Better Water Management in the Middle East and North Africa*. World Bank, Washington, D.C.

World Bank. (2004c). *Kingdom of Saudi Arabia: Assessment of the Current Water Resources Management Situation*. Unpublished manuscript, Rural Development Water and Environment Department, Middle East and North Africa Region, World Bank, Washington, D.C.

World Bank. (2018). *Securing Water for Development in West Bank and Gaza: Sector Note (English)*. Water Global Practice. Washington, DC: World Bank.

World Health Organization, United Nations Environment Programme & Global Environment Monitoring System. (1991). *WHO/UNEP report on water quality:*

progress in the implementation of the Mar del Plata Action Plan and a strategy for the 1990s. Nairobi: UNEP.

WWAP (World Water Assessment Programme). (2012). *The United Nations World Water Development Report 4: Managing Water under Uncertainty and Risk*. Paris, UNESCO.

Yang, C. Y., Cheng, M. F., Tsai, S. S., and Hsieh, Y. L. (1998). Calcium, magnesium, and nitrate in drinking water and gastric cancer mortality, *Cancer Sci.*, 89, 124–130.

Yonggua, Chang, Seip, Hans Martin, Vennemo, Haakon, 2001. The environmental costs of water pollution in Chongqingm, China. *Environment and Development Economics* 6 (Part 3).

Zanist Q. Hama-Aziz, Kevin M. Hiscock, and Richard J. Cooper. (2017). Indirect Nitrous Oxide Emission Factors for Agricultural Field Drains and Headwater Streams *Environmental Science & Technology*. 51 (1), 301-307. doi: 10.1021/acs.est.6b05094.

Zghibi, A., Tarhouni, J. and Zouhri, Lahcen. (2013). Assessment of seawater intrusion and nitrate contamination on the groundwater quality in the Korba coastal plain of Cap-Bon (North-east of Tunisia). *Journal of African Earth Sciences*. 87. 1–12. 10.1016/ j.jafrearsci.2013.07.009.

Zhao, H., Li, X., Jiang, Y. (2019). Response of Nitrogen Losses to Excessive Nitrogen Fertilizer Application in Intensive Greenhouse Vegetable Production. *Sustainability,* 11(6):1513.

Zubari, W. K., Madany, I. M., Al-Junaid, S. S., Al-Manaii, S. (1994). Trends in the quality of groundwater in Bahrain with respect to salinity, 1941–1992, *Environment International,* 20 (6), 739-746, https://doi.org/10.1016/0160-4120(94)90306-9.

Zubari, W. K., Mubarak, M. A. and Madany, I. M. (1993). Development impacts on groundwater resources in Bahrain, *International Journal of Water Resources Development,* 9:3, 263-279, doi: 10.1080/07900629308722588.

Zyoud, S. H. and Fuchs-Hanusch, D. (2017). *Appl Water Sci,* 7: 1255. https://doi.org/ 10.1007/s13201-016-0520-2.

In: Advances in Hydrogeochemistry Research
Editors: S. Saeid Eslamian and Faezeh Eslamian
ISBN: 978-1-53617-829-6

Chapter 4

GROUNDWATER MODELING AND ITS CONCEPTS, CLASSIFICATIONS, AND APPLICATIONS FOR SOLUTE TRANSPORT SIMULATION IN SATURATED POROUS MEDIA

Masoud Saatsaz[1*] and Saeid Eslamian[2, 3]
[1]Department of Earth Sciences, Institute for Advanced Studies in Basic Sciences (IASBS), Zanjan, Iran
[2]Department of Water Engineering, Collage of Agriculture, Isfahan University of Technology, 8415683111, Isfahan, Iran
[3]Center of Excellence in Risk Management and Natural Hazards, Isfahan University of Technology, 8415683111, Isfahan, Iran

ABSTRACT

Numerical modeling of flow and solute transport has received increased attention during the last three decades. Groundwater models can offer the most accurate representation, characterization, and prediction of the hydrogeologic processes. This chapter discusses (i) the basic concept of groundwater modeling, (ii) the classification of groundwater models (general view), (iii) the application of solute transport models, (iv) the identification of hydrogeochemical processes, (v) the governing equation of solute transport, (vi) the guide for the development of coupled flow and solute transport numerical models, (vii) the applicability of various kinds of models to simulate the fate and transport of the major ions, (viii) the inverse estimation of solute transport properties, and (ix) the evaluation of approaches for modeling groundwater contamination (specific view). One clear conclusion of this study is that groundwater models should consider a series of hypothesizes, assumptions, constraints, and simplifications based on the model objective, modeling techniques, model reliability, solution complexity, data availability, and

* Corresponding Author's Email: saatsaz@iasbs.ac.ir.

characteristics of the modeled area. Although numerical models provide a more comprehensive vision of the groundwater systems under complicated conditions, these models are not suitable when the input data are not updated, reliable, sufficient, and precise. As the groundwater flow velocity mainly controls the magnitude of the movement, storage, and change in the solute concentration, an accurate definition of the flow system is of great significance. This chapter highlights the advection-dispersion equation suffers from several conceptual limitations, especially in large scale heterogeneous systems. To overcome these difficulties, advanced groundwater models can solve complex problems, either one- or multidimensional, that often involve multiple chemical species, diverse transport processes, material heterogeneity, complex boundary conditions, and time-varying stresses.

Keywords: finite difference method, finite element method, groundwater, hydrogeochemical processes, saturated porous media, numerical solute transport model

INTRODUCTION

Men have always tried to improve their abilities in presenting the truth. Depending on system complexity, techniques, sources (data, budget, and time), user experience, and needs, the task may be easy or difficult. In this way, modeling is a practical way used to define, represent, quantify, analyze, test, document, teach or advertise something else using physical objects, texts, symbols, graphs, equations, structures, computer codes, and a variety of other things. Scientific modeling is a scientific process, the aim of which is to generate an abstract, conceptual framework, and graphical or mathematical representation of an object under examination. According to Schwarz et al. (2009), a scientific model is a representation of a particular phenomenon (states, objectives, and events) that abstracts and simplifies it to allow one to make explanations and predictions. Modern science offers a growing collection of methods, techniques, theories, tools, and facilities about all kinds of specialized scientific modeling (Prior 2017).

Generally, scientific models at best are not exact replicas; these are approximations of the systems and objects that they represent (Pallant and Lee 2015). To make models more realistic, the modeling process has recently become a problematic task for model designers. They routinely follow several formalized and standardized steps, allowing models to be faster to set up, more secure to run, more accurate to present, more comfortable to understand, and more flexible to meet the needs. Practical issues in developing a scientific model consist of (i) the establishment of model objectives, (ii) data gathering, classification, and preparation, (iii) selection of the system's main components, functions, behaviors, and problems, (iv) identification and documentation of constraints, limitations, and assumptions within the simulation model, and (v) validation of the simulation results (Gray and Rumpe 2016). This chapter aims to provide a brief but comprehensive explanation of principles, concepts, theories, and applications of flow and solute transport modeling. This research answers (i) what are the main processes controlling solute transport; (ii) how the mathematical equations describe these processes; (iii) how numerical methods solve the governing equation; and (iv) how the accuracy of the model is measured. It is important to note that the focus is on non-reactive miscible solutes that pass through a saturated porous medium.

Groundwater Modeling

Groundwater resources are hidden systems and dynamic; their quality and quantity are often hard to measure and understand. The state of any hydrogeological system can change over time. In this regard, model designers and operators are repetitively developing new scientific models and solution techniques to evaluate these systems. Groundwater models can help hydrogeologists explain the behavior of the whole or a part of a groundwater system, understand the processes, make predictions, make management decisions, and test a hypothesis (Figure 1). Some of the typical groundwater model applications are:

I. To improve hydrogeological understanding by analysis of data and existing literature (characterization and planning);
II. To evaluate the aquifer behavior through simulation (prediction);
III. To identify practical solutions to meet objectives (engineering construction and design);
IV. To optimize designs for maximum performance, minimum costs and environment conservation (control and optimization);
V. To assess input, output, and water balance factors (water availability assessment);
VI. To estimate the predevelopment conditions of an aquifer system (development and enhancement);
VII. To predict the behavior of the groundwater system under different hydrogeological stresses (management and decision-making);
VIII. To quantify the sustainable and safe yield (water allocation policies);
IX. To analyze the sensitivity and optimization of parameters (uncertainty or risk analysis);
X. To visualize the behaviors, processes and ongoing and upcoming events of a groundwater system (scaling and representation);
XI. To point out the strengths and weaknesses of other research and evaluating other studies and theories through modeling (learning and teaching).

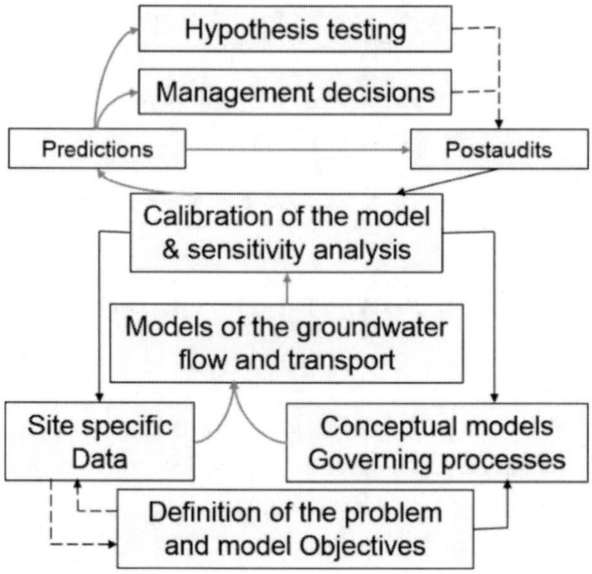

Figure 1. The development and application of the modeling in the analysis of groundwater problems, decision making, hypothesis testing (Adapted from Konikow et al. 1996).

Groundwater Model Simplification

Similar to other models, the development of groundwater models requires the consideration of a series of hypotheses, assumptions, constraints, and simplifications to ensure that an idealization or abstraction sufficiently expresses an idea. Simplifications may decrease the accuracy of the model. For example, when the migration of chemicals from a point-source is simulated and predicted by a simple model that assumes homogeneous groundwater conditions, the results may be less accurate than those of a heterogeneous groundwater system. As another example, a regional scale model with a 100 × 100 m cell size would not account for preferential solute transport through the relatively small open framework gravel lenses (Dommisse 2015). The range of assumptions and constraints of a groundwater model depends upon the model objective, modeling techniques, model reliability, solution complexity, data availability, and characteristics of the modeled area. According to Beven (2009), a model should not be more complicated than is needed to make sufficiently useful predicts of the observations. In more complex groundwater models, a more detailed understanding of the groundwater system is necessary. Hence, a logical relation between the data availability and model complexity should be considered for better presentation. Selecting complex systems, when there is a shortage of information, would not end in desirable results and would not show the real world. In such situations, operators should revise their expectations of the model's ability based on data availability and accessibility. The trade-offs between model complexity and simplicity are discussed by Konikow (2011) and Hill (2006).

Groundwater models can be classified based on the result's quality into the board, detailed, and detailed, coupled with predictive modeling classes (Middlemis et al. 2001). The first is basic models and low-cost models, fit for a baseline evaluation, limited planning, and data gap identification. As these models do not need significant information to develop, they are not appropriate for evaluating complex conditions. The second models are relatively complex, requiring more data and a better understanding of the groundwater system. These models are generally suitable for characterizing suspected groundwater issues. The third models (also known as aquifer simulators) are too specific, time-consuming, and costly. These models can predict the changes in groundwater bodies and their interactions with surface water on a large scale (aquifer or watershed). Their outputs can be used to support comprehensive groundwater management and protection plans (Payne and Woessner 2010).

Groundwater Model Classification

Groundwater models can be classified using a variety of independent methods. Based on structure, groundwater models are routinely classified into three categories, including physical, analog, and mathematical. Mathematical models are divided into two main categories, including deterministic and stochastic (statistical models). The deterministic models can be analytical or numerical.

Physical Groundwater Models

Physical models (also known as porous media models or mocked-up models) such as sand-box, bench-scale, parallel-plate, and viscous fluid models are an abstracted and scaled-down laboratory representation of a groundwater system. The earliest works on groundwater modeling were physical models that appeared in European literature. The Darcy's experiments (1855) describing the flow of fluids through porous media can be regarded as the first works in groundwater modeling (Schweizer 2015; Landmeyer 2011; Simmons 2008). By applying this pioneering model, Darcy could experimentally demonstrate the law of water flow through saturated sand (now known as Darcy's law). Physical models can be currently applied to simulate seepage, artificial recharge, dispersion, and seawater intrusion, in situations that little is known about the processes and dynamics that act in a groundwater system. The most significant problem in physical models is finding a proper scale, so that be able to represent a real groundwater system in the most effective way possible. Besides, a small number of situations can be described by these models, mostly supporting the hydraulic properties of the porous media. Another shortcoming associated with physical groundwater models is that the setup of such models is relatively costly and time-consuming. Despite all the disadvantages, these models can be used as an appropriate educational tool in the hands-on learning of the basic concepts of groundwater hydrology. A description of the pioneering works related to numerical modeling of the groundwater in porous media is provided by Pinder and Bredehoeft (1968).

Analog Groundwater Models

Analog models utilize the correspondence between Darcy's Law and other laws of physics, such as Fourier's law of heat transfer, Ohm's law of electricity, Navier-Stokes equation of fluid motion, and Fick's law of solute diffusion to describe the dynamics of a groundwater system (Delleur 2010).

Membrane, thermal, conductive liquid, and electrical models are common forms of groundwater analog models. For example, to simulate aquifer storage (capacitance) and groundwater flow (resistance), electric analog models used a network of capacitors and resistors, arranged in a grid to approximate aquifer geometry. Here, the analogy between the electrical flow in a material conductor (Ohm's law) and the water flow in a porous media (Darcy's law) can be stated mathematically as (Thangarajan 2007):

$$I = -\sigma \frac{dV}{dx} \text{ (Ohm's law)} \tag{1}$$

where I is the electrical current per unit cross-sectional area, σ is the specific electrical conductivity of a substance, and $\frac{dV}{dx}$ is the voltage (potential) gradient V.

$$q = -K \frac{dh}{dx} \text{ (Darcy's law)} \tag{2}$$

where I is the Darcy velocity and is similar to the electrical current stated in equation (1). The hydraulic conductivity K is similar to the electrical specific conductivity σ, and hydraulic head h corresponds to the electric potential.

Mathematical Groundwater Models

The behavior of a groundwater system can be mathematically formulated and represented by a set of equations, variables, and numbers constituting a mathematical model. These equations and related parameters are formulated based on the geometrical and hydrogeological characteristics of the groundwater system, spatial and temporal boundary conditions, initial conditions, and stresses of the system. In Figure 2, a schematic shows the connections between the elements of a simplified solute transport numerical model.

As mentioned, mathematical models are divided into deterministic and stochastic (statistical) models. Stochastic models are provided with possible outcomes and solutions based on spatial variabilities and probabilities of occurrence. Deterministic models are established based on the cause and effect relationships (one process, a cause, contributes to the production of another state, an effect) of known systems and processes. These types of models are typical for identifying local and regional groundwater systems. The governing equations of deterministic models can be solved with analytical (if the model is relatively simple), numerical, or a mixture of analytical and numerical techniques.

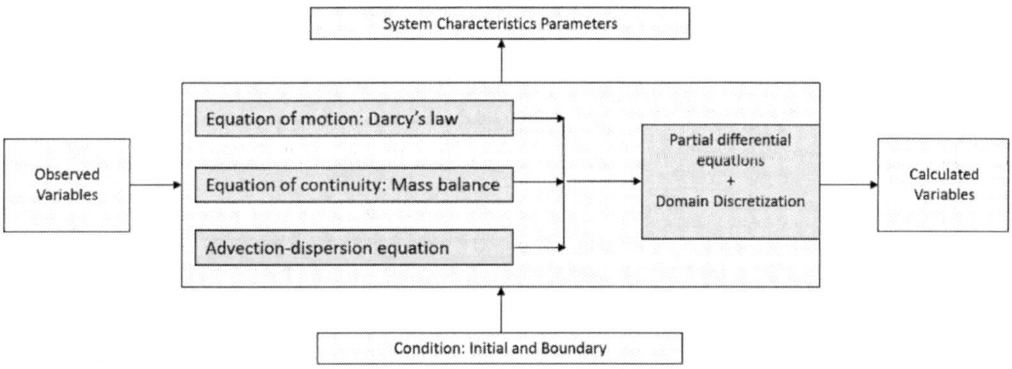

Figure 2. Schematic showing a simplified solute transport numerical mathematical model.

Analytical Groundwater Models

The first analytical solution is provided by Dupuit (1863) to describe the steady-state groundwater flow to pumping wells in confined or unconfined aquifers (Renard 2005). Twenty-three years later, in 1886, Dupuit's approach was extended and generalized by an Austrian engineer, Philipp Forchheimer. The Theis solution (1935), as one of the commonly used analytical solutions in groundwater hydrology that considers different types of aquifers, flow regimes, stresses, and boundary conditions. Through the ages, analytical methods have become standard tools for explaining the dynamics of groundwater quantity and quality. Also, these models are occasionally applied before a numerical model starts. These models have their disadvantages concerning mathematical assumptions that may not occur naturally. Another problem is that they do not allow for simulating spatial and temporal variations in various parameters simultaneously. Hence, these models are often inadequate to mimic a heterogeneous subsurface system closely.

Numerical Models

In the late 1960s, following the development of computers and digital technology, numerical models were widely adapted to provide a more comprehensive vision of the groundwater systems under complicated conditions. With numerical modeling, hydrogeologists were directly able to approximate the solutions of complicated differential equations of the system by converting them to discrete equations and dissecting the domain into meshes or grids. In numerical models, dissection of a region and discretization of differential equations are generally accomplished by approximation approaches like finite difference, finite element, and finite volume boundary. Each of these approaches can be applied solely or combined with other methods to reduce the computational complexity of numerical models. Considering the applicability, ease of use, performance, and computational speed, each method has its advantages and disadvantages. Among them, however, finite difference and finite element methods are particularly the most preferred approaches for solving groundwater flow and transport equations (Wang et al. 2019).

Finite Difference Method

The finite difference methods are perhaps the best known among the numerical techniques used to solve a set of partial differential equations based on the geometry of the system, boundary conditions, and initial conditions. This method was first introduced by Richardson in 1910 and was improved later in 1927 (Raats and Knight 2018). In this method, a structured grid is divided into regular-shaped small cells or grid points (nodes), the derivatives of the partial differential equations are approximated in space, concerning to the differences between values of variables at adjacent cells or nodes. The nodes represent the specific values of all computational properties of the system (e.g., hydraulic head, solute concentration, effective porosity, etc.) as functions of time. All nodes are defined in a body-fitted coordinate system and labeled consecutively by a set of indices (i, j, and k along the x, y, and z directions, respectively). The distance between neighboring cells or nodes can be varied. The continuous medium can be discretized either into a block or mesh centered node (point centered) domain. To solve a finite difference equation, it is necessary to start with the initial distribution of heads or solute concentration and compute their values at later time instants. The relation between initial (observed) and calculated locations of particles in a finite-difference grid is shown in Figure 3.

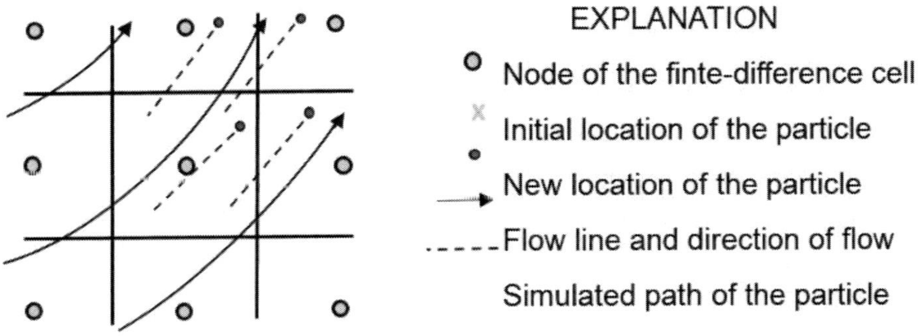

Figure 3. A hypothetical finite-difference grid is depicting the relation between flowlines and computed flow paths (Adapted from Konikow and Bredehoeft 1978).

In the finite difference method, the solution is obtained by an iterative process. An iterative process is an on-going process conducted for calculating the desired result through a repeated cycle of operations. An iterative process should come closer to the ideal outcome as the number of iterations increases, i.e., it should be convergent. Some of the prevalent iterative numerical methods have been referred by Thangarajan (2007), that are (i) Alternating Direction Implicit (ADI) Procedure (Peaceman and Rachford 1955); (ii) Successive Over-Relaxation (SOR) method (Young 1971); (iii) Modified Iterative Alternating Direction Implicit Method (Prickett and Lonnquist 1971); (iv) Strongly Implicit Procedure (SIP) (Stone 1968).

Finite Element Method

The finite element method, in its presently known form, was first theorized by Turner et al. (1956) and used to solve civil engineering and construction problems (Zeinkiewicz 1971). In this method, the grid is discretized into a network of nodes that form the vertices of irregularly/regularly-sized triangular elements. The elements and nodes are used for the interpolation of field parameters such as the hydraulic head and solute concentration. The elements may have different spatial dimensions, orientations, and sizes. Where the size of triangles decreases (i.e., a dense array of fine elements), more details is required. In the finite element method, systematic approaches, such as the Weighted Residual Method, are needed to generate the finite element equations. The Galerkin method is the most common method of the finite element method for solving groundwater problems. First, an approximate solution based on the behavior of the dependent variable of the system is assumed. The obtained solution is often selected to satisfy the boundary conditions of the model (Alexandrov 2004). The residual is the difference between the actual and assumed solution. The Galerkin method minimizes the weighted sum of the residuals of the differential equation for each of the finite elements until it reaches the best approximate solution over the domain.

The main disadvantages of the finite element methods over finite difference methods are their complexity to formulate, difficulty to solve, and time required to use. Conversely, these methods have more flexibility than finite difference methods to approximate a geometrically irregular mesh. However, a large amount of calculation and storage memory for software simulation is necessary to approximate an unconditionally irregular scheme.

Finite Volume Method

In the finite volume models, the small volume surrounding each node point on a mesh is measured based on the differential form of the conservation equations a control volume. This method relies primarily on the concept that the mass of a dependent variable that diverges out from a control volume (or a finite volume) is equal to the mass that passes across the boundary. When the entire flux is integrated over the area and time, the total variation of the mass in each control volume can be approximated (Rapp 2016). It means the change of the mass inside a control volume plus the total mass fluxes through the boundary should be zero (Neill and Hashemi 2018). Like in the case of the finite element method, this method is very flexible, it can be easily implemented for irregular and complex geometries. Another advantage is that the approach can enjoy an axisymmetric non-body fitted coordinate system that allows the grid lines to be non-orthogonal, where edges are not aligned with the horizontal and vertical axes. Further details on these methods may be found in the following references: Scesi and Gattinoni (2012); Constanda (2016); Howell (2018); Argyros et al. (2018); Dobrushkin (2017); Moatamedi and Khawaja (2018).

Flow and Solute Transport Models

All mentioned above have confirmed the practice of groundwater modeling has a lengthy history and comes in many different forms. At first, however, groundwater models focused on the quantitative aspects of groundwater. Groundwater flow models have been routinely used to calculate the rate, velocity, and direction of groundwater flow within and across the boundaries of an aquifer system (Mandle 2002). The outputs of a groundwater flow model are the hydraulic heads and groundwater fluxes that are in equilibrium with the physical and hydrogeological conditions defined for the model. Over time, the focus of hydrogeologists has shifted from quantitate to qualitative issues. Groundwater solute transport models have been extensively developed to simulate the distribution and fate of various chemical species in different hydrogeological settings since the mid-1960s. These models could simulate the flux and concentration of solutes concerning the hydrogeological, geochemical, and spatiotemporal conditions. Since the magnitude of the movement, storage, and change in the concentration (transformation) of solutes are mainly controlled by the groundwater flow velocity (the flow distance for a given time), an accurate definition of the flow system is crucial to simulate solute transport problems.

In general, the term "solute transport model" is used to describe models that simulate the 'movement, storage, and transformations' of a soluble chemical species (contaminant). In a solution, a solute constitutes only a small portion of a solvent (Bear and Cheng 2016). Since the most accessible groundwater resources are in the alluvial plains, the modeling of solute transport through porous media has received a lot of attention. According to Lage and Narasimhan (2000), a porous medium is "a region in space comprising of at least two homogeneous material constituents, presenting identifiable interfaces between them at a resolution level, with at least one of the constituent remaining fixed or slightly deformable."

Application of Solute Transport Models

Solute transport modeling in porous media deals with hydrogeological, hydrological, climatological, agricultural, industrial, geoecological, geographical, environmental, hydraulic engineering, water economics, and health issues. The specific and regulatory purposes of these models in hydrogeology include:

I. To analyze the mass balance of chemicals;
II. To interpret concentration data;
III. To describe the bulk and large scale motion of chemicals (advection or convection); or random, small scale motion of chemicals (diffusion);
IV. To assess the degree of mixing or dilution along the flow path (longitudinal dispersion) and perpendicular to the flow path (transverse dispersion);
V. To predict chemical processes and the changes in chemical concentration taking place by dissolution, retardation, and degradation (e.g., adsorption, precipitation, hydrolysis, cation exchange, bioaccumulation, biodegradation, redox reaction);
VI. To develop a monitoring strategy and risk assessment;
VII. To estimate the cleanup time using remediation plans;
VIII. To manage and protect groundwater quality

Migration of Solutes in Groundwater

Contaminants can be mobilized and released into the groundwater from various sources. The movement of soluble chemicals in the saturated zone of a groundwater system represents a miscible displacement process, where a distinct interface doesn't exist in the fluid. The solubility of a chemical species is the most significant solute characteristic (Ebrahim 2013), defined as the maximum mass of the chemical species that can be dissolved in a unit quantity of the solvent under specified conditions. The solubility of a substance depends on the nature and amount of solute and solvent, presence of other ions, as well as the pH of the solution, temperature, and pressure (Saatsaz et al. 2013_a). Chemical components with low solubility are often referred to as immiscible with water. Dense and light non-aqueous phase liquids (DNAPLs and LNAPLs) are examples of liquid solution contaminants that do not dissolve in or easily mix with water. A more detailed description of NAPLs can be obtained from, for example, Müller and Sedláčková (2003); Mayer and Hassanizadeh (2005); Rubin et al. (2013); Bharagava (2017).

Typical solute transport models are advection, advection-dispersion, and advection-dispersion-chemical/biological reaction models (Bobba 2012). The movement of solutes through groundwater flow, where there is no change in the solute concentration with distance, is mainly described by advection models. According to Darcy's law of groundwater flow, advection is attributed mostly to the effective porosity, hydraulic conductivity, and hydraulic gradient of the medium.

Hydrogeochemical Processes and their Models

Typical solute transport models are advection, advection-dispersion, and advection-dispersion-chemical/biological reaction models (Bobba 2012). Advection-dispersion models describe the transport of solutes when solute concentration is changed in time and space due to diffusion and dispersion. These models have been developed based on the diffusion dispersion concept. The dispersion term describes a mechanical mixing process caused due to the heterogeneity of the medium (Pandey et al. 2018). This process allows solutes to spread in both longitudinal and transverse directions. The spreading rate depends upon the saturation degree and the value of water velocity in porous media, the rate of solute mixing because of convergence and divergence of flow paths, and molecular diffusion (Leij and van Genuchten 2002). Diffusion is a gradual movement of individual solutes from one part with higher concentration to another section with lower concentration. This process is carried by random molecular motions due to the concentration gradient, with or without movement of water (Demirel 2013). In the aquitard sediments, in which the hydraulic conductivity values are relatively low, the advective proportions are negligible. In such a condition, the diffusion effect is usually modeled because it becomes the controlling factor of mass transport.

Advection-dispersion-chemical/biological reaction models are inherently more complex. These models are frequently used to quantify the change in the concentration of contaminants when chemical or biological reactions incorporate into advection-dispersion processes. Other main mechanisms within the groundwater systems that affect the transport, persistence, and fate of chemicals include:

i. Retardation is the act or result of delaying by which a solute becomes attached to sediments and the geologic rack material and removed from the water. Often the sorption is a reversible process. When solutes desorb, contamination plumes are retarded, rather than solutes being permanently retained by the aquifer solids (Rivett et al. 2006).
ii. Filtration is a process to separate components based on their relative particle sizes. This process affects particulate contaminants (contaminants consist of particles suspended in the water) rather than dissolved solutes.
iii. Precipitation is the removal of dissolved and suspended solids from solution by the formation of insoluble compounds through chemical means.
iv. Cation exchange is the interchanging between cations in the solution and cations on the surfaces of clay particles or organic colloids.
v. Biodegradation is the transformation of a substance into new compounds through biochemical reactions or the actions of microorganisms (Insam and Bertoldi 2007). Biodegradation is a complex process by which an aquatic system can depollute chemicals (Ney et al. 1981). In general, soluble solutes are more biodegradable than those with low water solubility (Powell 2017).
vi. Hydrolysis occurs in saturated and unsaturated zones of the groundwater systems so that a molecule of water is added to a substance to alter the chemical's structure and break down its bonds. The rate of hydrolysis reaction s depends on many factors such as temperature, pH, solubility, adsorption or absorption, and the redox (oxidation-reduction) potential (Richens 1997).
vii. Dissociation or ionization is the separation of a compound into separated anion and cation when the chemical becomes electrically charged through the redox reaction (IUPAC 2014).
viii. Bioaccumulation is the gradual accumulation of chemicals into the living tissue of an organism, such as plants or animals (Alexander 1999). As living organisms are not generally found in groundwater, the bioaccumulation process in groundwater is negligible. A simplified graphic of the described hydrogeochemical processes for the one-dimensional case is given in Figure 4.

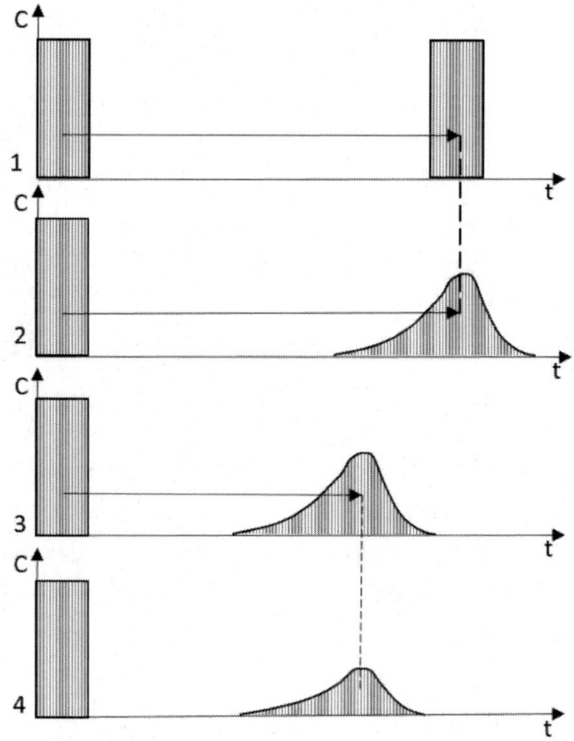

Figure 4. The effect of the advection (A), advection and dispersion/diffusion (B) advection, dispersion/diffusion, and retardation (C), and advection, dispersion/diffusion, retardation and degradation (D), on the concentration of chemical species along a flow path (Adapted from Merkel et al. 2005).

Further information on physicochemical processes that control the fate and transport of chemicals in groundwater is provided by Freeze and Cherry (1979); Stumm and Morgan (1996); Bedient et al. (1999); Fetter (2018); Kovarik (2000); Schwartz and Zhang (2003); Appelo and Postma (2004); Merkel et al. 2005; Clark (2015); Scozzari and Dotsika (2016); Tikhomirov (2016); Dassargues (2018).

Governing Equations

The form of the governing equations for three-dimensional flow in saturated porous media is a function of the properties of the water (i.e., density and viscosity), the features of the porous media (i.e., intrinsic permeability, grain size, shape factor), and the gradient of the hydraulic head. Darcy's law represents a generalized form as follows:

$$\frac{\partial}{\partial x}\left(K_{xx}\frac{\partial_h}{\partial_x}\right)+\frac{\partial}{\partial y}\left(K_{yy}\frac{\partial_h}{\partial_y}\right)+\frac{\partial}{\partial z}\left(K_{zz}\frac{\partial_h}{\partial_z}\right)-Q=S_s\frac{dh}{dt} \quad (3)$$

where h is the hydraulic head [L], S_s is specific storage that is equal to the volume of water released from the storage per unit change in head per unit volume of porous material (flow area × Thickness) [L^{-1}], and K_{xx}, K_{yy}, and K_{zz} [LT] are hydraulic conductivities along the *x*, *y*, and *z* axes. The term Q is the volumetric water flux (recharge/discharge) per unit volume of porous media (sources and sinks) [T^{-1}] (Karatzas 2017):

$$Q = \sum_{i=1}^{r} Q_i \delta\,(x - x_i)(y - y_i)(z - z_i) \tag{4}$$

where δ is the number of sources and sinks along the *x*, *y*, and *z* axes, and Q_i is the volumetric injection/discharge rate [L^3T^{-1}] at point (x_i, y_i, and z_i). The value of Q_i in withdrawal/discharging points is positive, and for injection/recharging points is negative.

Darcy's velocity is the volumetric flow rate (volume per unit time) through a unit surface area of porous media that is related to hydraulic conductivity and hydraulic gradient, as follows:

$$q_x = -\left(K_{xx}\frac{\partial_h}{\partial_x}\right) \tag{5}$$

$$q_y = -\left(K_{yy}\frac{\partial_h}{\partial_y}\right) \tag{6}$$

$$q_z = -\left(K_{zz}\frac{\partial_h}{\partial_z}\right) \tag{7}$$

where $-\left(\frac{\partial_h}{\partial_x}\right)$, $-\left(\frac{\partial_h}{\partial_y}\right)$, and $-\left(\frac{\partial_h}{\partial_z}\right)$ (dimensionless) are the hydraulic gradients along the x, y, and z axes.

Darcy's velocity is not the actual velocity of groundwater through the voids. Hence, to calculate the actual velocity (seepage velocity), Darcy's velocity is divided by porosity (Bear and Verruijt 2012):

$$q_x = -\left(\frac{K_{xx}}{\varphi}\frac{\partial_h}{\partial_x}\right) \tag{8}$$

$$q_y = -\left(\frac{K_{yy}}{\varphi}\frac{\partial_h}{\partial_y}\right) \tag{9}$$

$$q_z = -\left(\frac{K_{zz}}{\varphi}\frac{\partial_h}{\partial_z}\right) \tag{10}$$

where φ is the porosity of the medium (dimensionless).

The specific storage in Equation (3) can be written as:

$$S_s = \frac{S}{d} \tag{11}$$

where S is the storage coefficient (dimensionless). Hence, Equation (3) can be rewritten as:

$$\frac{\partial}{\partial x}\left(K_{xx}\frac{\partial_h}{\partial_x}\right) + \frac{\partial}{\partial y}\left(K_{yy}\frac{\partial_h}{\partial_y}\right) + \frac{\partial}{\partial z}\left(K_{zz}\frac{\partial_h}{\partial_z}\right) - Q = \frac{S}{d}\frac{dh}{dt} \tag{12}$$

The above equation is equivalent to:

$$\frac{\partial}{\partial x}\left(T_{xx}\frac{\partial_h}{\partial_x}\right) + \frac{\partial}{\partial y}\left(T_{yy}\frac{\partial_h}{\partial_y}\right) + \frac{\partial}{\partial z}\left(T_{zz}\frac{\partial_h}{\partial_z}\right) - (Q') = S\frac{dh}{dt} \tag{13}$$

where T is the transmissivity (L^2T^{-1}), and Q' is the volumetric water flux per unit area [LT^{-1}].

Advection and dispersion are the typical mass transport mechanisms for non-reactive solutes. The advective transport of a chemical through porous media is given by (Bear 1988, Freeze and Cherry 1979, Scheidegger 1961):

$$J_A = q.c \tag{14}$$

where J_A is the advective mass flux of chemical species [$ML^{-2}T^{-1}$], c is the solute concentration per unit volume [ML^{-3}], and q is Darcy's velocity [LT^{-1}].

For dispersive transport, the dispersive mass flux is considered equal to Fick's first law:

$$J_D = -D\nabla c \tag{15}$$

where J_D is the dispersive mass flux of a solute [$ML^{-2}T^{-1}$], and D is the molecular diffusion tensor [L^2T^{-1}]. As stated by this equation, the dispersive mass flux is a function of the solute concentration gradient. The dispersion tensor can be indicated based on the following components (Karatzas 2017):

$$D_{xx} = \frac{\alpha_L {q_x}^2 + \alpha_T {q_y}^2 + \alpha_V {q_z}^2}{|q|} + D \tag{16}$$

$$D_{yy} = \frac{\alpha_T {q_x}^2 + \alpha_L {q_y}^2 + \alpha_V {q_z}^2}{|q|} + D \tag{17}$$

$$D_{zz} = \frac{\alpha_V {q_x}^2 + \alpha_T {q_y}^2 + \alpha_L {q_z}^2}{|q|} + D \tag{18}$$

$$D_{xy} = D_{yx} = \frac{(\alpha_{T}-\alpha_{V})(q_x \times q_y)}{|q|} \tag{19}$$

$$D_{zy} = D_{yz} = \frac{(\alpha_{L}-\alpha_{V})(q_y \times q_z)}{|q|} \tag{20}$$

$$D_{xz} = D_{zx} = \frac{(\alpha_{L}-\alpha_{V})(q_x \times q_z)}{|q|} \tag{21}$$

where α_L is longitudinal dispersivity [L], α_T is transversal dispersivity [L]; α_V is vertical dispersivity [L], D_m is the molecular diffusion coefficient in a saturated porous medium [$L^{-2}T^{-1}$], and $|q|$ is the average velocity of groundwater flow [LT^{-1}]. The value of $|q|$ is given by (Burnett and Frind 1987):

$$|q| \equiv \sqrt{{q_x}^2 + {q_y}^2 + {q_z}^2} \tag{22}$$

The convection-diffusion equation (transport equation) on the domain Ω is a partial differential equation as follows (Babu et al. 1984):

$$\left\{\frac{\partial}{\partial x}\left(D_{xx}\frac{\partial c}{\partial_x} + D_{xy}\frac{\partial c}{\partial_y} + D_{xz}\frac{\partial c}{\partial_z}\right) + \frac{\partial}{\partial y}\left(D_{yy}\frac{\partial c}{\partial_y} + D_{yx}\frac{\partial c}{\partial_x} + D_{yz}\frac{\partial c}{\partial_z}\right) + \frac{\partial}{\partial z}\left(D_{zz}\frac{\partial c}{\partial_z} + D_{zy}\frac{\partial c}{\partial_y} + D_{zx}\frac{\partial c}{\partial_x}\right) - \left(q_x\frac{\partial c}{\partial_x} + q_y\frac{\partial c}{\partial_y} + q_z\frac{\partial c}{\partial_z}\right) + Q(C_W - C) - \varphi[1 + E(C)]\frac{dC}{dt} = 0\right\} \tag{23}$$

where C is the concentration of contaminant [ML^{-3}] at the point (x, y, z) at time t, C_W is the concentration of water entering the system [ML^{-3}], $E(C)$ is a function representing chemical adsorption properties.

The concentration of the withdrawn water (C_W) at the discharging (pumping) point of the groundwater system is assumed to be the same as the concentration of the surrounding ambient water (C). Thus, the term $Q(C_W - C)$ at all withdrawal points becomes equal to zero. When the point is injecting (recharging), the concentration of the injection water is considered equal to the C_W component.

The conventional form of the advection-dispersion equation suffers from conceptual shortcomings, the most known of which is that dispersivity is scale-dependent and describes the heterogeneity. (Fetter 2018). Dispersivity is a function of the domain size; it increases with increasing solute transport distance. Thus, the dispersivity derived from small scale laboratory experiments may not be valid at large scales (Willmann et al. 2008). Solute transport can be varied by the viscosity and density differences between groundwater and leachate. Transport through a homogeneous groundwater system can also be quite complex where horizontal hydraulic gradient frequently changes by variation of hydrological stresses (e.g., extraction of groundwater by pumping wells or surface recharge), boundary conditions, and the physical and chemical properties of the solute. In these cases, analytical models and laboratory studied using sank tank models can improve the assessment of the effect of changes in horizontal hydraulic gradient, and groundwater draft and recharge on solute transport (Atlabachew et al. 2018). For more detailed information about solute transport equations and their supporting literature see, e.g., McDowell-Boyer et al. (1986); Batu (2005); Woessner (2007); Bear and Bachmat (2012); Selim (2014); Adelana (2015); Kelbaliyev et al. (2019).

Development of Solute Transport Numerical Models

The development of solute transport models in saturated porous media should be performed in a logical sequence (Figure 5). The fundamental steps required for the model development are: (i) the identification of the model objectives: (ii) data collection: (iii) hydrogeological characterization: (iv) model conceptualization: (v) code selection: (vi) model calibration: (vii) model verification: (viii) performing post-processing tasks, and (ix) presentation of results (Saatsaz et al. 2013_b). At first, a conceptual model which outlines components and processes of the system should be constructed based on available topographic, climatological, geological, hydrogeological, hydrogeochemical, and land-use data. The next step is code selecting, input parameters, and model design. Subsequently, calibration and adjustment between field data (observed data) and model outputs (calculated data) should be performed. Following the calibration, to make sure about the reliability of the model, the verification process (history matching) should be initiated. This process analyzes the ability of the calibrated model to behave like the real system. After completing calibration and validation, the model can be used for predicting the future under applied scenarios. Upon performing post-processing tasks and predictive simulations, the results of the model should be described clearly and concisely.

Inverse Estimation of Solute Transport Properties

Proper estimation of parameter values in a solute transport model of a groundwater system is often tricky, needs a detailed analysis of field measurements. The task difficulty is likely because field measurements are often complicated, time-consuming, expensive, and technique-dependent. The hydrogeochemical properties of an aquifer can vary due to the spatial heterogeneity and complex chemical processes taking place in a groundwater system. Besides, field measurements can be affected by data lacking, data quality, and data analysis error. For instance, when multiple solutes with different diffusion coefficients are used to assess the effects of diffusion and hydrodynamic dispersion in a porous medium, the quality of data is so important, especially for models whose cell size is too small. The majority of the problems discussed can be fixed by both forward and inverse modeling techniques. Forward models mathematically optimize the system processes based on specific parameters. Inverse models optimize the system's parameters based on the system's operations. Inverse models are routinely applied to quantify parameters which are hard, challenging, and time-consuming to measure directly. The significant advantages of inverse modeling are (i) the precise calculation of parameter values that yield the best fit model; (ii) evaluation of the uncertainty of measurements, (iii) identifying data gaps, and needs; (iv) checking the reliability of the model; and (v) documentation of matters such as model limits, ignored by the manual calibration methods (Poeter and Hill 1997).

The practice of inverse modeling techniques along with numerical solutions for estimating solute transport parameter values is well-acknowledged (Mishra and Parker 1989; Abbasi et al. 2003; Giudici 2004; Vrugt 2008; Kirkham et al. 2019). A robust approach for parameter value estimation is the nonlinear regression method. In recent decades, the regression approach has been applied effectively to optimize groundwater flow and solute transport models. In the regression analysis, an objective function is applied to make an analogy between simulated values (e.g., hydraulic heads and solute concentrations), and observed values (prior values). Parameter values that provide the best fit between observed and simulated data are defined as those which yield the smallest amount of the objective function (Tiedeman and Hill 2007). It means the goal, during regression analysis, is the estimation of parameter values that minimize the objective function in a given error tolerance. Preferably, a better fit is found as regression moves forward. One of the widely used methods for minimizing the objective function is the weighted least-squares objective function. This method reduces the objective function as follows (Hill 2000):

$$S = \sum_{i=1}^{ND+NPR} \omega_i \,(y_i - \hat{y}_i)^2 \tag{24}$$

where *ND* is the total number of observation, *NPR* is the total number of prior data equations, y_i is the ith observed value (prior value), $\hat{y}_i$ is the ith simulated value, ω_i is the weight for the ith observation, $(y_i - \hat{y}_i)$ is the residual for the ith observed value, and $\omega_i\,(y_i - \hat{y}_i)^2$ is the square weighted residual, showing the regression fitting as a matter of how the residuals are weighted. In most cases, weights are determined with respect to the standard deviation of the error measured for the difference between observation and simulation values. Further detailed information on inverse modeling can be found in Rastogi (2007), Rastogi et al. (2010), Sun (2013), and Bear and Cheng (2016).

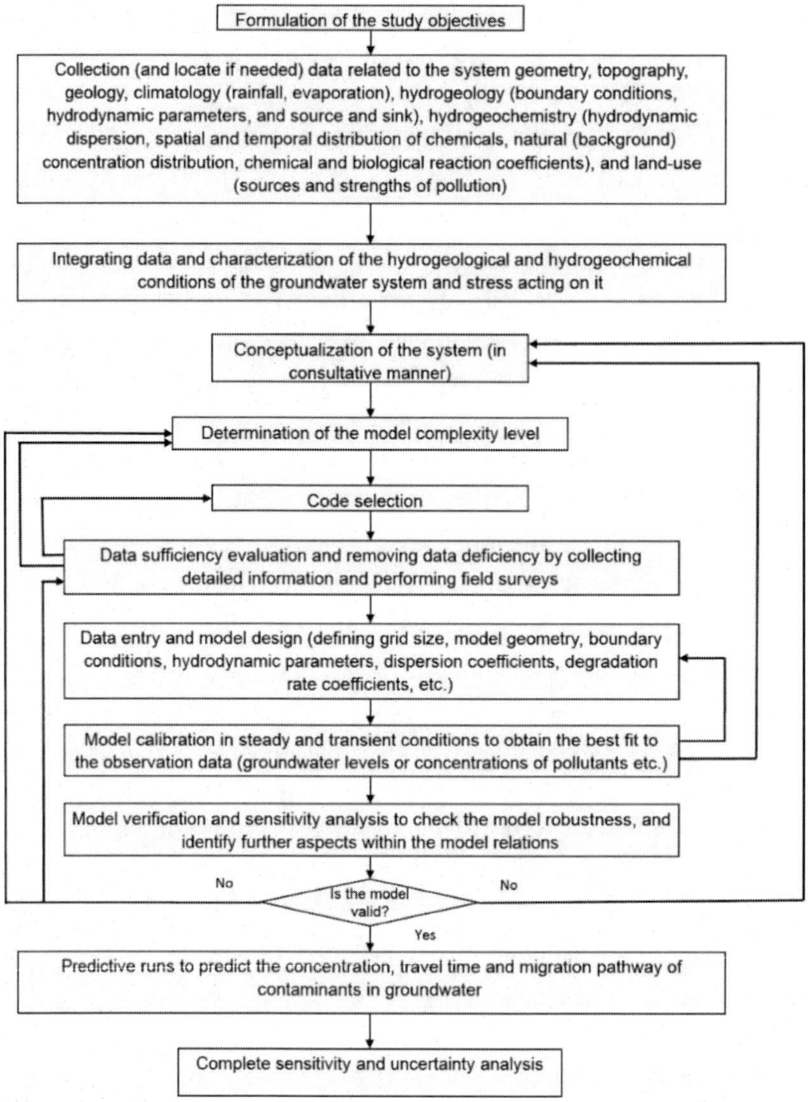

Figure 5. Flow chart classifying the main steps involved in the development of a solute transport numerical model.

Classification of Solute Transport Numerical Models

Solute transport models can simulate simple, complex, or highly sophisticated processes. The earliest works on numerical solutions for solute transport were evidence-based but straightforward in their design and performance, providing a solution to the advection-dispersion equation for one-dimensional transport of a single, nonreactive solute under isotropic and homogenous conditions in a steady-state regime (Lewis et al. 1986). Later, several model designers have incorporated solute-porous media interactions through the sources/sinks term in the governing mass-balance equations. Consequently, mechanisms such as ion exchange, degradation, and adsorption have been received closer attention, presenting new insights into the simulation of single-solute transport. Since the early 1980s, numerical solution techniques have been widely used to deal with multiple solutes involving in various chemical reactions. At the regional scale, solute transport models can be categorized based on components of the model, and their relations are categorized as a transfer function, compartment, and advanced advection-dispersion models (Orban 2009).

Transfer Function Models

Transfer function models are particularly useful when no sufficient data exist in the structure and functioning of the system. These models are mainly controlled by a limited number of parameters describing mass fluxes and concentration values only at the system input and output. In such models, transfer function (also called impulse response) is the ratio of the output of a system to the input of a system. The relation between the input and output of the system can be expressed as follows:

$$Y(t) = F(t) * X(t) \tag{25}$$

where $X(t)$ is the input function, $F(t)$ is the transfer function, and $Y(t)$ is the output (response of the system) function. A non-parametric transfer function is not explicitly described by a function or a mathematical model. It is obtained by deconvolution of the system response $Y(t)$ by the solicitation of $X(t)$ applied to the system (Orban 2009). Some of transfer function models are constructed based on more parameters describing the model stresses and responses. Such models need time series with enough history to adjust the parameters. Further information about the application of these types of models is provided by Maloszewski (2000), Carter et al. (2002), and Rumynin (2011).

Advanced Solute Transport Numerical Models

The advanced solute transport models, also known as extended solute transport models or fully coupled (hybrid) models, are applied to solve complex problems, either one- or multidimensional, that often involve multiple chemical species, diverse transport processes, material heterogeneity, complex boundary conditions, and time-varying stresses. HYDRUS (2D/3D), FEHM (Finite Element Heat and Mass Transfer), MOC (Method of Characteristics), MOFAT (Multiphase Flow and Multicomponent Transport), MIKE SHE, MODFLOW-2000 (Modular Three-dimensional Flow Model), NAPL Simulator, SUTRA, MODPATH (Particle-tracking for MODFLOW), and MT3DMS

(Modular Transport, 3-Dimensional, Multi-Species model) are among the most widely used advanced solute transport models. Table 1 provides some examples of the advanced solute transport models.

Table 1. Simulator software for groundwater flow and transport modeling

Model	Features
MODFLOW	This model simulates water flow in the saturated zone; it can be linked with chemical transport modules and particle tracking modules (for flow path simulations)
MODPATH, path3D	the model is a 3-dimensional particle tracking software for flow path and pattern simulation; the model can be integrated with MODFLOW;
MT3D	MT3D simulates 3-D advection, dispersion, sorption, and reaction; it can be linked to MODFLOW, the model solves equations based on the finite-difference method; it is suitable for both saturated and unsaturated zones.
BIOPLUME 1, 2, 3	Family of models that couples chemical transport with biodegradation. BIOPLUME 2 focuses on the aerobic reactions that occur under oxygen-limiting conditions. BIOPLUME 3 simulates the aerobic and anaerobic reactions with multispecies biodegradation kinetics. The model can be used to simulate the fate and transport of hydrocarbons and electron acceptors.
3DFEMFAT	A 3-Dimensional flow and solute transport model that simulates both saturated and unsaturated media. The solution is based on the finite element method; it is suitable for density-dependent flow and transport.
BIOF&T	A 2- and 3-Dimensional software for both fractured media and porous media; it is suitable for simulation under heterogeneity; the model can consider advection, dispersion, sorption, reaction, and biodegradation.
CHEMFLO	CHEMFLO is a one-dimensional vadose zone screening model; its simulation is based on the finite difference method.
CHEMFLUX	This model is a finite element model useful to solve advection, dispersion, sorption, and decay problems; it has the capability for automatic mesh size and time step refinement.
MOFAT	MOFAT can Simulate the LNAPL movement from the surface spill to the water table.
UTCHEM	This software can simulate multiphase fluid flow as well as formation in vadose and saturated zones.
FEFLOW	A 2- and 3-D software that is used for both saturated and unsaturated media, the model is suitable for salinity and heat-dependent transport.
FLONET/ TRANS	This model is a 2-D flow and solute transport-retardation model that simulates based on the finite element method.
HST3D	A 3-D groundwater flow and solute transport software that used for a single solute with advection, dispersion, linear sorption, and first-order decay.
MOC	A 2-D finite-difference software that simulates groundwater flow and solute transport under the influence of advection and dispersion, along with first-order decay. This software is also useful to simulate the reversible equilibrium-controlled sorption and reversible equilibrium-controlled ion exchange.
SWMS 2D	A 2-D software that is practical for simulating flow and solute transport in variably saturated media with irregular boundaries under the influence of dispersion, linear sorption, zero-order production, and first-order decay.
VSAFT2	A 2-D finite element program that is applied for solving flow and reactive solute transport in variably saturated porous media.
SUTRA	A multidimensional software that is used for variable-density flow and transport modeling of energy or dissolved substances in a saturated and unsaturated media

CONCLUSION

Groundwater modeling is a powerful tool for evaluating the quantity and quality of groundwater resources over time. The comparison among different types of groundwater models shows numerical models are often more versatile, practical, and reliable to cope with sophisticated solutions that rarely have simple or straightforward solutions. In this regard, selecting model complexity is a challenging task that controls the performance of a numerical model. The performance of a groundwater numerical model is also dependent on so many criteria such as the definition of the conceptual model, model structure, quality of input data, solution techniques, characteristics of the modeled area (spatial and temporal resolutions), operator proficiency, etc.). Comparing with flow models, the performance of solute transport models can be affected by more factors such as the domain size, viscosity and density variations, changes in hydrological stresses (e.g., extraction of groundwater by pumping wells or surface recharge), heterogeneity of the medium, physical and chemical properties of the solute, and complex chemical processes taking place in the groundwater system. Although most solute transport models provide solutions for one- or two-dimensional transport of a single, nonreactive solute, under isotropic and homogenous conditions, in a steady-state regime, the advanced ones can be used to solve complex problems, either one- or multidimensional, involving multiple chemical species, diverse transport processes, material heterogeneity, complex boundary conditions, and time-varying stresses. However, more investigations and computations are required to increase their robustness and to decrease uncertainty in the results.

REFERENCES

Abbasi, F., Simunek, J., Feyen, J., Van Genuchten, M. T., & Shouse, P. J. (2003). Simultaneous inverse estimation of soil hydraulic and solute transport parameters from transient field experiments: Homogeneous soil. *Transactions of the ASAE,* 46(4), 1085.

Adelana, S. (2015). *Groundwater: Hydrogeochemistry, Environmental Impacts, and Management Practices*. Nova Publishers, USA.

Alexander (1999). Bioaccumulation, bioconcentration, biomagnification. Environmental Geology, *Encyclopedia of Earth Science*. pp. 43–44. doi: 10.1007/1-4020-4494-1_31.

Alexandrov, A. (2004). *Computer-aided design of electrical apparatus*. Avangard Prima, Sofia, Bulgaria.

Appelo, C. A. J., & Postma, D. (2004). *Geochemistry, groundwater, and pollution*. CRC press. New York, USA.

Argyros, I. K., George, S., & Thapa, N. (eds.). (2018). *Mathematical modeling for the solution of equations and systems of equations with applications*. Nova Science Publishers, USA.

Atlabachew, A., Shu, L., Wu, P., Zhang, Y., & Xu, Y. (2018). Numerical modeling of solute transport in a sand tank physical model under varying hydraulic gradient and hydrological stresses. *Hydrogeology Journal*, 26(6): 2089-2113.

Babu, D. K., Pinder, G. F., Niemi, A., Ahlfeld, D. P., & Stothoff, S. A. (1984*). Chemical Transport by Three Dimensional Ground Water Flows*. Princeton University, New Jersey, USA.

Batu, V. (2005). *Applied flow and solute transport modeling in aquifers: fundamental principles and analytical and numerical methods*. CRC Press. New York, USA.

Bear, J. (1988). *Dynamics of Fluids in Porous Media*, Dover, New York, USA.

Bear, J., & Verruijt, A. (2012). *Modeling groundwater flow and pollution* (Vol. 2). Springer Science & Business Media. Berlin, Germany.

Bear, J., & Bachmat, Y. (2012). *Introduction to modeling of transport phenomena in porous media (Vol. 4)*. Springer Science & Business Media, Germany.

Bear, J., & Cheng, A. H. D. (2016). *Modeling groundwater flow and contaminant transport*. Springer, New York, USA.

Bedient, P. B., Holder, A. W., Enfield, C. G., & Wood, A. L. (1999). *Enhanced remediation demonstrations at Hill air force base: introduction*. Department of Environmental Science and Engineering, Rice University, Houston, USA.

Beven, K. (2009). *Environmental modeling: An uncertain future?*. The Cromwell Press. London, UK. ISBN 978-0-415-45759-0.

Bharagava, R. N. (2017). *Environmental Pollutants and their Bioremediation Approaches*. CRC Press. New York, USA.

Bobba, A. G. (2012). Ground Water-Surface Water Interface (GWSWI) modeling: recent advances and future challenges. *Water resources management*, 26(14): 4105-4131.

Burnett, R. D., & Frind, E. O. (1987). Simulation of contaminant transport in three dimensions: 2. Dimensionality effects. *Water Resources Research*, 23(4), 695-705.

Clark, I. (2015). *Groundwater geochemistry and isotopes*. CRC press. New York, USA.

Constanda, C. (2016). *Solution techniques for elementary partial differential equations*. Chapman and Hall/CRC. New York, USA.

Dassargues, A. (2018). *Hydrogeology: groundwater science and engineering*. CRC Press. New York, USA.

Delleur, J. W. (2010). *The handbook of groundwater engineering*. CRC press. Boca Raton, FL, USA, 992 pp.

Demirel, Y. (2013). *Nonequilibrium thermodynamics: transport and rate processes in physical, chemical, and biological systems*. Elsevier Science. pp. 792. https://doi.org/10.1016/C2012-0-00459-0.

Dobrushkin, V. (2017). *Applied Differential Equations with Boundary Value Problems*. Chapman and Hall/CRC. New York, USA.

Dommisse, J. P. (2015). *Mathematical modeling of solute transport in a heterogeneous aquifer. Dissertation for the degree of Master of Science*. Victoria University of Wellington, New Zealand.

Ebrahim, G. Y. (2013). *Modeling groundwater systems: Understanding and improving groundwater quantity and quality management*. IHE Delft Institute for Water Education. Delft, Netherlands.

Fetter, C. W. (2018). *Applied hydrogeology*. Waveland Press. Illinois, USA.

Freeze, R. A. & Cherry, J. A. (1979). *Groundwater*. Prentice-Hall Inc, USA.

Giudici, M. (2004). *Inverse modeling for flow and transport in porous media*. Department of Earth Sciences, Geophysics Section, University of Milan, Milan, Italy.

Gray, J., & Rumpe, B. (2016) *Models in Simulation Software & Systems Modeling*. Retrieved from https://doi.org/10.1007/s10270-016-0544-y.

Hill, M. (2006). The practical use of simplicity in developing groundwater models. *Ground Water*, 44(6): 775-781.

Hill, M. C. (2000). Methods and guidelines for effective model calibration. In *Joint Conference on Water Resource Engineering and Water Resources Planning and Management*, Minneapolis, Minnesota, USA. (pp. 1-10).

Howell, K. B. (2018). *Ordinary Differential Equations: An Introduction to the Fundamentals*. CRC Press. New York, USA.

Insam, H., & De Bertoldi, M. (2007). Microbiology of the composting process. In Compost Science and Technology, In LF. Diaz, M de Bertoldi, W Bidlingmaier, E Stentiford (eds.). *Waste management series* (Vol. 8, pp. 25-48). Elsevier.

IUPAC. (2014). *Compendium of Chemical Terminology, Gold Book Version* 2.3.3. International Union of Pure and Applied Chemistry. Available at: http://goldbook.iupac.org/PDF/D01817.pdf. Accessed 22 May 2018.

Karatzas, G. P. (2017). Developments on the modeling of groundwater flow and contaminant transport. *Water Resources Management*, 31(10), 3235-3244.

Kelbaliyev, G. I., Tagiyev, D. B., & Rasulov, S. R. (2019). *Transport Phenomena in Dispersed Media*. CRC Press. New York, USA.

Kirkham, J. M., Smith, C. J., Doyle, R. B., & Brown, P. H. (2019). *Inverse modeling for predicting both water and nitrate movement in a structured-clay soil (Red Ferrosol)*. PeerJ, 6, e6002.

Konikow, L. (2011). The secret to successful solute transport modeling. *Ground Water* 49(2): 144-159.

Konikow, L. F., & Bredehoeft, J. D. (1978). Computer Model of Two-Dimensional Solute Transport and Dispersion in Ground Water. *U.S. Geological Survey Techniques of Water-Resources Investigations*.

Konikow, L. F., Goode, D. J., & Hornberger, G. Z. (1996). A three-dimensional method-of-characteristics solute-transport model (MOC3D). *Water-Resources Investigations Report,* 96, 4267.

Kovarik, K. (2000). *Numerical models in groundwater pollution*. Springer Science & Business Media, Germany.

Lage, J. L., & Narasimhan, A. (2000). Porous Media Enhanced Forced Convection: Fundamentals and Applications. In: Vafai, K. (ed.) *Handbook of Porous Media*, 8, 357-394.

Landmeyer, J. E. (2011). *Introduction to phytoremediation of contaminated groundwater: historical foundation, hydrologic control, and contaminant remediation*. Springer Science & Business Media.

Leij, F. J., & van Genuchten, M. T. (2002). Solute transport. *Soil physics companion*, 189-248.

Lewis, F. M., Voss, C. I., & Rubin, J. (1986). Numerical simulation of advective-dispersive multi-solute transport with sorption, ion exchange, and equilibrium chemistry. *Water-Resources Investigations Report*, 86, 4022.

Mandle, R. J. (2002). *Groundwater modeling guidance. Michigan Department of Environmental Quality*, GMP, draft, 1, USA.

Mayer, A. S., & Hassanizadeh, S. M. (2005). *Soil and groundwater contamination: Nonaqueous phase liquids* (No. 17). American Geophysical Union, USA.

McDowell-Boyer, L. M., Hunt, J. R., & Sitar, N. (1986). Particle transport through porous media. *Water Resources Research*, 22(13), 1901-1921.

Merkel, B. J., Planer-Friedrich, B., & Nordstrom, D. K. (2005). *Groundwater geochemistry. A practical guide to modeling of natural and contaminated aquatic systems, 2*. Spring-Verlag Berlin Heidelberg, Germany. pp. 200.

Middlemis, H., Merrick, N., & Ross, J. (2001) *Groundwater flow modeling guideline*. Murray Darling Basin Commission (MDBC) Report, Australia.

Mishra, S., & Parker, J. C. (1989). Parameter estimation for coupled unsaturated flow and transport. *Water Resources Research*. 25:385–396.

Moatamedi, M., & Khawaja, H. A. (2018). *Finite Element Analysis*. CRC Press. New York, USA.

Müller, F. K. Z. B. P., & Sedláčková, M. K. I. (2003). Contamination of soils and groundwater by petroleum hydrocarbons and volatile organic compounds–Case study: ELSLAV BRNO. *Bulletin of Geosciences*, 78(3), 225-239.

Neill, S. P., & Hashemi, M. R. (2018). *Fundamentals of Ocean Renewable Energy: Generating Electricity from the Sea*. Academic Press. Cambridge, Massachusetts, USA.

Ney, U., Schoberth, S. M., & Sahm, H. (1991). Anaerobic degradation of sulfite evaporator condensate in a fixed-bed loop reactor by a defined bacterial consortium. *Applied microbiology and biotechnology*, 34(6), 818-822.

Orban, P. (2009). *Solute transport modeling at the groundwater body scale: Nitrate trends assessment in the Geer basin (Belgium)*. Doctoral Dissertation. University of Liège, Belgium.

Pallant, A., & Lee, H. S. (2015). Constructing scientific arguments using evidence from dynamic computational climate models. *Journal of Science Education and Technology*, 24(2-3): 378-395.

Pandey, A. K., Kumar, R., & Singh, M. K. (2018). A Solution to Advection–Dispersion Equation for the Heterogeneous Medium Using Duhamel's Principle. In Amit Kumar Pandey, Rohit Kumar, Mritunjay Kumar Singh (eds.). *Applications of Fluid Dynamics* (pp. 559-571). Springer, Singapore.

Payne, S. M., & Woessner, W. W. (2010). An Aquifer Classification System and Geographical Information System-Based Analysis Tool for Watershed Managers in the Western US 1. *JAWRA Journal of the American Water Resources Association*, 46(5): 1003-1023.

Peaceman, D. W. & Rachford Jr., H. H. (1955). Numerical solution of the parabolic and elliptic differential equations. *Journal of the Society for Industrial and Applied Mathematics*. 3: 28-41.

Pinder, G. F., & Bredehoeft, J. D. (1968) Application of the digital computer for aquifer evaluation. *Water Resource Research*, 4: 1069–1093. doi: 10.1029/WR004i005p01069.

Poeter, E. P., & Hill, M. C. (1997). Inverse modeling: A necessary next step in groundwater modeling. *Ground Water*: 35(2): 250-260.

Powell, A. L. (2017). *Water Resource Planning, Development, and Management*. Nova Publishers, USA.

Prickett, T. A. & Lonnquist, C. G. (1971). A selected digital computer for groundwater resource evaluation. *Illinois State Water Survey Bull.*, 55: 62.

Prior, J. M. (2018). *Contemporary Mission Theology: Engaging the Nations (Essays in Honor of Charles E. Van Engen)*. Robert Gallagher & Paul Hertig (eds.) American Society of Missiology Series No. 53. Maryknoll, New York, Orbis Books, USA.

Raats, P. A., & Knight, J. H. (2018). The Contributions of Lewis Fry Richardson to Drainage Theory, Soil Physics, and the Soil-Plant-Atmosphere Continuum. *Frontiers in Environmental Science*, 6, 13.

Rapp, B. E. (2016). *Microfluidics: Modeling, Mechanics, and Mathematics*. William Andrew. Norwich, NY, USA.

Rastogi, A. K. (2007). *Numerical groundwater hydrology*. Penram International Publication, India. pp. 981.

Rastogi, A. K., Aparna, B., & Sanjeeta, K. (2010). *Modeling and Inverse modeling in Groundwater Hydrology: Application of FEM and soft computing optimization.* Omniscriptum Publishing Group, Germany. pp. 412.

Renard, P. (2005). The future of hydraulic tests. *Hydrogeology Journal*, 13: 259–262.

Richens, D. T. (1997). *The chemistry of aqua ions: synthesis, structure, and reactivity: a tour through the periodic table of the elements*. Wiley. ISBN 0-471-97058-1.

Rivett, M., Drewes, J., Barrett, M., Chilton, J., Appleyard, S., Dieter, H. H., Wauchope, D. & Fastner, J., (2006). Chemicals: Health relevance, transport, and attenuation, in protecting groundwater for health managing the quality of drinking-water source. In Schmon, O., Howard, G., Chilton, J., Chorus, I., & World Health Organization (eds.). *Protecting groundwater for health: Managing the quality of drinking-water sources.* pp. 81-137. Recuperado de http://www. WHO. int/iris/handle/10665/43186.

Rubin, H., Narkis, N., & Carberry, J. (Eds.). (2013). *Soil and aquifer pollution: non-aqueous phase liquids-contamination and reclamation.* Springer Science & Business Media.

Saatsaz, M., Sulaiman, W. N. A., Eslamian, S., & Mohammadi, K. (2013)$_a$. Hydrogeochemistry and groundwater quality assessment of Astaneh-Kouchesfahan Plain, Northern Iran. *International journal of water*, 7(1-2), 44-65.

Saatsaz, M., Sulaiman, W. N. A., Eslamian, S., & Javadi, S. (2013)$_b$. Development of a coupled flow and solute transport modeling for Astaneh-Kouchesfahan groundwater resources, North of Iran. *International journal of water*, 7(1-2), 80-103.

Scesi, L. T. G., & Gattinoni, P. (2012). *Methods and Models to Determine the Groundwater Flow in Rock Masses: Review and Examples.*

Scheidegger, A. E. (1961). On the statistical properties of some transport equations. *Canadian Journal of Physics*, 39(11), 1573-1582.

Schwartz, F. W., & Zhang, H. (2003). *Fundamentals of Groundwater.* John Wiley & Sons. New York, USA. pp. 583.

Schwarz, C. V., Reiser, B. J., Davis, E. A., Kenyon, L., Acher, A., Fortus, D., Shwartz, Y., Hug, B., & Krajcik, J. (2009). Developing a learning progression for scientific modeling: Making scientific modeling accessible and meaningful for learners. *Journal of Research in Science Teaching*, 46(6): 632–654.

Schweizer, B. (2015). Darcy's law and groundwater flow modeling Snapshots of modern. *Mathematics from Oberwolfach*, no 7, Germany.

Scozzari, A., & Dotsika, E. (2016). *Threats to the Quality of Groundwater Resources.* Springer Berlin Heidelberg, Germany.

Selim, H. M. (2014). *Transport & fate of chemicals in soils: principles & applications.* CRC Press. New York, USA.

Simmons, C. T. (2008). Henry Darcy (1803–1858): Immortalized by his scientific legacy. *Hydrogeology Journal*, 16(6), 1023. doi: 10.1007/s10040-008-0304-3.

Stone, H. L. (1968). Iterative Solution of Implicit Approximations of Multidimensional Partial Differential Equations. *SIAM Journal on Numerical Analysis.* 5 (3): 530-538. doi: 10.1137/0705044.

Stumm, W., & Morgan, J. J. (1996). Aquatic Chemistry: Chemical Equilibria and Rates in Natural Waters. *Environmental Science and Technology.* John Wiley & Sons. New York, USA.

Sun, N. Z. (2013). *Inverse problems in groundwater modeling* (Vol. 6). Springer Science & Business Media, Germany.

Thangarajan, M., (ed.). (2007). *Groundwater: Resource evaluation, augmentation, contamination, restoration, modeling, and management*. Springer Science & Business Media.

Tiedeman, C. R., & Hill, M. C. (2007). Model calibration and issues related to validation, sensitivity analysis, post-audit, uncertainty evaluation, and assessment of prediction data needs. In *Groundwater* (pp. 237-282). Springer, Dordrecht, Germany.

Tikhomirov, V. V. (2016). *Hydrogeochemistry fundamentals and advances: groundwater composition and chemistry, volume 1 for designers, owners, and operators*. John Wiley & Sons Inc; ISBN-10: 1119160391.

Turner, M. J., Clogh, R. W., Martin, H. C., & Topp, L. J. (1956). Stiffness and deflection analysis of complex structures. *Journal of Aeronautical Science*, 23(9):805–823, 854.

Vrugt, J. A., Stauffer, P. H., Wöhling, T., Robinson, B. A., & Vesselinov, V. V. (2008). Inverse modeling of subsurface flow and transport properties: A review with new developments. *Vadose Zone Journal*, 7(2), 843-864.

Wang, D., Wu, C., Huang, W., & Zhang, Y. (2019). Vibration investigation on fluid-structure interaction of AP1000 shield building subjected to multi earthquake excitations. *Annals of Nuclear Energy*, 126: 312-329.

Willmann, M. Carrera, J. Sanchez-Vila, X. Willmann, M., Carrera, J., & Sánchez-Vila, X. (2008). Transport upscaling in heterogeneous aquifers: What physical parameters control memory functions?. *Water Resources Research*, 44(12).

Woessner, W. W. (2007). *Applied Flow and Solute Transport Modeling in Aquifers: Fundamental Principles and Analytical and Numerical Methods*. CRC Press Book. New York, USA.

Young, D. M. (1971). *Iterative Solution of Large Linear Systems*. Academic Press, New York, USA.

Zienkiewicz, O. C. (1971). *The Finite Element Method in Engineering Science*. McGraw-Hill, London, UK.

In: Advances in Hydrogeochemistry Research
Editors: S. Saeid Eslamian and Faezeh Eslamian
ISBN: 978-1-53617-829-6

Chapter 5

Assessing the Microbiological Quality of Potable Groundwater from Selected Protected and Unprotected Wells in Murehwa District, Zimbabwe

Fiona Munyebvu[1], Never Mujere[2,*], Rajendra Kumar Isaac[3] and Saeid Eslamian[4,5]

[1]Department of Geography and Environmental Science,
University of Zimbabwe, Harare, Zimbabwe
[2]Department of Geography and Environmental Science,
University of Zimbabwe, Harare, Zimbabwe
[3]Department of Soil Water Land Engineering and Irrigation and Drainage Engineering, Deemed University, Allahabad, India
[4]Department of Water Engineering, Collage of Agriculture,
Isfahan University of Technology, 8415683111, Isfahan, Iran
[5]Center of Excellence in Risk Management and Natural Hazards,
Isfahan University of Technology, 8415683111, Isfahan, Iran

Abstract

Groundwater is one of the most available source of freshwater in arid and semi-arid regions. However, the quantity and quality of groundwater are under threat from overexploitation, climate change, pollution, land use and, land cover changes. The purpose of this paper is to highlight the spatial variations of microbiological groundwater quality in three rural wards of Murehwa District in Zimbabwe. The study was conducted from February to March in 2011. Water samples from 60 randomly selected wells comprising 30 protected and 30 unprotected hand dug wells were analyzed for faecal coliforms using

* Corresponding Author's Email: mujere@arts.uz.ac.zw.

a portable Oxfam Delagua Water Testing Kit. An isotropic kriging was used to map the spatial variations of faecal coliform levels in the wells. Results indicated significant ($p = 0.00$) differences in coliform counts between the protected and unprotected wells. Coliforms in 67% of the protected wells and all unprotected wells failed to meet the Standard Association of Zimbabwe (SAZ) guideline of zero CFU per 100 ml. The distance from pit latrine was significantly negatively correlated ($R^2 = 0.064$, $p = 0.04$) with coliform counts in unprotected wells. However, the relationship between the distance from pit latrine and coliforms counts in protected wells was positive, but not significant ($R^2 = 0.0005$, $p = 0.628$). Well depth was significantly related to coliform counts in both protected (R2 = 0.1986, $p = 0.028$) and unprotected wells ($R^2 = 0.39$, $p = 0.001$) in a negative sense. Although slope angle increased with increasing coliform counts protected and unprotected wells, the relationships were not significant ($p>0.05$). Overall, the research findings showed both protected and unprotected wells were vulnerable to bacterial contamination. The study also highlights an important aspect of the application of GIS in wellhead protection which is relevant in water resource management.

Keywords: spatial variations, thermotolerant coliforms, statistical significance, water wells, murehwa district

INTRODUCTION

Access to adequate safe drinking water is vital for primary health care hence a fundamental human right. It is the basic component of human development as well as a precondition for man's ability to deal with hunger, poverty, disease, and death. Regrettably, surface water has become a scarce commodity as only a small proportion of the world population has access to quality drinking water. Groundwater has become a major alternative source of water for socio-economic activities (e.g., domestic, agricultural, industrial and drinking) and, environmental processes in arid and semi-arid regions. In southern Africa, almost 70% of the people rely on groundwater as a major source of drinking water and, in most cases, consumed without receiving treatment to improve quality (Rosa and Clasen, 2010).

Groundwater is widely distributed, ubiquitous and represents 97% of the world's freshwater resources. It is a primary source of drinking water for half of the world's population, including over 1.5 billion urban dwellers. It also constitutes over 40% of the total annual volume of irrigated water (Ndoziya et al., 2019). In arid and semi-arid regions, groundwater is vital in satisfying domestic, industrial, municipal, agricultural and recreational water needs. Its relative pervasiveness, reliability, and drought resistance reduce livelihoods sensitivity to rainfall and risk of crop failure. It is often preferred over surface water because of its relative stability in terms of both quality and quantity. It is relatively a safe water source due to natural purification processes that take place because water is filtered by soil particles as it seeps underground.

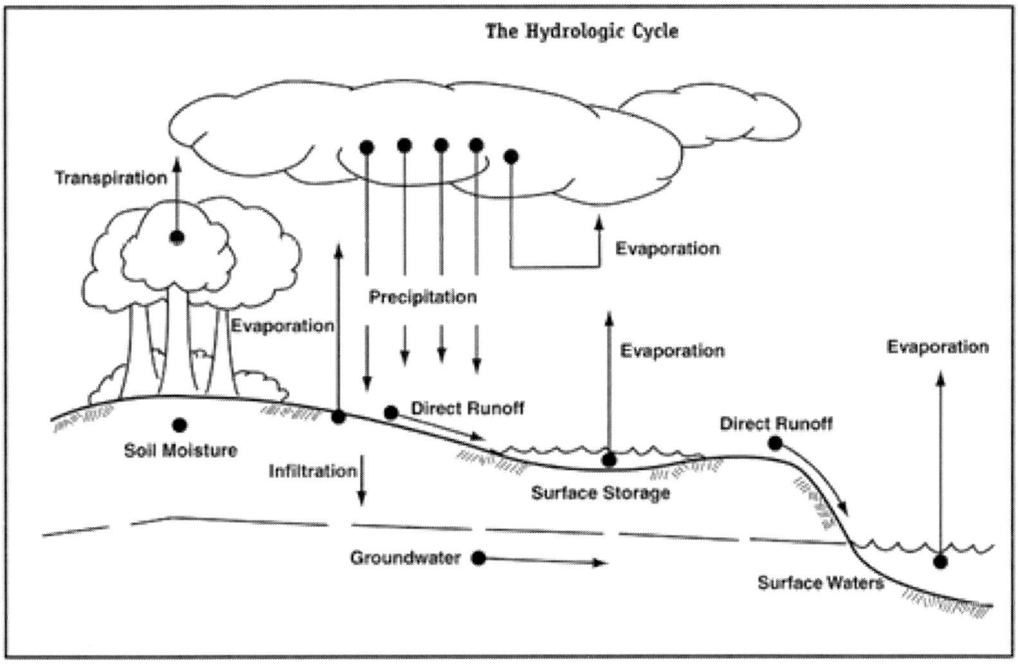

Source: Https://www.ques10.com/p/28761/diagram-of-hydrological-cycle-and-aquifers-1/?

Figure 1. Hydrological cycle showing the groundwater movement.

Partial development of groundwater resources is due to inaccessible, remote locations, unaffordable costs of drilling and the maintenance of wells or boreholes. Thus, although groundwater resources are pervasive, it is often the poor who are less likely to benefit from groundwater resources because of the high investment costs needed to access it. While most low-income households may not have land rights or capital necessary to install private well, middle and high-income households can afford to do so with or without permits thus, resulting in intensive and unregulated use.

Urban and rural settings provide slightly different challenges with regard to groundwater management. In urban settings, large numbers of private wells complicate equitable groundwater use and sustainable management. The quality of groundwater is often compromised by contamination from urban sewerage or waste due to poorly designed collection and treatment systems. Also, peri-urban areas are often characterised by heavily compromised groundwater, with excess levels of nitrate, chloride and microbial pathogens. There is a concern that chemical and microbial contaminants in pit latrines can leach into groundwater sources thereby threatening human health (Dzwairo et al. 2006; Graham and Polizzotto 2013). Previous studies on the impacts of pit latrines on groundwater quality have demonstrated deterioration in groundwater quality (Haruna et al., 2005; Graham and Polizzotto 2013). Thus, the protection of groundwater sources from pollution by pit latrines is critical. There is strong evidence that access to improved sanitation and safe drinking water can reduce diarrhoea morbidity and mortality and soil-transmitted helminths.

In rural settings, groundwater-related challenges for drinking water, sanitation and hygiene take different forms. Given the sparse infrastructure and degree of remoteness, groundwater is often the only alternative to scarce or polluted surface water. It has the potential to significantly improve the livelihoods of women, children, the elderly and people with disabilities. It reduces their burden in water collection and makes time available for education, economically productive activities and reduces the burden of physical challenges they face. Yet achieving equity within and among rural communities remains logistically challenging. Direct connections to water and sanitation services are not always possible or economically viable and distances from central community water points vary greatly. Thus, ensuring the sustainable use of groundwater resources through appropriate allocation and pollution prevention practices is critical.

Occurrence of Groundwater

Groundwater exists whenever water penetrates beneath surface rocks which are permeable enough to transmit the water (Figure 1). The rate of infiltration should be sufficient so that the rocks are saturated with an appreciable thickness. Groundwater becomes a useful resource when the rocks in the zone of saturation are permeable enough to yield useful supplies of water to springs, streams, rivers, reservoirs, boreholes, and wells.

The Role of Groundwater in Achieving Sustainable Development Goals

The United Nations Sustainable Development Goals (SDGs) also known as the Global Goals are a universal sets of goals, targets, and indicators that UN member states are be expected to use to frame their agendas and political policies over a period of 15 years, from January 2016 to January 2030. There are 17 goals with a focus on key areas including poverty alleviation, democratic governance, and peace building, climate change, disaster risk, and economic inequality. UNDP provides support to governments to integrate SDGs into their national development plans and policies.

The critical role of groundwater to successfully implement the SDGs is widely acknowledged. Groundwater is explicitly linked to ensuring availability and sustainable management of water and sanitation for all (Goal 6); contribute towards poverty eradication (Goal 1); food security (Goal 2); gender equality (Goal 5); sustainability of cities and human settlement (Goal 11); combating climate change (Goal 13) and protecting terrestrial ecosystems (Goal 15) among others. In providing water for domestic use, enabling food production and sustaining critical ecosystems function, groundwater is central in relating to human development, human dignity, and well-being. Increased water availability can also free resources for other investments.

Groundwater Contamination

Water quality determines the degree of water suitability for particular purposes. It is a measure of the amount and type of substances contained in water with respect to its application (Moyo and Mtetwa, 2002). This definition takes into account the fact that even when water is polluted with sewage, depending on the level of the contaminant, the water can be used for other purposes such as agriculture though is not suitable for use in homes. Levels of water quality are assessed in terms of physical (for example, temperature, taste, turbidity, hardness, and smell) chemical (conductivity, dissolved salts, viruses, heavy metals, nitrates, and oxygen) biological (amount of micro-organisms present such as bacteria, viruses, protozoa, and algae) parameters. In its natural form, water has a quality suitable for drinking and sustaining aquatic life. The deterioration of water quality often has negative effects on animal health. However, substances introduced through human activities and natural processes deteriorate the water quality. Water pollutants can originate from point and/or non-point sources. Point source pollution include mine dumps, oil spills or pit latrines in the case of groundwater, industrial processes, mining activities, and sewage effluent. Non-point or diffuse sources of water pollution reduce the quality of water by introducing excessive levels of nutrients such as nitrogen and phosphorus, organic matter, and pathogens into the environment. Pathogens contained in manure can affect human health either through the ingestion of unwashed crops or through the ingestion of polluted groundwater used as a source of drinking water (Paidamoyo, 2016).

Levels of groundwater contamination vary from time to time and, from place to place. During the wet seasons, contaminants find their way into groundwater through infiltration and lateral sub-surface movement. Water pollution will result if the contamination plume spreads faster than the rate of contaminant attenuation by the soil. The mineral substances dissolved by water during percolation through the soil and rocks can reach high concentrations making groundwater unfit for the desired use. Thus, the quality of groundwater is compromised due to both natural processes and socio-economic activities including agriculture, mining, dumping waste and industrial activities.

The depth to the water table and lateral separation between onsite sanitation facilities and the groundwater source are some of the key parameters affecting groundwater pollution. The level of contamination may also be influenced by the depth of the water source, and where hydraulic loads are high and exceed natural attenuation potential in the subsurface. Latrine siting guidelines have been developed taking into account transport distances for microbiological and chemical contaminants originating from pit latrines.

WHO (2011)suggests that the distance of latrines from water sources should be at least 30 m and the bottom of the pits should be at least 1.5 m above the groundwater table.

Contaminants find their way into groundwater through infiltration and lateral groundwater movement. This is due to the movement of pollutants by rainwater and underground seepage. However, for longer distances of up to 200 m, underground travel of total and fecal coliform have also been reported in Nigeria (Adekunle and Adetunji, 2007). Usually, these long distances are associated with macropores found in gravel or other coarse materials, structured or cracked clay soils, fractured rock, and cavernous limestone (FAO, 1992).

Coliforms in Groundwater

Coliform bacteria are commonly live in soil, on vegetation, in surface water, and in the intestines of warm-blooded creatures. Most coliform bacteria do not cause illness. However, since coliform bacteria are commonly associated with sewage or surface waters, their presence in drinking water indicates the potential for disease-causing organisms or pathogens such as strains of bacteria, viruses, and protozoa. Waterborne diseases from pathogens include flu-like symptoms such as nausea, vomiting, fever, and diarrhea. In Zimbabwe, bacterial contamination in domestic water poses great health risks to people and animals (Ward, 1989). There are three different groups of coliform bacteria including the total coliform, *Escherichia coli* (E. coli) and fecal or thermotolerant coliform (Paidamoyo, 2018). Each category contain a different level of risk. High coliform counts of up to 240 CFU/100 ml in groundwater were observed in the rainy season in Zimbabwe (Nyati, 2004).

Total coliform bacteria can persist within slime or biofilm formed by naturally occurring groundwater micro-organisms (Paidamoyo, 2018). The slime clings to the well screen, casing, drop pipe and pump and can also harbor bacteria in protected wells. Disturbances during pumping or well maintenance can cause the slime to dislodge, releasing the coliform bacteria.

Escherichia coli, (E.coli) *is* a common waterborne pollutant that causes death and illnesse when ingested in contaminated food and water (Crump and Suka, 2002). Humans have a large and harmless population of *E. coli* in lower large intestines. When released into drinking water sources, *E coli* can be ingested and enter the upper small intestine causing diarrhea.

Thermotolerant or fecal coliform bacteria occur in water containing fecal material from humans and livestock. The presence of thermotolerant coliform in drinking water indicates the presence of faecal material and that intestinal pathogen could be present (WHO 2011). Major surface sources fecal coliforms in groundwater include seepage from on-site sanitation systems such as pit latrines. Overflow and leakages from a pit latrine can percolate (seep) down to the water table and contaminate groundwater.

Studies in some parts of Zimbabwe have revealed that pit latrines were microbiologically contaminating groundwater up to 25 m distance (Dzwairo et al., 2006). In in eastern Zimbabwe total coliform, *E.coli*, and thermotolerant coliforms were posing a threat to human health. The bacterial counts exceeded the WHO standards of zero counts/100 ml in untreated surface and groundwater used for drinking (WHO, 1993). Research conducted in north-eastern Zimbabwe, studies have observed coliform counts up to 240 CFU/100 ml in a boreholes during the wet seasons (Nyati, 2004). The presence of the coliforms in protected wells can be attributed to contamination through groundwater seepage.

Protected and Unprotected Water Wells

A well (as opposed to a borehole) is a shallow and large diameter hand-dug hole into the groundwater body using simple equipment. Wells are highly susceptible to contamination in areas with intensive agriculture and pollutant discharge from shallow aquifers can add significant amounts of contamination (Rodwang, 2000).

A protected well (Figure 2a) comprises a completely sealed wellhead brickwork lining, heavy concrete slab and in-built or housed lifting device such as rope and washer, wellhead protection and an apron a concrete slab and an apron. Protected water wells include sources that, by nature of their construction are protected from outside contamination, particularly fecal matter.

Unprotected water wells comprise of unlined wells, with an opening on the surface without proper wellhead protection and mobile water abstraction devices which include bucket and rope device (UNICEF and WHO, 2004) (Figure 2b). Unprotected water wells have inadequate sanitary completion which is usually characterized by a poor wellhead protection, and no concrete slab or apron.

Contamination of wells is caused by surface water flowing along the well casing or from a loose fitting or absent of the well cap that allows insects, animals, or surface water to enter. Bacteria can also be introduced during the well construction and can remain if the water system is not thoroughly disinfected and flushed. The well construction defects such as insufficient well casing depth, improper sealing, corroded or cracked well casings, and poor well seals or caps can facilitate contamination of water. Poor design of aprons allow waste water to wash back into wells.

Photo by Fiona Munyebvu (2012)

Figure 2. Schematic view of (a) protected well and (b) an unprotected well in Murehwa District.

Groundwater monitoring in both protected and unprotected water wells has become an important component in water and sanitation programs. This issue suggests groundwater exploitation should be viewed as part of water resources management in which socio-economic, legal and institutional aspects are integrated. Studies have shown that outbreaks of waterborne disease are more prominent during the wet season than the dry season because pollutants are washed by rainwater into groundwater (Adekunle and Adetunji., 2007; Pritchard and Mkandawire, 2007; Omo-Irabor and Olobaniyi, 2008, Muzenda et al., 2019). Hardly any work has been undertaken to monitor the spatial variations in water quality in shallow wells.

Most studies in Zimbabwe have been conducted to assess seasonal variations in the groundwater quality focusing on boreholes. Hardly any work has been undertaken to investigate water in hand dug wells. This study, therefore, assesses the microbiological water quality in selected protected and unprotected water wells during the wet season. The study focuses in three wards of Murehwa District in Zimbabwe. Specifically, the study aimed at: (i) testing for differences in thermotolerant coliform levels in protected and unprotected wells, (ii) determine the relationship between thermotolerant coliform counts and, slope, distance from pit latrine and well depth (iii) determine the spatial variations of fecal coliform levels, within the vicinity of protected and unprotected wells. Understanding levels of groundwater quality helps in providing useful information for rural water supply planning and, putting in place measures for protection wells, handling and purifying contaminated groundwater.

Case Study: Murehwa District

Murehwa district lies in the north-eastern side of Zimbabwe. Three wards, namely ward 12, 13 and 28 were considered for analysis in this study. During the 2012 national population census, Ward 12 had a population on 9,353 people (2,376 households), Ward

13 had 5,987 people (1,442 households) and Ward 28 had 5399 people (1330 households) (ZimStat, 2012). Figure 3 shows the location of Murehwa District in Zimbabwe and the relative location of Ward 12, 13, and 28.

The relief consists of undulating slopes, distinct hills and granitic rock outcrops with few portions of dolerite uplands and basalt in low lying areas. Soils in the district are predominantly sandy loams and clay loams from the weathering of granite rocks. The presence of these fertile soils make the area good for crop farming. However, frequent droughts inhibit rainfed farming activities.

Rainfall is highly variable and erratic. Most of the rains are received during the rainy season which spans from October to April. The 30 year (1981-2010) mean annual rainfall received in the area is 850 mm (Katsi, 2006). Thus, most households depend on groundwater for their water needs.

Groundwater occurs in fractured and weathered greenstones and granite rocks. Although greenstones and granite have very low permeability, fracturing and weathering creates adequate porosity and permeability, thus, high groundwater development potential. The transmissivity of the rocks is low (1 – 10 m^2/d), The depth of weathering are shallow and varies from 10 m to 30 m. Aquifer thickness is generally less than 15 m. The rock aquifers support yields of 1.2 - 2.9 l/s. Boreholes are typically 30 m to 40 m deep and have low to moderately low specific capacity of the order 2 – 20 $m^3/d/m$. The water level is typically from 5 m to 20 m below ground level.

Figure 3. Location of the Wards 12, 13 and 28 in the Murehwa District and sampled water wells *(Coordinates are in meters UTM Zone 36 South based on the WGS84 reference spheroid).*

The main form of sanitation systems in Murehwa district are pit latrines that are often near to shallow wells. Most of the pit latrines are brick lined, but there no mechanism of desludging the filled up pits. In 2010, sanitation coverage was 17.5% meaning that 82.5% of the population had no proper sanitation facilities to ensure hygienic separation of human excreta from human contact. Almost 42% of the households were practizing open defecation (UNICEF et al., 2010).

Domestic water supply is predominantly from community or privately owned boreholes and wells. In 2002, 24% of the households got water from boreholes and wells on their premises, 46% of the households got water from sources located within 500 m away from their premises, whilst 4% of the households relied on water from sources located 1,000 m or more away from their homesteads (Katsi, 2006).

Data Collection Methods

Mapping of wells and collection of water samples were done during the field survey conducted from February to March 2011 when the research team observed locations of pit latrines and wells and collected water samples. A Global Positioning System (GPS) was used to take the coordinates of the wells and pit latrines. The locations of wells and pit latrines were as mapped in a GIS. An inventory of the water wells in the three wards was made by recording the well number, well depth, distance from pit latrine and whether protected or unprotected. From a total of 75 wells comprising 41 protected wells and 34 unprotected wells were identified in the three wards. 30 protected wells and 30 unprotected wells were randomly selected for analyzing the presence of coliforms. Thus, 60 water samples were collected in 500 ml sterilized plastic bottles and placed in a cooler box and then, analyzed for thermotolerant coliform presence using the potable OXFAM DELAGUA water testing kit. This was done within 24 hours after samples were collected in the field.

Preparation of testing kit and determination of thermotolerant coliforms were done following standard methods. The temperature of the incubator was calibrated using a thermometer and a testing incubator lid. The temperature was monitored over a period of 30 minutes until it had stabilized at 44.5°C. Calibrating to 44.5°C was done to ensure that only thermo-tolerant bacteria grew. Filtration apparatus and Petri dishes were sterilized using boiling water (Oxfam, 2004).

The membrane filtration method was used to determine coliform levels in the water samples using the Oxfam Delagua Water Testing Kit (Oxfam, 2004). A culture medium was prepared using 38.1 g of Membrane Laurel Sulphate Broth (MLSB) and 500 ml distilled water. The analysis of water samples for thermotolerant coliform was then carried out by passing 100 ml and 50 ml of water samples through a sterile filter. The filter membrane was then placed onto a paper pad soaked in the culture medium and incubated for 16 hours at 44.5°C. Physical colony count of all yellow colonies with a diameter of between 1 and 3 mm on Petri dishes was done using a hand lens within 15 minutes of removing the Petri dishes from the incubator. The procedure was repeated for all 60 water samples and results were recorded.

A digital elevation model (DEM) of the study area was used in calculating the slope in Integrated Land and Water Information System (ILWIS), a GIS-based software using the linear filter technique, dFdX (slope angle), which is the slope difference in the X-direction and dFdY which is the slope difference in the Y-direction. Slope percentages (*Slopepct*) were converted to slope degrees (Slopedeg) using the command:

$$Slopedeg = Atan(\frac{Slopepct}{100}) \quad (1)$$

The slope value of each of the sampled water well was determined using the coordinates of the water point and the calculated slope in degrees using the equation:

$$WaterwellSlope = mapvalue(\text{Slopedeg, coordinate}) \quad (2)$$

Spatial and statistical analyses were performed to determine the spatial variations of coliforms in selected wells. Isotropic kriging of coliform counts was done in ILWIS GIS. Isotropic kriging was used with the assumption that spatial autocorrelation of coliform counts was equal in all directions from an output pixel up to a limiting distance or range. Statistical analyses were applied to whether there were significant differences in coliforms counts between protected and unprotected water wells. The analysis also focused on identifying the relationships between coliform levels and slope, well depth as well as the distance from the pit latrine.

RESULTS

Variation of Thermotolerant Coliform Counts

The mean coliform count in protected wells was 1.57 CFU/100 ml while in unprotected as wells was 120.83 CFU/100 ml. Almost 16.67% of all the wells had coliform levels of zero CFU/100 ml whilst 83.3% of the wells had coliform levels above zero. The Mann Whitney-U test statistic for independent samples indicated a statistically significant difference ($p = 0.000$) in thermotolerant coliforms between protected and unprotected water wells.

The thermotolerant coliform colonies in 33.3% of protected wells were within the Standard Association of Zimbabwe (SAZ) guidelines of zero CFU/100 ml whilst thermotolerant coliform levels 66.6% failed to meet the guidelines. Figure 4 shows that

80% of the protected wells failed to meet the guidelines were in Ward 12, 58% in ward 13, and 62.5% in the Ward 28. All the 30 sampled unprotected wells from the Wards 12, 13, and 28 did not meet the WHO guidelines of 0 CFU/100 ml (WHO, 2004).

Relationship between Coliform Levels and Distance from Pit Latrine

Ten percent of the sampled wells had no pit latrines and in all these homesteads, water wells were not protected. The minimum distance of a water source from a pit latrine which was the likely source of contamination was 12 m while the maximum was 37 m. The distance from a pit latrine was significantly negatively correlated ($p = 0.036$, $R^2 = 0.062$) with thermotolerant coliform count in unprotected wells. In protected wells, an insignificant positive relationship ($R^2 = 0.0005$, $p = 0.6278$) existed between thermotolerant coliform counts and of the distance from the pit latrine. With an increase in the distance from the pit latrine, thermotolerant coliform colonies were decreasing in unprotected wells and the reverse was true for protected wells (Figure 5).

Variation of Thermotolerant Coliform Counts with Well Depth

The average depth of groundwater in protected wells was 4 m while in unprotected wells it was 10 m. Significant negative relationships between the well depth and thermotolerant coliform level in both protected wells ($R^2 = 0.1986$, $p = 0.028$) and unprotected wells ($R^2 = 0.39$, $p = 0.001$). Thus, as well depth increasds, the coliform counts significantly decreased (Figure 6).

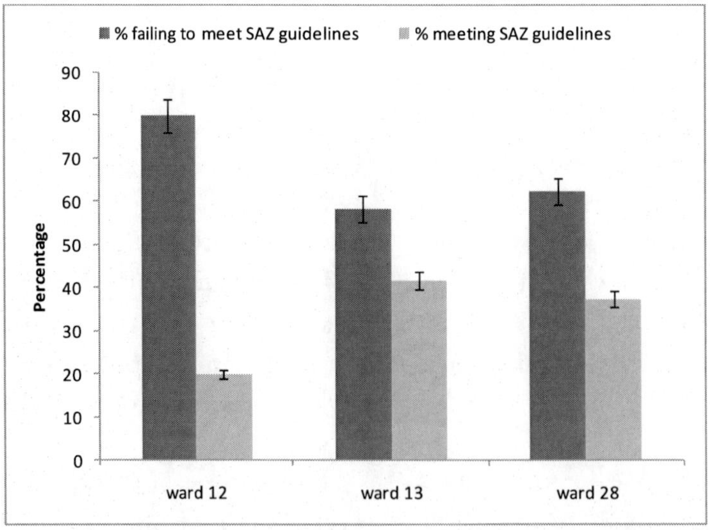

Figure 4. Thermotolerant coliform counts in protected wells compared to SAZ guidelines

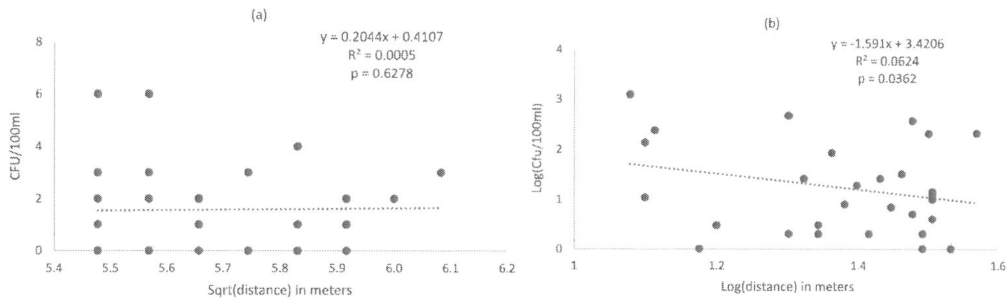

Figure 5. Correlation between distance from pit latrine and thermotolerant coliform levels in (a) protected and (b) unprotected wells.

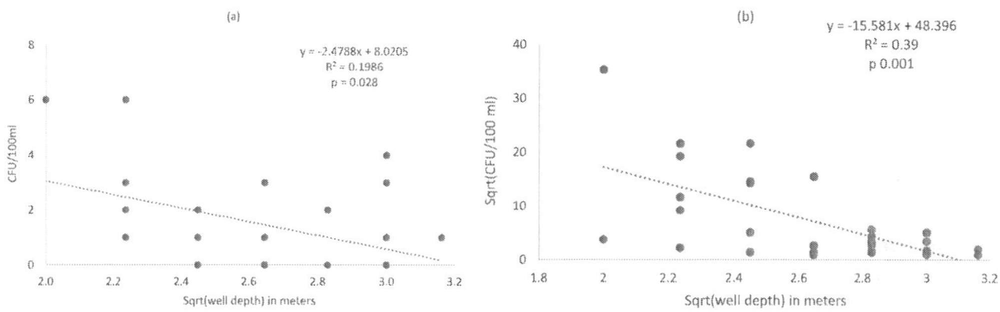

Figure 6. Significant linear negative relationships between well depth and thermotolerant coliform counts in (a) protected and (b) unprotected wells.

Variation of Coliform Counts with Slope

Elevated thermotolerant coliform counts between 400 and 1248 CFU/100 ml were recorded aton slopes measuring from zero to one degree in unprotected wells. Low thermotolerant coliform levels of 19 CFU/100 ml were also recorded on slopes measuring 14.7 degrees. However, the results indicated no significant relationships between slope angles and thermotolerant coliform counts in protected (R^2 = 0.0093, p = 0.756) and unprotected wells (R^2 = 0.0144, p = 0.109). Slope increased with increasing thermotolerant coliforms in protected wells while in protected wells coliforms counts decreased as slopes increased. Thus, thermotolerant coliform levels did not significantly vary with an increase or decrease in slope angle in both protected and unprotected wells.

Figure 8 shows that wells in areas to the south east and north had high number of coliform colonies. Low concentration of coliform colonies were observed in the central and southern parts of the area and in most of these areas, the pit latrines were present.

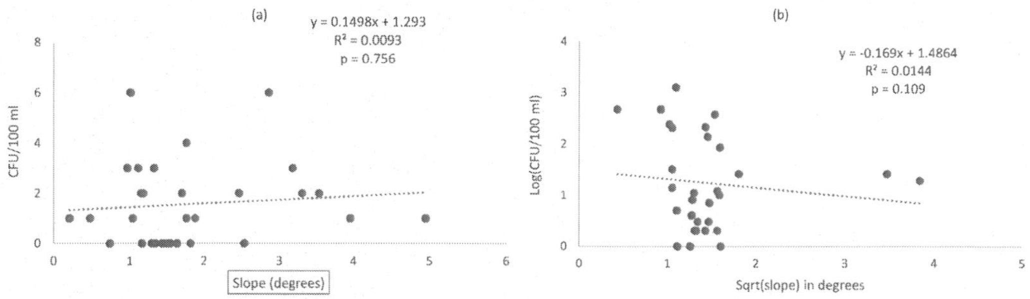

Figure 7. Correlation between slope and coliform levels in (a) protected and (b) unprotected wells.

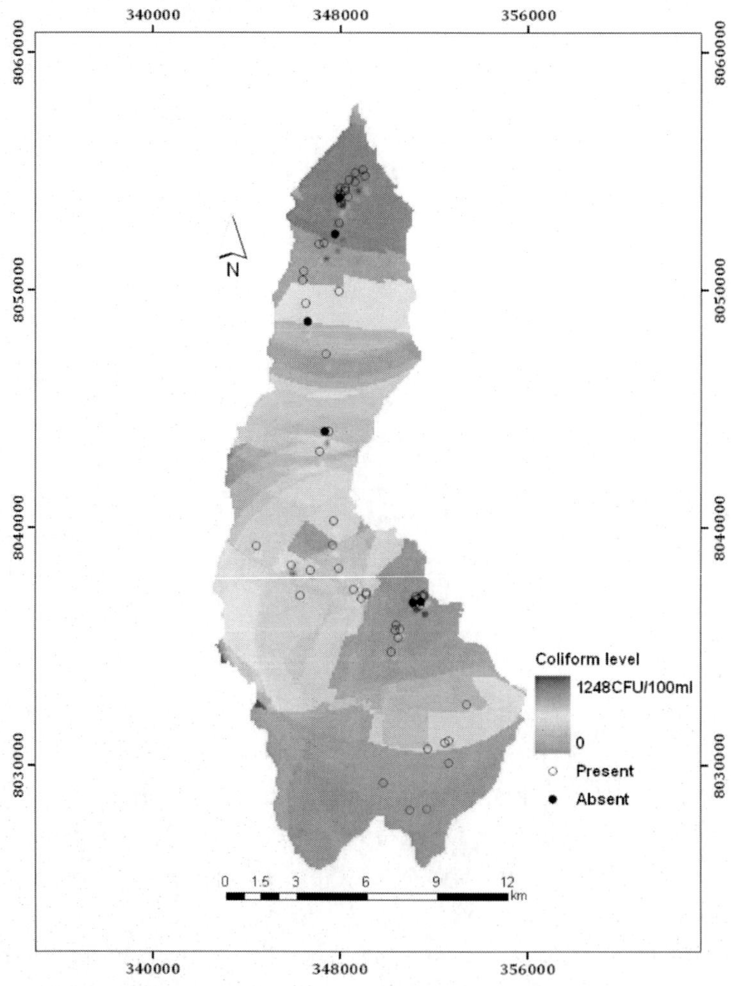

Figure 8. Spatial variations of thermotolerant coliform levels in relation to presence and absence of pit latrines *(Map coordinates are in meters UTM zone 36 south based on the WGS84 reference spheroid).*

CONCLUSION

Contamination of groundwater result in poor water quality, loss of water supply, high clean-up costs, high costs for alternative water supplies, and/or potential health problems. This paper highlights findings from a study conducted to map and determine the spatial variations of groundwater quality levels in protected and unprotected wells in three wards of Murehwa District of Zimbabwe during the wet season of 2011. The quality of groundwater was determined by taking samples from 60 wells within the study area.

The study revealed the presence of thermotolerant coliforms in all the unprotected wells while only two-thirds of the protected wells had coliforms. Significant negative correlations existed between coliform counts in wells and, the distance from to the pit latrines in unprotected wells and depth of wells in protected and unprotected wells. These findings agree with those found in a study in Iran by Dobaradaran et al. 2009, where total coliform and thermotolerant coliform decreased with increasing the depth of well (r=-0.5572).

The study results showed no significant relationship between slope and coliform counts in both protected and unprotected wells. Contrary to the research findings, a study conducted by Boyer (2008) concluded that the mean splash distance for fecal coliforms was less than 50 mm in all directions at a zero slope angle and more than 500 mm in the down slope direction on a 40.8 slope angle, meaning slope angle influenced coliform movement.

Therefore the protected and unprotected shallow wells were all vulnerable to the contamination than boreholes. Thus, integrated approaches in rural water supply and sanitation schemes need to be developed in areas prone to groundwater contamination like Murehwa district. These approaches can include policies that address both protection of the wells and provision of improved sanitation facilities. Routine water quality monitoring before and after implementation of these water supply schemes (e.g., drilling or protection of a well) at each specific water point is also important to ensure the public health safety. Monitoring and maintenance of abstraction devices are also important.

REFERENCES

Adekunle, I. M., Adetunji, M. T., Gbadebo, A. M. & Banjoko, O. P. (2007). Assessment of Groundwater quality in a typical rural settlement in southwest Nigeria. *International Journal of Environmental Research and Public Health*, *4*(4), 307-18.

Boyer, D. G. (2008). Fecal Coliform dispersal by rain splash on slopes. *Agricultural and Forest Meteorology*, *148*, 1395-1400.

Crump, J. A. & Sulka, A. A. (2002). An outbreak of Escherichia Coli O157:H7 Infections among visitors to a dairy farm. *The New England Journal of Medicine*, *347*(8), 555-60.

Dzwairo, B., Hoko, Z., Love, D. & Guzha, E. (2006). Assessment of the impacts of pit latrines on groundwater quality in rural areas: a case study from Marondera district, Zimbabwe. *Physics and Chemistry of the Earth*, *Parts A/B/C*, *31*, (15–16), 779–788.

Dobaradaran, S., Bina, B. & Mahmoudian, M. H. (2009). 'The Effects of well depth on water Quality in Neighboring Zone of Zayandahroud, Iran. *World Applied Sciences Journal*, *5*, 80-83.

Graham, J. P. & Polizzotto, M. L. (2013). Pit latrines and their impacts on groundwater quality: a systematic review. *Environmental Health Perspectives*, *121*, 521–530.

Katsi, L. (2006). *Assessment of factors which affect multiple uses of water and their impacts on the sustainability of rural water supply sources*. Department of Civil Engineering, University of Zimbabwe.

Moyo, N. A. G. & Mtetwa S. (2002). Water quality management and pollution control. In: Hirji, R., Johnson, P. Muzenda, F., Masocha, M., Misi, S.N. (2019). Groundwater quality assessment using a water quality index and GIS: A case of Ushewokunze Settlement, Harare, Zimbabwe. *Physics and Chemistry of the Earth*, *Parts A/B/C*, Volume *112*, 134-14

Ndoziya, A. T., Hoko, Z. & Gumindoga, W. (2019). Assessment of the impact of pit latrines on groundwater contamination in Hopley Settlement, Harare, Zimbabwe. *Journal of Water*, *Sanitation and Hygiene for Development*, 9 (3), 464-476.

Nyati, H. (2004). "Evaluation of the microbial quality of water supplies to municipal, mining and squatter communities in the Bindura urban area of Zimbabwe, in relation to incidence of specific enteric diseases" *Zimbabwe Journal of Science and Technology*, *4*, 26-30.

Omo-Irabor, O. O. & Olobaniyi, S. B. (2008). Surface and groundwater water quality assessment using multivariate analytical methods: A case study of the Western Niger Delta, Nigeria. *Physics and Chemistry of the Earth*, *33*, 666-673.

Oxfam. (2004). *Oxfam Delagua water testing kit Users' Manual*, Oxfam, UK.

Paidamoyo, M., (2016). *Assessment of groundwater vulnerability in Norton Town, Zimbabwe*. University Of Zimbabwe. MSc theses. Civil Engineering, University of Zimbabwe, Harare.

Pritchard, M. & Mkandawire, T. (2007). Biological, chemical and pyhsical drinking water quality from shallow wells in Malawi: case study of Blantyre, Chiradzulu and Mulanje. *Physics and Chemistry of the Earth Journal*, *32*, 1167-1177.

Rodvang, J. (2000). *Agricultural land resources Atlas of Alberta- Aquifer vulnerability index for the Agricultural area of Albert*. Government of Alberta, Agriculture and Rural Development.

Rosa, G. & Clasen, T. (2010). Estimating the scope of household water treatment in low- and medium-income countries. *The American Journal of Tropical Medicine and Hygiene*, *82*(2), 89–300.

UNICEF, CASS and Government of Zimbabwe. (2010). *A Situational Analysis on the Status of Women's and Children's Rights in Zimbabwe, 2005-2010*, UNICEF, Zimbabwe.

WHO. (2004). *Meeting the MDG Drinking Water and Sanitation Target, A mid-term Assessment of Progress.*

WHO. (2011). *Guidelines for Drinking-Water Quality* (4th Edn) World Health Organization, Geneva, Switzerland.

Zvidzai, C., Mutukirwa, T., Mundembe, R. & Sithole-Niang, I., (2007). Microbial Community analysis of drinking water sources from rural areas of Zimbabwe. *African Journal of Microbial Research*, *1*, 100-103.

ZimStat, (2012). Zimbabwe National population census, *ZimStat*, Harare. Https:// www.ques10.com/p/28761/diagram-of-hydrological-cycle-and-aquifers-1/? (Accessed on 10 December 2019).

In: Advances in Hydrogeochemistry Research
Editors: S. Saeid Eslamian and Faezeh Eslamian
ISBN: 978-1-53617-829-6

Chapter 6

^{36}CL CONTRIBUTION FOR DATING GROUNDWATERS FROM THE GUARANI AQUIFER SYSTEM

Daniel Marcos Bonotto[1,*], Leslie Keith Fifield[2] and Saeid Eslamian[3,4]

[1]Departamento de Petrologia e Metalogenia,
Universidade Estadual Paulista (UNESP), Câmpus de Rio Claro, Av. 24-A No. 1515,
C.P. 178, CEP 13506-900, Rio Claro, São Paulo, Brasil
[2]Department of Nuclear Physics, Research School of Physics and Engineering,
Australian National University, Canberra, ACT 2601, Australia
[3]Department of Water Engineering, Collage of Agriculture,
Isfahan University of Technology, 8415683111, Isfahan, Iran
[4]Center of Excellence in Risk Management and Natural Hazards,
Isfahan University of Technology, 8415683111, Isfahan, Iran

ABSTRACT

The field of groundwater age dating nowadays considers a groundwater sample not as water that recharged the flow regime at a point in the past, but as a mixture of waters that have resided in the subsurface for varying lengths of time. This recognition resolves longstanding inconsistencies encountered in age dating and suggests new ways to carry out age dating studies. Groundwater age dating takes advantage of the known decay rates of radioactive isotopes, the timing of the introduction into the atmosphere of isotopes from nuclear testing or reactors, or the history of the release of manufactured gases to estimate the age of a groundwater sample. ^{36}Cl has properties which make it suitable for the study of confined groundwaters. Its long half-life (301 ka) and the fact that chloride is not removed from solution by mineral interaction or secondary mineral formation would, in principle, permit the estimation of very long groundwater residence times. This chapter reports the use of the ^{36}Cl method for estimating the groundwater residence time in the

* Corresponding Author's Email: danielbonotto@yahoo.com.br.

northern part of the Guarani Aquifer System (GAS) in Brazil. It underlies 1.2M km^2 in the Paraná sedimentary basin of South America and is an important source of water for industry, agriculture and domestic supplies. The isotopic database was set after collecting and analyzing rainfall and groundwater samples coming from São Paulo and Mato Grosso do Sul states, helping to understand the dynamics of this groundwater system and its sustainability. The ages varied between 43.8 ka and 2.0 Ma that exceed the range of 30-40 ka from the conventional ^{14}C dating, thus, expanding the results obtained from previous work in the region, including other radio-isotope analyses.

Keywords: ^{36}Cl measurements, AMS-accelerator mass spectrometry, groundwater, GAS-Guarani aquifer system, dating, residence time

INTRODUCTION

^{36}Cl has properties which make it suitable for the study of confined groundwaters in addition to other hydrogeochemical tools, for instance, those reported by Saatsaz et al. (2013a, 2013b). Its long half-life (301 ka) and the fact that chloride is not removed from solution by mineral interaction or secondary mineral formation would, in principle, permit the estimation of very long groundwater residence times. It is produced naturally in the atmosphere and lithosphere via various reactions. In the atmosphere, the principal production reactions are proton- and neutron-induced spallation of ^{40}Ar. The primary cosmic-ray protons will certainly produce ^{36}Cl, but most of the isotope will be produced by secondary particles, many of which are neutrons. Further, the probability of ^{36}Ar absorbing a neutron to make ^{36}Cl is very low. An upper limit for the cross-section (<1.5 mbarn) was measured by Jiang et al. (1990). For comparison, the equivalent reaction on ^{14}N to make ^{14}C has a cross-section that is 3 orders of magnitude higher at 1.75 barns. On this basis, and the facts that the isotopic abundance of ^{36}Ar is only 0.33% and argon is only ~ 1% of the atmosphere, the formation of ^{36}Cl via the $^{36}Ar(n,p)^{36}Cl$ reaction is negligible. Although the atmosphere attenuates cosmic-ray intensity, the residual fluxes of fast secondary cosmic rays (mostly neutrons) produce ^{36}Cl by spallation of potassium and calcium in continental rocks. Cosmic-ray induced reactions are negligible at depths greater than about 50m of rock, below which reactions caused by the in-situ neutron flux continue to generate ^{36}Cl. This flux is the result of (α,n) reactions due to irradiation of light nuclei by α-particles emitted by the natural radioelements within the rock (Bentley et al., 1986a, 1986b). Because of the large activation cross-section of ^{35}Cl, the reaction $^{35}Cl(n,\gamma)^{36}Cl$ is particularly important in high Cl^- environments (Fabryka-Martin, 1988).

The huge transboundary Guarani Aquifer System (GAS) consists of Triassic-Jurassic eolian-fluvio-lacustrine sandstones confined by thick (up to 1,500 m) Cretaceous basalt flows of the Serra Geral Formation. It has continental dimensions, extending over some 1.2 million km^2 within the Paraná sedimentary basin, South America, and comprising southern Brazil (states of Mato Grosso, Mato Grosso do Sul, Goiás, Minas Gerais, São

Paulo, Paraná, Santa Catarina and Rio Grande do Sul), eastern Paraguay, NW Uruguay and the northeastern extreme corner of Argentina. The GAS encompasses all of the Paraná basin and part of the Chaco-Paraná Basin and is one of the world's largest (Araújo et al., 1999).

The GAS has been the focus of different approaches by many investigators since the 1970s. The number of studies increased dramatically, however, after the GEF (Global Environment Fund)-funded Guarani Aquifer Program for Groundwater Resource Sustainability and Environmental Protection (Foster et al., 2006). Since then, hydrochemical, stable isotopes and ^{14}C data have been obtained (see, for instance, Gastmans et al., 2010a, 2010b and Hirata et al., 2011, among other), but the Brazilian agencies have also supported several radiometric investigations in GAS groundwaters (see, for instance, Bonotto and Caprioglio, 2002; Bonotto, 2004, 2006, 2011, 2012; Bonotto and Mello, 2006; Bonotto and Bueno, 2008; Bonotto and Armada, 2008; Bonotto et al., 2009a, 2009b; Soler i Gil and Bonotto, 2015). The groundwater residence time in the GAS is an essential information for a proper understanding and some approaches have been realized, principally based and ^{3}H and ^{14}C data (Silva, 1983; Kimmelmann e Silva et al., 1989; Kimmelmann et al., 1995). The large size of the Paraná basin requires the evaluation of very long groundwater residence times and, for such purpose, ^{36}Cl and the U-isotopes ^{234}U and ^{238}U have also allowed to perform some age determination (Bonotto, 2006; Creswell and Bonotto, 2008).

This chapter reports a novel ^{36}Cl database in rainwater and groundwater of the GAS and discusses the major implications on the age estimate in groundwater occurring in the aquifer.

Materials and Methods

Multiaquifer systems mainly comprising sandstones and basalts plus sediments from the Passa Dois Group behaving as aquitards have been proposed to represent the hydrostratigraphy of the Paraná basin (Campos, 2000). Groundwater occurs within the interflow zones and along cooling joints in basalts and diabases from the Serra Geral Formation, where interbedded sediments greatly increase the average porosity of large volumes of rocks. The sandstones of Cretaceous age (Bauru Group), which have an average thickness of 100 m, are moderately cemented, and exhibit adequate properties to store water. The Paleozoic sediments (Devonian-Permian age) also provide water, which is highly mineralized in the central parts of the basin and also contains H_2S in some wells.

The Guarani Aquifer System (GAS) has an average thickness of 300-400 m, and is composed of silty and shaly sandstones of fluvial-lacustrine origin (the Pirambóia and Rosário do Sul formations in Brazil and Buena Vista formation in Uruguay), and variegated quartzitic sandstones accumulated by eolian processes under desert conditions (the

Botucatu formation in Brazil, Misiones formation in Paraguay, and Tacuarembó formation in Uruguay and Argentina) (Araújo et al., 1999). A thick basaltic group of the Serra Geral formation (up to 1,500 m) overlies this aquifer, reducing its exposed areas in non-continuous elongated strips, 10-100 km wide, along the edges of the Paraná basin, where the longest strip stretches between the states of São Paulo and Santa Catarina, in Brazil. The aquifer overlies previous formations ranging from the igneous basement to the Paleozoic sediments of the Passa-Dois and Tubarão Groups, being covered by Cretaceous sediments of the Bauru-Caiuá formations. The percolating water moves from the phreatic exposed areas that surround the entire basin towards its central part.

The rainwater and groundwater samples in this study were collected at different cities of the Brazilian states of São Paulo and Mato Grosso do Sul (Figure 1). The wet season in the area generally occurs between October and March, while the dry period is between April and September. Two sampling campaigns were conducted, in the end of March 2006 and November 2007. The rainfall samples were collected with bulk (dry and wet deposition) collectors that were installed in São Paulo State at the stations described by Creswell and Bonotto (2008) and in Mato Grosso do Sul State at the offices of SANESUL (Mato Grosso do Sul State agency responsible for water collection, treatment and supply) situated at Amambai (AMB), Sidrolândia (SDL), Maracaju (MRJ) and Camapuã (CMP) municipalities. The samplers consisted of 80 cm-long and 50 cm-wide stainless steel manufactured trays coupled to polyethylene flasks (20 L) that allowed the collection of enough rainwater volume within short time periods (~3 days).

The groundwater samples (40 L) were stored in polyethylene vessels and taken (a) along the SE-NW traverse in São Paulo State described by Silva (1983), Sracek and Hirata (2002), Bonotto (2006), Creswell and Bonotto (2008), among other, and (b) from five deep bores drilled at the following localities of Mato Grosso do Sul State: Três Lagoas (TLG), Amambai, Maracaju, Sidrolândia and Ribas do Rio Pardo (RRP) (Table 1).

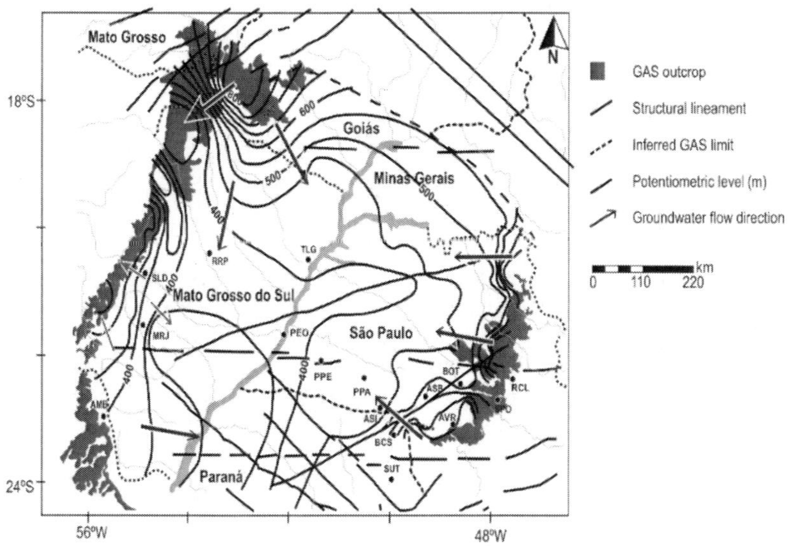

Figure 1. A simplified map showing the location of the sampling points, as well the outcrop and groundwater flow direction in the Brazilian northern portion of the GAS. The symbols for the cities are: RCL- Rio Claro, SPO- São Pedro, BOT- Botucatu, ASB- Águas de Santa Bárbara, AVR- Avaré, SUT- Sarutaiá, BCS- Bernardino de Campos, ASI- Assis, PPA- Paraguaçu Paulista, PPE- Presidente Prudente, PEO- Presidente Epitácio, AMB- Amambai, MRJ- Maracaju, SDL- Sidrolândia, RRP- Ribas do Rio Pardo, CMP- Camapuã, and TLG- Três Lagoas. Adapted from Silva (1983), Araújo et al. (1999) and Gastmans (2007).

Table 1. Description of the tube wells drilled at the aquifers of this study. Data reported by Silva (1983) and Gastmans (2007)

Sample Code	City	State[1]	Latitude	Longitude	Altitude (m asl)	Depth (m)
AVR	Avaré	SP	23°06´46´´S	48°54´33´´W	640	150
SUT	Sarutaiá	SP	23°16´03´´S	49°29´05´´W	750	152
BCS	Bernardino de Campos	SP	23°01´41´´S	49°29´05´´W	660	509
PPA	Paraguaçu Paulista	SP	22°25´21´´S	50°33´38´"W	474	3663
PPE	Presidente Prudente	SP	22°06´45´´S	51°22´43´´W	407	1800
PEO	Presidente Epitácio	SP	21°46´29´´S	52°05´27´´´´W	258	3953
AMB	Amambai	MS	23°03´50´´S	55°15´21´´W	460	150
MRJ	Maracaju	MS	21°39´34´´S	55°09´46´´W	380	150
SDL	Sidrolândia	MS	20°56´09´´S	54°58´07´´W	400	95
RRP	Ribas do Rio Pardo	MS	20°26´42´´S	53°45´36´´W	500	136
TLG	Três Lagoas	MS	20°47´19´´S	51°41´49´´W	315	4582

[1] SP = São Paulo, MS = Mato Grosso do Sul

The temperature, pH, redox potential Eh, and conductivity readings of the water samples were done *in situ* using potentiometric methods (Bonotto, 2006, 2012). The filled flasks were transported to LABIDRO-Isotopes and Hydrochemistry Laboratory of UNESP at Rio Claro city, São Paulo State. The suspended solids were separated on filtering each sample through 0.45-μm Millipore membranes. The dry residue (DR) (~TDS, total dissolved solids) content was evaluated on evaporating the filtrate to dryness in a weighed flask that was dried to constant weight at 180°C, with the increase in flask weight representing TDS (APHA, 1989).

A 100-mL aliquot of each water sample was separated and utilized for measuring alkalinity by titration with 0.02 N H_2SO_4 (Hach, 1992). Another 100-mL aliquot was also separated and sent to CENA-Center for Nuclear Energy in Agriculture, Piracicaba, for analysis of the major cations (Na^+, K^+, Ca^{2+}, and Mg^{2+}) by atomic absorption spectrometry, and the major anions (Cl^-, SO_4^{2-}, NO_3^-) by FIA-flow injection analysis. The bromide data were obtained by ICP-MS analysis performed at Nicholas School of the Environment, Duke University, USA.

The remaining water volume was used for ^{36}Cl analysis. The measured Cl^- concentrations were used to calculate the volume needed for the ^{36}Cl analysis. Evaporative concentration was needed in some cases, in which the original water volume (3.1-19.5 L) was slowly evaporated in a clean environment, reducing to final volumes of 250 mL to 1 L. Dissolved chloride was precipitated as silver chloride (AgCl) in the clean laboratories of the Department of Nuclear Physics at the ANU and purified by re-dissolution in ammonia, precipitation of barium sulfate, followed by re-precipitation of AgCl from the supernatant to remove unwanted sulfur (Conard et al., 1986). The AgCl sources were pressed into silver bromide-masked copper holders and the $^{36}Cl/Cl$ ratio measured using the 14UD Tandem Accelerator the Department of Nuclear Physics at the ANU. Processing blanks yielded $^{36}Cl/Cl$ ratios of $1.0\pm0.5\times10^{-15}$.

HYDROCHEMICAL HIGHLIGHTS

Table 2 reports the pH, conductivity, ionic concentrations, and ^{36}Cl analytical results in the bulk composition of the rainwater samples of this study. The pH values indicate slightly acid conditions as also reported elsewhere (e.g., Berner and Berner, 1987; Vieira et al., 1988; Moreira-Nordemann et al., 1988; Lara et al., 2001; Pelicho et al., 2006). The conductivity is typically low (mean = 90 μS/cm). The rainwater data treatment according to the Aquachem 4.0 software (Waterloo Hydrogeologic, 2003) showed they are dominantly mixed in terms of dissolved cations, whereas HCO_3^- dominated the dissolved anions, despite Ca^{2+} and Cl^- were identified as major ions in some samples.

Table 2. The pH, ionic concentrations, and ^{36}Cl analytical results in the bulk composition of rainwater in São Paulo and Mato Grosso do Sul states, Brazil

Parameter	Unit	Minimum	Maximum	Average	CV* (%)
pH	-	5.8	6.9	6.3	5.1
Conductivity	μS/cm	60	120	90	25.1
Sodium	mg/L	0.17	1.90	0.87	66.6
Potassium	mg/L	0.13	1.40	0.53	79.4
Calcium	mg/L	0.31	2.60	1.37	62.6
Magnesium	mg/L	0.02	0.40	0.13	89.2
Chloride	mg/L	0.53	4.10	1.5	70.5
Bromide	mg/L	0.01	0.03	0.02	55.8
Sulfate	mg/L	0.27	1.34	0.86	42.6
Nitrate	mg/L	0.40	4.82	2.10	61.3
Bicarbonate	mg/L	1.0	7.0	3.4	73.6
Cl^-/Br^- ratio	-	44	111	72	36.6
TDS	mg/L	5.0	18.0	10.8	35.5

* Coefficient of variation

Location	Sample code	Original volume	Volume after evaporation	Chloride content	^{36}Cl/Cl	^{36}Cl/Cl error	^{36}Cl atoms/L
		(L)	(L)	(mg/L)	$\times 10^{-15}$	$\times 10^{-15}$	$\times 10^{6}$
Rio Claro	RCL	19.5	0.25	1.11	24.0	2.5	0.4
São Pedro	SPO	4.5	0.25	0.87	123.1	8.2	1.8
Botucatu	BOT	3.1	0.25	1.76	25.9	4.9	0.2
Maracaju	MRJ-RW	1.0	-	1.60	29.6	2.2	0.8
Sidrolândia	SDL-RW	1.0	-	2.00	162.6	8.3	5.5

Na^+ and Cl^- are the dominant ions in coastal areas and in rain over the oceans and oceanic islands. Keywood et al. (1997) and Faure (1998) pointed out a general trend of decreasing Na^+ and Cl^- concentrations with increasing distance from the coast and such relationship was found by Creswell and Bonotto (2008) considering the rainfall data in São Paulo State. However, the dataset reported in Table 2 did not confirm such finding, showing an opposite trend between the sodium content and coast distance (Pearson correlation coefficient $r = -0.70$) as the Na^+ levels were slightly higher in the monitoring stations situated at Mato Grosso do Sul State. One possible explanation for such finding could be related to the wind directions that tend dominantly to follow the SE-NW direction in São Paulo State but other directions in Mato Grosso do Sul State, thus, bringing air masses from distinct Brazilian continental and coastal zones (Amarante et al., 2001).

Table 3. The pH, ionic concentrations, and ^{36}Cl analytical results in the bulk composition of the groundwater samples collected at São Paulo and Mato Grosso do Sul states, Brazil

Parameter	Unit	Minimum	Maximum	Average
Temperature	°C	23	70	36
pH	-	5.9	9.6	7.6
Eh	mV	-72	545	140
Conductivity	μS/cm	60	1340	473
Sodium	mg/L	2.2	208.4	64.3
Potassium	mg/L	0.06	3.5	1.5
Calcium	mg/L	1.1	33.0	10.9
Magnesium	mg/L	0.01	8.1	2.2
Bromide	mg/L	0.01	0.17	0.05
Sulfate	mg/L	1.0	79.3	18.2
Nitrate	mg/L	0.5	6.5	1.1
Bicarbonate	mg/L	23	337	135
TDS	mg/L	37.6	771.9	280.6

Parameter	Sample Code	Original volume	Volume after evaporation	Chloride concentration	Chloride/ Bromide	$^{36}Cl/Cl$	$^{36}Cl/Cl$ error	^{36}Cl atoms/L
Unit	-	(L)	(L)	(mg/L)	ratio	($\times 10^{-15}$)	($\times 10^{-15}$)	($\times 10^{6}$)
Avaré	AVR	11.35	1	0.65	32	109.5	5.7	1.2
Sarutaiá	SUT	11.35	1	0.49	49	83.3	5.1	0.7
Bernardino de Campos	BCS	11.30	0.25	0.63	63	145.4	7.3	1.6
Paraguaçu Paulista	PPA	0.25	-	9.36	234	9.6	1.4	1.5
Presidente Prudente	PPE	0.10	-	102.90	605	7.1	1.1	12.4
Presidente Epitácio	PEO	0.10	-	24.60	410	3.7	0.8	1.5
Três Lagoas	TLG	0.25	-	84.0	-	4.7	1.2	6.7
Amambai	AMB	10.75	0.25	0.15	-	369.0	15.0	0.9
Maracaju	MRJ	-	-	0.50	-	-	-	-
Sidrolândia	SDL	8.60	0.25	0.15	-	147.0	8.0	0.4
Ribas do Rio Pardo	RRP	7.80	0.25	8.6	-	284.0	13.0	41.5

Chlorine-36 in precipitation is a parameter needed for groundwater dating based on the ^{36}Cl-method. Because chloride in groundwater may come from various sources, reliable ^{36}Cl values in rainfall are required for adequate age determination. The Cl^-/Br^- mass ratio is an useful chemical criterion in order to select representative initial values of ^{36}Cl that have their origin in atmospherically produced radionuclides and, for precipitation, the ratio most commonly ranges from 80-160 (Davis et al., 1998). Table 2 shows a range of 44-111 for this parameter that is not correlated with the coast distance. The same happened for the ^{36}Cl values in rainwater (range of 0.2-5.5 × 10^6 atoms/L, Table 2) that were compatible with other reported elsewhere, for instance, a range of 0.1-10 × 10^6 atoms/L for wet deposition collected near West Lafayette, Indiana (USA) (Davis et al., 1998).

Table 3 reports the physicochemical data, ionic concentrations, and ^{36}Cl analytical results obtained in the groundwater samples of this study. The insertion of the acquired data in an Eh-pH diagram (Krauskopf and Bird, 1995; Brownlow, 1996) indicates that the analyzed groundwaters are oxidizing, transitional/reducing and reducing in character, depending on the sampling site. According to temperature, the groundwaters analyzed are cold (< 25°C), hypothermal (25-33°C), and hyperthermal (> 38°C) (DFPM, 1966). The groundwater temperature depends on the water flow from the border of the Paraná basin towards its central part, in the direction of the dip of the geological units, according to the natural geothermal gradient of ~ 1°C/35m (Teissedre and Barner, 1981). There was an excellent linear relationship between the groundwater temperature and geostatic (lithostatic) pressure (Castany, 1982) in the bores, indicating pressures above 200 bar in the hyperthermal waters and values lower than 100 bar in the cold/hypothermal waters. Temperature was also useful for identifying some trends of the hydrochemical data in the groundwaters studied. For instance, the cold/hypothermal waters exhibited lower values of conductivity, TDS, sodium, (bi)carbonate, chloride, and sulfate relative to the hyperthermal waters.

The results of the groundwater data treatment according to the Aquachem 4.0 software (Waterloo Hydrogeologic, 2003) showed that the cold/hypothermal waters were Ca-dominated in terms of dissolved cations, whereas the hyperthermal waters were Na-dominated. In terms of dissolved anions, (bi)carbonate dominated the composition of all groundwater samples. The chemistry of the hyperthermal waters in this study was equivalent to others occurring at the Paraná basin as reported by Gastmans et al. (2012).

The mean dissolved chloride content in the rainwater samples analyzed in this study corresponded to 1.5 mg/L (Table 2). The comparison of this value with those of the groundwaters indicates molar ratios (groundwater/rainwater) below 1 for the GAS cold/hypothermal waters, possibly indicating Cl^- removal by vegetation and superficial soil covers from the infiltrating rainfall during the water-soil/rock interactions. Chlorine is an essential element for plants and is thus taken up by their roots and utilized in plant tissues and fluids (Sheppard et al., 1999). Such processes could affect the choice of the chlorine-36 input values for the groundwater dating based on the ^{36}Cl-method.

On other hand, groundwater/rainwater Cl^- molar ratios exceeding unity were found for the hyperthermal waters, indicating chloride enrichment relative to the precipitation in the GAS groundwater of the deep zones, possibly favored by increased dissolution processes caused by higher temperatures occurring there. One cold water exhibited a groundwater/rainwater Cl^- molar ratio above 1, but that bore (RRP) was exploiting water from Serra Geral Aquifer (SGA). Bonotto (2012) reported Cl^- levels generally lower than 3 m/L in SGA groundwaters and the higher value found for RRP was accompanied by the more elevated nitrate content found in it relative to the other groundwater samples. Such values suggested anthropogenic inputs occurring there perhaps due to imperfect well sealing and poor sanitary conditions close to the well that is located in the urban area of the municipality. Thus, the chlorine-36 value in the groundwater sample RRP must be used with caution for dating purposes based on the ^{36}Cl-method.

^{36}Cl Data and the Standard Decay Equation

In general, the ^{36}Cl/Cl ratio rather than the ^{36}Cl concentration is a more robust parameter for determining groundwater ages. This is because the ^{36}Cl concentration is strongly influenced by evapo-transpiration, which, in turn, depends upon climate, and hence is likely to vary over time (Cresswell et al., 1999). Such aspects may be illustrated in Figure 2 that shows no clear variability of the ^{36}Cl data according to the Cl^- content but an evident tendency of decreasing of the ^{36}Cl/Cl ratio according to the increase of the Cl^- content in precipitation and groundwater in three different scenarios.

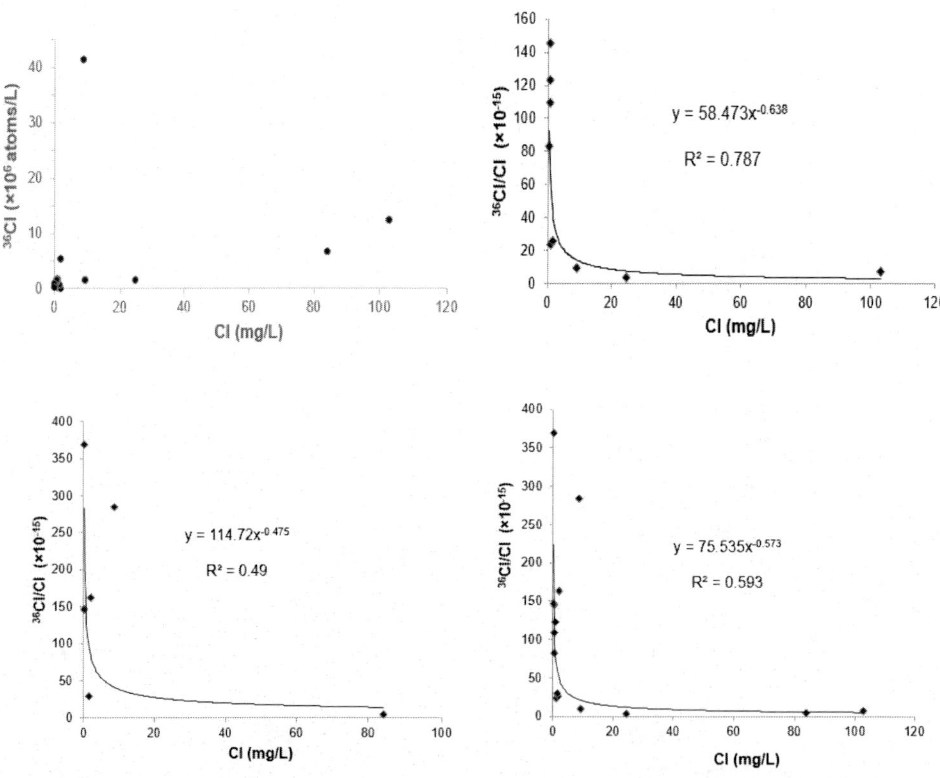

Figure 2. The number of ^{36}Cl atoms per litre (top left), ^{36}Cl/Cl ratio in rainwater and groundwater samples from São Paulo State (top right), ^{36}Cl/Cl ratio in rainwater and groundwater samples from Mato Grosso do Sul State (except sample RRP) (bottom left), and all ^{36}Cl/Cl ratio in rainwater and groundwater samples (except sample RRP) (bottom right) plotted against the chloride content.

Cresswell and Bonotto (2008) utilized ^{36}Cl data in precipitation and groundwater sampled at São Paulo State to suggest travel times of over 1 Ma for waters to reach the centre of the Paraná sedimentary basin. They adopted diverse approaches of those considered here for proposing possible evolutionary scenarios taking place at GAS from ^{36}Cl values. Davis et al. (1998) considered three aspects of the water chemistry in order to select representative initial values of ^{36}Cl that have their origin in atmospherically produced radionuclides. They are: a) anthropogenic chemicals must be absent in the samples; b) choice of water samples in the region with the lowest chloride concentration; c) Cl^-/Br^- mass ratio, which most commonly ranges from 80-160 for precipitation. The Cl^-/Br^- ratios of 44-111 and 32-63 found, respectively, in rainwater and cold/hypothermal waters of this study are compatible with the third criterion mentioned by Davis et al. (1998). Therefore, they permit choosing samples that would exhibit suitable initial ^{36}Cl-values derived from precipitation for applying the ^{36}Cl dating method in groundwaters. Davis et al. (1998) also pointed out that chloride from water having Cl^-/Br^- ratio greater ~ 200 contains abundant Cl^- from sources other than atmospheric ones. This could be the case of the hyperthermal waters of this study as they exhibit Cl^-/Br^- ratios between 234 and 605.

As a general working principle, ^{36}Cl dating of old groundwater is most straightforward and least ambiguous in situations where the meteoric ^{36}Cl/Cl ratio at the aquifer recharge area is high relative to the steady-state subsurface ratios (Phillips, 2000). Good examples of favorable situations for application of the method have been found in large sedimentary basin aquifers with recharge in the continental interior and permeabilities high enough to flush most connate fluids (Phillips, 2000), for instance, in the Australian interior at Palm Valley (Wischusen et al., 2004), the Amadeus basin (Cresswell et al., 1999), the Coonamble embayment in the Great Artesian Basin (Mahara et al., 2007), and, in North Africa, the Nubian aquifer (Sturchio et al., 2004). Wischusen et al. (2004) pointed out three major assumptions for the use of the ^{36}Cl/Cl ratio to estimate the groundwater age: a) the only sink for ^{36}Cl in the aquifer is radioactive decay; b) the only source for additional ^{36}Cl is normal deep subsurface production; c) the production rate for ^{36}Cl at the time of recharge would be equal to the present day value (Andrews and Fontes, 1992). The groundwater dating could be done using the standard decay equation:

$$t = -(\lambda)^{-1} \ln(R_{36}/R_0) \quad (1)$$

where: R_{36} is the ^{36}Cl/Cl ratio measured in the groundwater sample, R_0 is the ^{36}Cl/Cl ratio for the recharge (or input function) and $1/\lambda$ is the mean life of ^{36}Cl (434,251 yr).

Table 4. ^{36}Cl ages in the GAS cold/hypothermal waters analyzed in this study

Input function	R_0 value	Code sample	R_s value		Age (ka)[1]	
					$R_e = 1.9 \times 10^{-15}$	$R_e = 15.7 \times 10^{-15}$
Rainfall station RCL	24.0×10^{-15}	AVR	109.5×10^{-15}	nv (eq. 1)	nv (eq. 6) nv (eq. 7)	nv (eq. 6) nv (eq. 7)
		SUT	83.3×10^{-15}	nv (eq. 1)	nv (eq. 6) nv (eq. 7)	nv (eq. 6) nv (eq. 7)
		BCS	145.4×10^{-15}	nv (eq. 1)	nv (eq. 6) nv (eq. 7)	nv (eq. 6) nv (eq. 7)
		AMB	369.0×10^{-15}	nv (eq. 1)	nv (eq. 6) nv (eq. 7)	nv (eq. 6) nv (eq. 7)
		SDL	147.0×10^{-15}	nv (eq. 1)	nv (eq. 6) nv (eq. 7)	nv (eq. 6) nv (eq. 7)
		RRP	284.0×10^{-15}	nv (eq. 1)	nv (eq. 6) nv (eq. 7)	nv (eq. 6) nv (eq. 7)
Rainfall station SPO	123.1×10^{-15}	AVR	109.5×10^{-15}	50.8 (eq. 1)	51.7 (eq. 6) 178.3 (eq. 7)	58.8 (eq. 6) 185.4 (eq. 7)
		SUT	83.3×10^{-15}	169.6 (eq. 1)	172.9 (eq. 6) 422.2 (eq. 7)	201.1 (eq. 6) 450.4 (eq. 7)
		BCS	145.4×10^{-15}	nv (eq. 1)	nv (eq. 6) 66.8 (eq. 7)	nv (eq. 6) 58.2 (eq. 7)
		AMB	369.0×10^{-15}	nv (eq. 1)	nv (eq. 6) 282.2 (eq. 7)	nv (eq. 6) 246.3 (eq. 7)
		SDL	147.0×10^{-15}	nv (eq. 1)	nv (eq. 6) 685.3 (eq. 7)	nv (eq. 6) 676.2 (eq. 7)
		RRP	284.0×10^{-15}	nv (eq. 1)	nv (eq. 6) nv (eq. 7)	nv (eq. 6) nv (eq. 7)
Rainfall station BOT	25.9×10^{-15}	AVR	109.5×10^{-15}	nv (eq. 1)	nv (eq. 6) nv (eq. 7)	nv (eq. 6) nv (eq. 7)
		SUT	83.3×10^{-15}	nv (eq. 1)	nv (eq. 6) 24.9 (eq. 7)	nv (eq. 6) nv (eq. 7)
		BCS	145.4×10^{-15}	nv (eq. 1)	nv (eq. 6) nv (eq. 7)	nv (eq. 6) nv (eq. 7)
		AMB	369.0×10^{-15}	nv (eq. 1)	nv (eq. 6) nv (eq. 7)	nv (eq. 6) nv (eq. 7)
		SDL	147.0×10^{-15}	nv (eq. 1)	nv (eq. 6) 288.0 (eq. 7)	nv (eq. 6) nv (eq. 7)
		RRP	284.0×10^{-15}	nv (eq. 1)	nv (eq. 6) nv (eq. 7)	nv (eq. 6) nv (eq. 7)
Rainfall station MRJ-RW	29.6×10^{-15}	AVR	109.5×10^{-15}	nv (eq. 1)	nv (eq. 6) nv (eq. 7)	nv (eq. 6) nv (eq. 7)
		SUT	83.3×10^{-15}	nv (eq. 1)	nv (eq. 6) 45.8 (eq. 7)	nv (eq. 6) nv (eq. 7)
		BCS	145.4×10^{-15}	nv (eq. 1)	nv (eq. 6) nv (eq. 7)	nv (eq. 6) nv (eq. 7)
		AMB	369.0×10^{-15}	nv (eq. 1)	nv (eq. 6) nv (eq. 7)	nv (eq. 6) nv (eq. 7)
		SDL	147.0×10^{-15}	nv (eq. 1)	nv (eq. 6) 308.9 (eq. 7)	nv (eq. 6) 52.8 (eq. 7)
		RRP	284.0×10^{-15}	nv (eq. 1)	nv (eq. 6) nv (eq. 7)	nv (eq. 6) nv (eq. 7)
Rainfall station SDL-RW	162.6×10^{-15}	AVR	109.5×10^{-15}	171.7 (eq. 1)	174.2 (eq. 6) 662.4 (eq. 7)	194.8 (eq. 6) 683.0 (eq. 7)
		SUT	83.3×10^{-15}	290.5 (eq. 1)	295.4 (eq. 6) 906.3 (eq. 7)	337.1 (eq. 6) 948.0 (eq. 7)
		BCS	145.4×10^{-15}	48.3 (eq. 1)	49.2 (eq. 6) 550.9 (eq. 7)	54.1 (eq. 6) 555.8 (eq. 7)
		AMB	369.0×10^{-15}	nv (eq. 1)	nv (eq. 6) 766.2 (eq. 7)	nv (eq. 6) 743.9 (eq. 7)
		SDL	147.0×10^{-15}	43.8 (eq. 1)	44.4 (eq. 6) 1,169.4 (eq. 7)	48.8 (eq. 6) 1,173.8 (eq. 7)
		RRP	284.0×10^{-15}	nv (eq. 1)	nv (eq. 6) nv (eq. 7)	nv (eq. 6) nv (eq. 7)

[1] nv = negative value

The use of eq. 1 implies that the input rainfall values should give ^{36}Cl levels in excess of any of the observed groundwater samples. The cold/hypothermal groundwaters occur in the GAS recharge beds and exhibit ^{36}Cl/Cl ratios higher than those of some rainfall samples (RCL, BOT and MRJ-RW). The ratio in rainfall sample SPO is higher than in groundwater of bores AVR and SUT, whereas the ratio in rainwater SDL-RW is greater than that of groundwater from bores AVR, SUT, BCS and SDL. On the other hand, the ^{36}Cl/Cl level in groundwater from bores AMB and RRP was higher than any of the rainfalls. Table 4 records the estimated ages in the cold/hypothermal waters from using eq. 1 in the possible situations.

Some ^{14}C data by Kimmelmann e Silva et al. (1989) from bores near the SAG recharge areas have indicated a ^{14}C age of 1.1 ka for the groundwater sample SDL that is much lower than the value of 43.8 ka as estimated by the ^{36}Cl method. An initial explanation for such difference may be because of the difficulties on using ^{36}Cl at times that are short compared to its half-life. Additionally, uncertainties in the measured ^{36}Cl/Cl ratios should be considered. Thus, in that site, the values obtained with respective errors are: rainwater- $(162.6 \pm 8.3) \times 10^{-15}$; groundwater- $(147.0 \pm 8.0) \times 10^{-15}$. Therefore, they are equivalent within the analytical uncertainties of the measurements. The close correspondence between rainwater and groundwater at that site is a good evidence for young groundwater (and reasonably constant input ratio), as confirmed by the ^{14}C data.

^{36}Cl/Cl ratios of $3.7\text{-}9.6 \times 10^{-15}$ were found in the hyperthermal waters (bores PPA, PPE, PEO, TLG), which are lower than the range of $24\text{-}369 \times 10^{-15}$ in the precipitation and cold/hypothermal waters (AVR, SUT, BCS, AMB, SDL). Despite the rainfall values should represent pristine input waters, the ^{36}Cl/Cl ratios can modify for the region after mixing with any surface salts prior to infiltration (recharge). Therefore, the values recorded in the recharge beds of the aquifer exceeding those of the precipitation could also represent an input function for dating the deep hyperthermal waters. Nine different R_0 values were adopted in eq. 1 and reported in Table 5 in which R_s corresponded to three ^{36}Cl/Cl ratios of the hyperthermal waters. Tables 4 and 5 also show that, for a fixed R_s value, the increase on the R_0 values implies on older ages. Following such trend, younger ^{36}Cl groundwater ages were found adopting RRP ($R_0 = 284 \times 10^{-15}$) as the input function compared to AMB ($R_0 = 369 \times 10^{-15}$) (Table 5).

The ages reported in Table 5 (eq. 1) ranged from 0.40 Ma ($R_0 = 24 \times 10^{-15}$; $R_s = 9.6 \times 10^{-15}$) to 2.00 Ma ($R_0 = 369 \times 10^{-15}$; $R_s = 3.7 \times 10^{-15}$). Such residence times are much older than those obtained by Silva (1983) for the hyperthermal waters TLG (9.6 ka), PPA (18.9 ka), and PPE (30 ka). In that study, Silva (1983) utilized the ^{14}C method for groundwater dating, but without any "^{14}C correction procedure" such as proposed by Fontes and Garnier (1979). The hydraulic gradient of 0.0012 and hydraulic conductivity of 2.7×10^{-5} m/s reported by Silva (1983) from hydrogeological tests involving bores BCS and PPA allow estimate a velocity of the groundwater flow corresponding to 3.2×10^{-8} m/s after applying the Darcy's law for the groundwater flow in the porous media along the BCS-PPA profile

(Bonotto, 2006). That distance is ~ 380 km long, thus, the flowrate of 3.2×10^{-8} m/s implies on a groundwater residence time equivalent to 380 ka. Such elapsed time is more compatible with the ^{36}Cl ages shown in Table 5 (eq. 1) relatively to the ^{14}C ages reported by Silva (1983).

Despite the quite variable ages estimate, one relevant aspect on the application of eq. 1 to the acquired dataset is that the groundwater residence time of 43.8 – 290.5 ka for the cold/hypothermal waters is shorter than the range of 0.40 – 2.00 Ma for the hyperthermal waters. This agrees with the expected travel times in GAS, as the deep (hyperthermal) groundwaters should exhibit higher values of the ^{36}Cl ages in relation to those of the cold/hypothermal ones.

Table 5. ^{36}Cl ages in the GAS hyperthermal waters analyzed in this study ($R_e = 1.9 \times 10^{-15}$). Equations 6 and 7 were unsolved for $R_e = 15.7 \times 10^{-15}$

Input function	Min. R_s value[1] = 3.7×10^{-15}		Max. R_s value[2] = 9.6×10^{-15}		Mean R_s value[3] = 6.3×10^{-15}	
	R_0 value	Age (Ma)[4]	R_0 value	Age (Ma)[4]	R_0 value	Age (Ma)[4]
Mean $^{36}Cl/Cl$ ratio of the rainfall at São Paulo State (stations RCL, SPO, BOT)	57.7×10^{-15}	1.19 (eq.1) 1.49 (eq.6) 1.96 (eq.7)	57.7×10^{-15}	0.78 (eq.1) 0.86 (eq.6) nv (eq.7)	57.7×10^{-15}	0.96 (eq.1) 1.10 (eq.6) nv (eq.7)
Mean $^{36}Cl/Cl$ ratio of the rainfall at Mato Grosso do Sul State (stations MRJ-RW, SDL-RW)	96.1×10^{-15}	1.41 (eq.1) 1.72 (eq.6) 0.58 (eq.7)	96.1×10^{-15}	1.00 (eq.1) 1.09 (eq.6) 0.37 (eq.7)	96.1×10^{-15}	1.18 (eq.1) 1.33 (eq.6) nv (eq.7)
Mean $^{36}Cl/Cl$ ratio of the rainfall at São Paulo and Mato Grosso do Sul states (stations RCL, SPO, BOT, MRJ-RW, SDL-RW)	73.0×10^{-15}	1.30 (eq.1) 1.60 (eq.6) 0.37 (eq.7)	73.0×10^{-15}	0.88 (eq.1) 0.96 (eq.6) 0.16 (eq.7)	73.0×10^{-15}	1.07 (eq.1) 1.21 (eq.6) nv (eq.7)
Mean $^{36}Cl/Cl$ ratio of the rainfall (stations RCL, SPO, BOT) and cold/hypothermal groundwaters (codes AVR, SUT, BCS)	85.2×10^{-15}	1.36 (eq.1) 1.66 (eq.6) 0.24 (eq.7)	85.2×10^{-15}	0.95 (eq.1) 1.03 (eq.6) 0.02 (eq.7)	85.2×10^{-15}	1.13 (eq.1) 1.28 (eq.6) nv (eq.7)
Mean $^{36}Cl/Cl$ ratio of the rainfall (stations MRJ-RW, SDL-RW) and cold/hypothermal groundwaters (codes AMB, SDL)	177.0×10^{-15}	1.68 (eq.1) 1.99 (eq.6) 0.59 (eq.7)	177.0×10^{-15}	1.26 (eq. 1) 1.36 (eq.6) 0.37 (eq.7)	177.0×10^{-15}	1.45 (eq.1) 1.60 (eq.6) nv (eq.7)
Mean $^{36}Cl/Cl$ ratio of the rainfall (stations RCL, SPO, BOT, MRJ-RW, SDL-RW) and cold/hypothermal groundwaters (codes AVR, SUT, BCS, AMB, SDL)	121.9×10^{-15}	1.52 (eq.1) 1.82 (eq.6) 0.41 (eq.7)	121.9×10^{-15}	1.10 (eq.1) 1.19 (eq.6) 0.20 (eq.7)	121.9×10^{-15}	1.29 (eq.1) 1.44 (eq.6) nv (eq.7)
$^{36}Cl/Cl$ ratio of the rainfall at station RCL (minimum value)	24.0×10^{-15}	0.81 (eq.1) 1.09 (eq.6) nv (eq.7)	24.0×10^{-15}	0.40 (eq.1) 0.46 (eq.6) nv (eq.7)	24.0×10^{-15}	0.58 (eq.1) 0.70

						(eq.6) nv (eq.7)
$^{36}Cl/Cl$ ratio of the cold groundwater coded AMB (maximum value)	369×10^{-15}	2.00 (eq.1) 2.31 (eq.6) 0.09 (eq.7)	369×10^{-15}	1.58 (eq.1) 1.68 (eq.6) nv (eq.7)	369×10^{-15}	1.77 (eq.1) 1.92 (eq.6) nv (eq.7)
$^{36}Cl/Cl$ ratio of the cold groundwater RRP (contains anthropogenic chloride)	284×10^{-15}	1.88 (eq.1) 2.20 (eq.6) 1.74 (eq.7)	284×10^{-15}	1.47 (eq.1) 1.56 (eq.6) 1.53 (eq.7)	284×10^{-15}	1.66 (eq.1) 1.81 (eq.6) 1.00 (eq.7)

[1]Sample code PEO; [2]Sample code PPA; [3]Calculated from the values for sample codes PPA, PPE, PEO and TLG; [4]nv= negative value.

^{36}Cl Data and Subsurface Production

The major complications in using ^{36}Cl to estimate the residence time of groundwater are the introduction of subsurface Cl, subsurface production of ^{36}Cl, and variation in the input of ^{36}Cl due to changes in evapotranspirative concentration or changes in cosmogenic production (Phillips, 2000). ^{36}Cl is naturally produced by the reaction $^{35}Cl(n,\gamma)^{36}Cl$ within the geological environment whose amount is directly related to the neutron flux depending on the radioelement content of the rocks, mainly uranium and thorium (Andrews et al., 1986; Balderer and Synal, 1997; Phillips, 2000). Therefore, ^{36}Cl is produced at low levels (compared to meteoric chloride) in the subsurface by absorption by ^{35}Cl of thermal neutrons produced by U and Th series decay and U fission (Fabryka-Martin, 1988; Pearson et al., 1991; Lehmann et al., 1993; Phillips, 2000). Such production takes into account the secular equilibrium condition which is reached after approximately 1.5 million years (about five ^{36}Cl half-lives), depending on the local neutron flux.

If the groundwater forms a closed system with respect to chloride so that chloride is neither lost nor gained from the rock matrix, then ^{36}Cl can be used for the determination of groundwater residence times. The number of ^{36}Cl atoms, N_{36}, present in a groundwater at time t after recharge is given by (Andrews et al., 1986):

$$N_{36} = N_0 e^{-\lambda t} + N_e (1 - e^{-\lambda t}) \qquad (2)$$

where: N_0 is the initial ^{36}Cl concentration at recharge, N_e is the equilibrium ^{36}Cl concentration and λ is the ^{36}Cl decay constant. The number N_{36} of ^{36}Cl atoms per litre of water can be calculated using the measured $^{36}Cl/Cl$ ratio (R_{36}) and the chloride content in water (C, in mg/L) according to the equation (Pearson et al., 1991):

$$N_{36} = 1.7 \times 10^{19} CR_{36} \qquad (3)$$

Another way to express eq. 2 is to take into account the $^{36}Cl/Cl$ ratios rather than the number of ^{36}Cl atoms (Mook, 2000):

$$R_{36} = R_0 e^{-\lambda t} + R_e (1 - e^{-\lambda t}) \quad (4)$$

where: R_{36} is the $^{36}Cl/Cl$ ratio measured in the groundwater sample, R_0 is the cosmogenic $^{36}Cl/Cl$ ratio and R_e is the equilibrium $^{36}Cl/Cl$ ratio (underground ^{36}Cl production by neutron activation of ^{35}Cl), and λ is the ^{36}Cl decay constant.

Phillips et al. (1986) and Balderer and Synal (1997) utilized a three term mass balance approach to investigate processes occurring within the water cycle and influencing the ^{36}Cl concentration. The adopted equation accounts for decay of meteoric ^{36}Cl (first term), subsurface ^{36}Cl production (second term) and addition of chloride derived from the rock (by water-rock interaction) (third term). It is expressed by:

$$R_{36}C = R_0C_0 e^{-\lambda t} + R_eC_0 (1 - e^{-\lambda t}) + R_e (C-C_0) \quad (5)$$

where: C_0 is the recharge (input) concentration of chloride (in mg/L).

Equations 4 and 5 can be solved for time, yielding, respectively:

$$t = - (\lambda)^{-1} \ln [(R_{36} - R_e)/(R_0 - R_e)] \quad (6)$$

$$t = - (\lambda)^{-1} \ln [C(R_{36} - R_e)/C_0(R_0 - R_e)] \quad (7)$$

Phillips (2000) pointed out that eq. 7 does not account for situations in which subsurface chloride is introduced that is either markedly above or below the aquifer nucleogenic equilibrium. Andrews et al. (1986) remembered that if the in-situ neutron flux for the aquifer is small, eq. 2 reduces to the decay term only but if the in-situ neutron flux is high and the groundwater is closed to chloride, the ingrowth term controls the ^{36}Cl activity change with time for the dissolved chloride. Despite of all these difficulties inherent to the use of the ^{36}Cl method and of the small database reported in this paper, eqs. 6 and 7 have been used on the groundwater residence time evaluation.

The secular equilibrium of underground (*in situ*) production of the $^{36}Cl/Cl$ ratios has been determined by Fabryka-Martin (1988), Pearson et al. (1991) and Lehmann et al. (1993), varying according to depth (it increases towards ground surface), rock type, geochemical composition, petrophysical properties and radioelements (U and Th) content. The R_e values for some rocks are (Fabryka-Martin, 1988): clay/shale = 12×10^{-15} (U = 3.2 ppm; Th = 11 ppm; depth > 417m); basalt = 6.9×10^{-15} (U = 1 ppm; Th = 4 ppm; depth > 357m); sandstone = 1.9×10^{-15} (U = 0.45 ppm; Th = 1.7 ppm; depth > 435m). Pearson et al. (1991) reported the following R_e values for sandstones, siltstones, shales and marls from northern Switzerland: minimum = 8.7×10^{-15}, maximum = 32.4×10^{-15}, mean = $15.7 \times 10^{-}$

[15]. The U and Th content range in these rocks was 2-6 ppm (mean = 3ppm) and 5-27 ppm (mean = 11.2ppm), respectively.

Kimmelmann et al. (1995) reported ranges of 0.6-1.2 ppm (mean=0.8 ppm) and 0.9-4.9 ppm (mean = 2.9 ppm), respectively, for the U and Th content in basalts and sandstones of the Paraná sedimentary basin. Such values are closer of those reported by Fabryka-Martin (1988), which implied on a R_e estimate of 1.9×10^{-15} for sandstone. This ratio will be adopted in eqs. 6 and 7 for calculating the groundwater residence time, but the mean R_e value of 15.7×10^{-15} reported by Pearson et al. (1991) will also be utilized as increased U levels have been also reported in the sedimentary rocks of the Paraná basin (Bonotto, 2013).

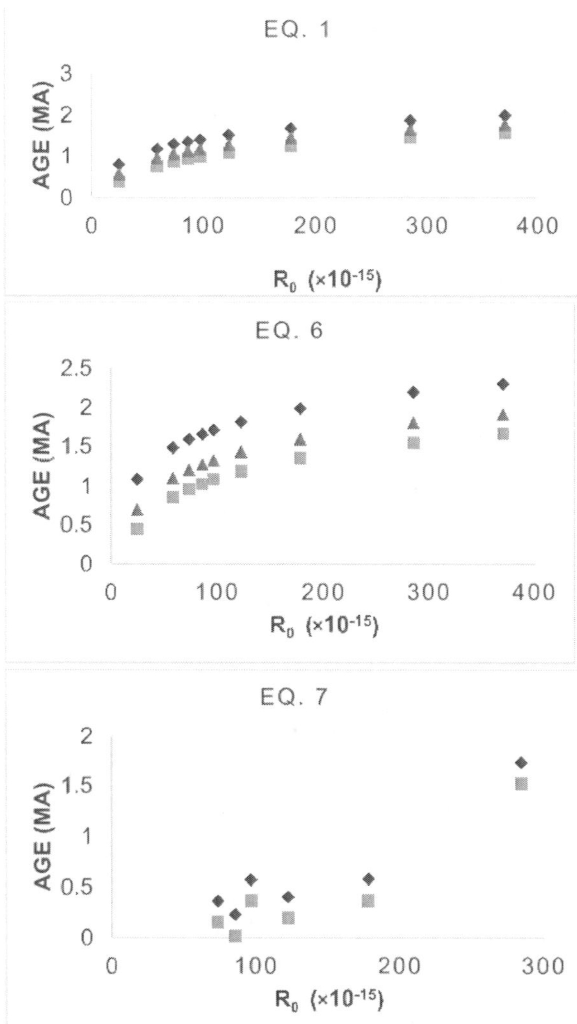

Figure 3. The relationships between the ^{36}Cl ages in the hyperthermal waters for $R_e = 1.9 \times 10^{-15}$ and input R_0 values from eqs. 1, 6, and 7, according to the following R_s values: 3.7×10^{-15} (diamonds), 9.6×10^{-15} (squares), and 6.3×10^{-15} (triangles).

Equations 6 and 7 were initially applied considering the cold/hypothermal waters (bores AVR, SUT, BCS, AMB, SDL, RRP) and all rainfall samples. Several "negative" ages were found like when eq. 1 was adopted (Table 4). None age was obtained for groundwaters AMB and RRP from eq. 6, whereas the use of eq. 7 to the cold/hypothermal waters indicated none age to the groundwater RRP. Equations 6 and 7 applied to the hyperthermal waters were unsolved for $R_e = 15.7 \times 10^{-15}$, whilst for $R_e = 1.9 \times 10^{-15}$ some "negative" ages were found when eq. 7 was adopted (Table 5).

Figure 3 illustrates the relationships between the ^{36}Cl ages in the hyperthermal waters for $R_e = 1.9 \times 10^{-15}$ and the input R_0 values from eqs. 1, 6, and 7, according to the lowest, highest and mean R_s values, respectively, 3.7×10^{-15}, 9.6×10^{-15}, and 6.3×10^{-15}. Equations 1 and 6 clearly show similar trends for the age increase according to the input R_0 value raising and also based on the R_s value decreasing. Considering the mean R_s value (6.3×10^{-15}), the ages estimated from eq. 1 were between 0.58 Ma ($R_0 = 24 \times 10^{-15}$) and 1.77 Ma ($R_0 = 369 \times 10^{-15}$), whereas eq. 6 allowed calculate ages from 0.70 Ma ($R_0 = 24 \times 10^{-15}$) to 1.92 Ma ($R_0 = 369 \times 10^{-15}$). Such differences are not so pronounced if considered the timescale of the processes involved and premises related to the groundwater residence time estimate.

Thus, the standard decay equation (eq. 1) yielded reliable results and its adoption to the reported GAS ^{36}Cl dataset may be recommended due to its easier use than eq. 6 that requires an appropriate choice of the R_e value. For $R_e = 15.7 \times 10^{-15}$, eqs. 6 and 7 were unsolved for the hyperthermal waters and also provided many "negative" ages for the cold/hypothermal waters (Table 4), suggesting no need to take into account some increased U levels in the sedimentary rocks of Paraná basin as the mean U and Th concentration in basalts and sandstones were valuable for using the GAS ^{36}Cl database. The "negative" ages in Table 5 from eq. 7 (hyperthermal waters) for practically all input R_0 values, $R_s = 6.3\times10^{-15}$ and $R_e = 1.9 \times 10^{-15}$ as well the unclear trend of the ages distribution from its use (Figure 3) indicate it is not suitable for applying to the GAS ^{36}Cl database reported in this research.

CONCLUSION

The combined use of isotope methods and groundwater flow modelling has been a common approach used for developing the required level of knowledge in the case of groundwater pumped from deep aquifers. Radiocarbon and tritium have been the common tools used in isotope hydrology studies to provide first estimates of groundwater age and dynamics. Over the last years, a number of long-live radionuclides like Chlorine-36, Uranium isotopes and Krypton-81 has been used as reliable age indicators in very old groundwaters. This chapter described the use of the ^{36}Cl method for estimating the groundwater residence time in the northern part of the Guarani Aquifer System (GAS) in Brazil. The standard decay equation yielded compatible Chlorine-36 ages and its adoption to the reported GAS ^{36}Cl dataset is appropriate due to its easier use than the equation accounting for decay of meteoric ^{36}Cl and subsurface ^{36}Cl production that requires an appropriate choice of the equilibrium value for the $^{36}Cl/Cl$ ratio, i.e., underground ^{36}Cl production by neutron activation of ^{35}Cl. Groundwater residence times of 43.8 – 290.5 ka were estimated from ^{36}Cl data in the cold/hypothermal waters, whereas higher values (0.40 – 2.00 Ma) in the hyperthermal waters. Such ranges agree with the expected travel times in GAS, as the deep (hyperthermal) groundwaters should exhibit higher values of the ^{36}Cl ages in relation to those of the cold/hypothermal ones. The mean U and Th concentration in basalts and sandstones of the sedimentary rocks of Paraná basin were valuable for using the GAS ^{36}Cl database as the equilibrium $^{36}Cl/Cl$ ratio representing increased U levels in them generates unsolved equations for the hyperthermal waters and many "negative" ages for the cold/hypothermal waters. The ^{36}Cl ages as estimated here were compatible with the radiokrypton ages of 70-834 ka reported elsewhere, despite the different premises related to each method.

ACKNOWLEDGMENTS

D.M.B. thanks CNPq (National Council for Scientific and Technologic Development) and FAPESP (Foundation for Supporting Research in São Paulo State), Brazil, for financial support of the projects that permitted the acquisition of the data reported in this chapter.

REFERENCES

Amarante, O. A. C., Brower, M., Zack, J. & Sá, A. L. (2001). *Atlas of the Brazilian eolian potential*. Brasilia: MME-Mines and Energy Ministry.

Andrews, J. N. & Fontes, J. C. (1992). "Importance of *in situ* production of ^{36}Cl, ^{36}Ar and ^{14}C in hydrology and hydrogeochemistry." In *Isotope Techniques in Water Resources Development*, edited by IAEA (International Atomic Energy Agency), 245-269. Vienna, Austria: IAEA.

Andrews, J. N., Fontes, J. C., Michelot, J. L. & Elmore, D. (1986). "*In-situ* ^{36}Cl production and groundwater evolution in crystalline rocks at Stripa, Sweden." *Earth and Planetary Science Letters*, *77*, 49-58.

APHA (American Public Health Association). (1989). *Standard methods for the examination of water and wastewater*. Washington: APHA.

Araújo, L. M., Franca, A. B. & Potter, P. E. (1999). "Hydrogeology of the Mercosul aquifer system in the Paraná and Chaco-Parana Basins, South America, and comparison with the Navajo-Nugget aquifer system, USA." *Hydrogeology Journal*, *7*, 317-336.

Balderer, W. & Synal, H. A. (1997). "Use of chlorine-36 as tracer for the evolution of waters in geothermal and tectonically active areas in Western Turkey." *Nuclear Instruments and Methods in Physics Research B, 123*, 387-393.

Bentley, H. W., Phillips, F. M. & Davis, S. N. (1986a). "^{36}Cl in the terrestrial environment." In *Handbook of Environmental Isotope Geochemistry*, edited by P. Fritz and J.-C. Fontes, 427-475. Amsterdam: Elsevier.

Bentley, H. W., Phillips, F. M., Davis, S. N., Habermehl, M. A., Airey, P. L., Calf, G. E., Elmore, D., Gove, H. E. & Torgersen, T. (1986b). "Chlorine 36 dating of very old groundwater. 1. The great artesian basin, Australia." *Water Resources Research*, *22*, 1991-2001.

Berner, E. K. & Berner, R. A. (1987). *The global water cycle*. Upper Saddle River: Prentice-Hall.

Bonotto, D. M. (2004). "Doses from ^{222}Rn, ^{226}Ra, and ^{228}Ra in groundwater from Guarani aquifer, South America." *Journal of Environmental Radioactivity*, *76*, 319-335.

Bonotto, D. M. (2006). "Hydro(radio)chemical relationships in the giant Guarani aquifer, Brazil." *Journal of Hydrology*, *323*, 353-386.

Bonotto, D. M. (2011). "Natural radionuclides in major aquifer systems of the Paraná sedimentary basin, Brazil." *Applied Radiation and Isotopes*, *69*, 1572-1584.

Bonotto, D. M. (2012). "A comparative study of aquifer systems occurring at the Paraná sedimentary basin: major hydrochemical trends." *Environmental Earth Sciences*, *67*, 2285-2300.

Bonotto, D. M. (2013). "A comparative study of aquifer systems occurring at the Paraná sedimentary basin: U-isotopes contribution." *Environmental Earth Sciences*, *68*, 1405-1418.

Bonotto, D. M. & Caprioglio, L. (2002). "Radon in groundwaters from Guarany aquifer, South America: environmental and exploration implications." *Applied Radiation and Isotopes*, *57*, 931-940.

Bonotto, D. M. & Mello, C. B. (2006). "A combined method for evaluating radon and progeny in waters and its use at Guarani aquifer, São Paulo State, Brazil." *Journal of Environmental Radioactivity*, *86*, 337-353.

Bonotto, D. M. & Armada, P. C. P. (2008). "Radon and progeny (^{214}Pb and ^{214}Bi) in urban water-supply systems at São Paulo State, Brazil." *Applied Geochemistry*, *23*, 2829-2844.

Bonotto, D. M. & Bueno, T. O. (2008). "The natural radioactivity in Guarani aquifer groundwater, Brazil." *Applied Radiation and Isotopes*, *66*, 1507-1522.

Bonotto, D. M., Bueno, T. O., Tessari, B. W. & Silva, A. (2009a). "The natural radioactivity in water by gross alpha and beta measurements." *Radiation Measurements*, *44*, 92-101.

Bonotto, D. M., Caprioglio, L., Bueno, T. O. & Lazarindo, J. R. (2009b). "Dissolved ^{210}Po and ^{210}Pb in Guarani aquifer groundwater, Brazil." *Radiation Measurements*, *44*, 311-324.

Brownlow, A. H. (1996). *Geochemistry*. Englewood Cliffs: Prentice-Hall.

Campos, H. C. N. S. (2000). "Hydrogeological map of Guarani aquifer." *Acta Geologica Leopoldensia*, *23*(4), 1-50.

Castany, G. (1982). *Principles and methods in hydrogeology*. Paris, France: Dunod.

Conard, N. J., Elmore, D., Kubik, P. W., Gove, H. E., Tubbs, L. E., Chrunyk, B. A. & Wahlen, M. (1986). "The chemical preparation of AgCl for measuring ^{36}Cl in polar ice with accelerator mass spectrometry." *Radiocarbon*, *28*(2A), 556-560.

Cresswell, R. G. & Bonotto, D. M. (2008). "Some possible evolutionary scenarios suggested by Cl-36 measurements in Guarani aquifer groundwater." *Applied Radiation and Isotopes*, *66*, 1160-1174.

Cresswell, R. G., Jacobson, G., Wischusen, J. & Fifield, L. K. (1999). "Ancient groundwaters in the Amadeus Basin, Central Australia: evidence from the radio-isotope Cl-36." *Journal of Hydrology*, *223*(3-4), 212-220.

Davis, S. N., Cecil, D. W., Zreda, M. & Sharma, P. (1998). "Chlorine-36 and the initial value problem." *Hydrogeology Journal*, *6*, 104-114.

DFPM (Division for Supporting the Mineral Production). (1966). *The mining code, the mineral waters code and how applying research in a mineral deposit*. Rio de Janeiro, Brazil: DFPM.

Fabryka-Martin, J. T. (1988). "*Production of radionuclides in the earth and their hydrogeologic significance, with emphasis on chlorine-36 and iodine-129*." PhD Thesis, University of Arizona.

Faure, G. (1998). *Principles and applications of geochemistry*. Upper Saddle River: Prentice-Hall.

Fontes, J. C. & Garnier, J. M. (1979). "Determination of the initial ^{14}C activity of the total dissolved inorganic carbon: a review of existing models and a new approach." *Water Resources Research*, *15*, 39-413.

Foster, S., Kemper, K., Garduño, H., Hirata, R. & Nanni, M. (2006). *The Guarani Aquifer Initiative for Transboundary Groundwater Management*. Case Profile Collection 9, Sustainable Groundwater Management - Lessons from practice. Washington: The World Bank.

Gastmans, D. (2007). "*Hydrogeology and hydrochemistry of the Guarany Aquifer System in the western portion of the Paraná sedimentary basin*." PhD Thesis, UNESP-São Paulo State University, Brazil.

Gastmans, D., Chang, H. K. & Hutcheon, I. (2010a). "Stable isotopes (^{2}H, ^{18}O and ^{13}C) in groundwaters from the northwestern portion of the Guarani Aquifer System (Brazil)." *Hydrogeology Journal*, *18*, 1497-1513.

Gastmans, D., Chang, H. K. & Hutcheon, I. (2010b). "Groundwater geochemical evolution in the northern portion of the Guarani Aquifer System (Brazil) and its relationship to diagenetic features." *Applied Geochemistry*, *25*, 16-33.

Gastmans, D., Veroslavsky, G., Chang, H. K., Caetano-Chang, M. R. & Pressinotti, M. M. N. (2012). "Hydrogeological conceptual model of the Guarany Aquifer System (GAS): a management tool." *Geological and Mining Bulletin*, *123*(3), 249-265.

Hach. (1992). *Water analysis handbook*. Loveland: Hach Company.

Hirata, R., Gesicki, A., Sracek, O., Bertolo, R., Giannini, P. C. & Aravena, R. (2011). "Relation between sedimentary framework and hydrogeology in the Guarani Aquifer System in São Paulo state, Brazil." *Journal of South American Earth Sciences*, *31*, 444-456.

Jiang, S. S., Hemmick, T. K., Kubik, P. W., Elmore, D., Gove, H. E., Tullai-Fitzpatrick, S. & Hossain, T. Z. (1990). "Measurement of the $^{36}Ar(n, p)^{36}Cl$ cross section at thermal energies using the AMS technique." *Nuclear Instruments and Methods in Physics Research Section B: Beam Interactions with Materials and Atoms*, *52*(3-4), 608-611.

Keywood, M. D., Chivas, A. R., Fifield, L. K., Cresswell, R. G. & Ayers, G. P. (1997). "The accession of chloride to the western half of the Australian continent." *Australian Journal of Soil Research*, *35*(5), 1177-1189.

Kimmelmann, A. A., Forster, M. & Coelho, R. (1995). "Environmental isotope and hydrogeochemical investigations of Bauru and Botucatu aquifers, Paraná basin, Brazil." In *Isotope Hydrology Investigations in Latin America (Report RLA/8/014 - ARCAL XIII; Application of isotope techniques in hydrology)*. Edited by IAEA (Internacional Atomic Energy Agency), IAEA-TECDOC-835, 57-74. Vienna, Austria: IAEA.

Kimmelmann e Silva, A. A., Rebouças, A. C. & Santiago, M. M. F. (1989). "^{14}C analyses of groundwater from the Botucatu aquifer system in Brazil." *Radiocarbon*, *31*, 926-933.

Krauskopf, K. B. & Bird, D. K. (1995). *Introduction to geochemistry*. New York: McGraw-Hill Inc.

Lara, L. B. L. S., Artaxo, P., Martinelli, L. A., Victoria, R. L., Camargo, P. B., Krusche, A., Ayers, G. P., Ferraz, E. S. B. & Ballester, M. V. (2001). "Chemical composition of rainwater and anthropogenic influences in the Piracicaba River basin, southeast Brazil." *Atmospheric Environment*, *35*, 4937-4945.

Lehmann, B. E., Davis, S. N. & Fabryka-Martin, J. T. (1993). "Atmospheric and subsurface sources of stable and radioactive nuclides used for groundwater dating." *Water Resources Research*, *29*(7), 2027-2040.

Mahara, Y., Habermehl, M. A., Miyakawa, K., Shimada, J. & Mizuochi, Y. (2007). "Can the ^{4}He clock be calibrated by ^{36}Cl for groundwater dating?" *Nuclear Instruments and Methods in Physics Research Section B: Beam Interactions with Materials and Atoms*, *259* (1), 536-546.

Mook, W. G. (2000). *Environmental isotopes in the hydrological cycle- Principles and Applications.* Technical Documents in Hydrology, No. 39. Paris, France: UNESCO.

Moreira-Nordeman, L. M. M., Forti, M. C., Dilascio, V. L., Espirito Santo, C. M. & Danelon, O. M. (1988). "Acidification in southeastern Brazil." In *Acidification in tropical countries*, edited by H. Rodhe and R. Herrera. *SCOPE*, *36*, 257-296.

Pearson, F. J., Balderer, W., Loosli, H. H., Lehmann, B. E., Matter, A., Peters, T., Schmassmann, H. & Gautschi, A. (1991). *Applied Isotope Hydrogeology – A Case Study in Northern Switzerland.* Studies in Environmental Sciences 43. New York: Elsevier.

Pelicho, A. F., Martins, L. D., Nomi, S. N. & Solci, M. C. (2006). "Integrated and sequential bulk and wet-only samplings of atmospheric precipitation in Londrina, South Brazil (1998-2002)." *Atmospheric Environment*, *40*, 6827-6835.

Phillips, F. M. (2000). "Chlorine-36." In *Environmental tracers in subsurface hydrology*, edited by P.G. Cook and A.L. Herczeg, 299-348. Dordrecht: Kluwer Academic.

Phillips, F. M., Bentley, H. W., Davis, S. N., Elmore, D. & Swannick, G. B. (1986). "Chlorine-36 dating of very old ground water II: Milk River aquifer, Alberta." *Water Resources Research*, *22*, 2003-2016.

Saatsaz, M., Azmin-Sulaiman, W. N., Eslamian, S. & Mohammadi, K. (2013a). "Hydrogeochemistry and groundwater quality assessment ofAstaneh-Kouchesfahan Plain, Northern Iran." *International Journal of Water*, *7*(1/2), 44-65.

Saatsaz, M., Azmin-Sulaiman, W. N., Eslamian, S. & Javadi, S. (2013b). "Development of a coupled flow and solute transport modelling for Astaneh-Kouchesfahan groundwater resources, North of Iran." *International Journal of Water*, *7*(1/2), 80-103.

Sheppard, S. C., Evendon, W. G. & MacDonald, C. R. (1999). "Variation among chlorine concentration ratios for native and agronomic plants." *Journal of Environmental Radioactivity*, *43*, 65-76.

Silva, R. B. G. (1983). "*Hydrochemical and isotopic study of groundwaters from Botucatu aquifer at São Paulo State*." PhD Thesis, USP-São Paulo University, Brazil.

Soler i Gil, A. & Bonotto, D. M. (2015). "Hydrochemical and stable isotopes (H, O, S) signatures in deep groundwaters of Paraná basin, Brazil." *Environmental Earth Sciences*, *73*, 95-113.

Sracek, O. & Hirata, R. (2002). "Geochemical and stable isotopic evolution of the Guarani Aquifer System in the state of São Paulo, Brazil." *Hydrogeology Journal*, *10*, 643-655.

Sturchio, N. C., Du, X., Purtschert, R., Lehmann, B. E., Sultan, M., Patterson, L. J., Lu, Z. T., Muller, P., Bigler, T., Bailey, K., O'Connor, T. P., Young, L., Lorenzo, R., Becker, R., El Alfy, Z., El Kaliouby, B., Dawood, Y. & Abdallah, A. M. A. (2004). "One million year old groundwater in the Sahara revealed by krypton-81 and chlorine-36." *Geophysical Research Letters*, *31*(5), L05503.

Teissedre, J. M. & Barner, U. (1981). "Geothermal and geochemical behavior of waters from Botucatu aquifer at Paraná basin." *Groundwaters Review*, *4*, 85-95.

Vieira, B. M., Forti, M. C., Tavares, M. F. M. & Moreira-Nordemann, L. M. (1988). "Atmospheric Na, Ca, K, Mg and Cl in the east region of São Paulo State." *Brazilian Review of Geophysics*, *6*(1-2), 17-23.

Waterloo Hydrogeologic. (2003). *AquaChem User´s Manual: Water Quality Data Analysis, Plotting & Modeling*. Waterloo, Canada: Waterloo Hydrogeologic.

Wischusen, J. D. H., Fifield, L. K. & Cresswell, R. G. (2004). "Hydrogeology of Palm Valley, central Australia; a Pleistocene flora refuge?" *Journal of Hydrology*, *293*, 20-46.

In: Advances in Hydrogeochemistry Research
Editors: S. Saeid Eslamian and Faezeh Eslamian
ISBN: 978-1-53617-829-6

Chapter 7

GEOSPATIAL TECHNIQUES IN HYDROGEOCHEMISTRY BY INTERSECTING THE LAYERS AND STATISTICAL ANALYSIS, MASHHAD, IRAN

Ali Akbar Jamali[1,*] and Saeid Eslamian[2,3]

[1]Department of GIS-RS and Watershed Management, Maybod Branch, Islamic Azad University, Maybod, Iran

[2]Department of Water Engineering, Collage of Agriculture, Isfahan University of Technology, Isfahan, Iran

[3]Center of Excellence in Risk Management and Natural Hazards, Isfahan University of Technology, Isfahan, Iran

ABSTRACT

Geospatial is a modern and useful tool in the precise and quick survey of hydrogeochemistry data of groundwater. The purpose of this study is to use geospatial and geochemical analysis of groundwater in Iran-Mashhad Plain. The GIS was used to provide different layers of water quality in three years. Statistical samples of piezometric wells were prepared in a coordinate system map layers. The layers were intersected with three western, middle and eastern sub-watersheds and their aquifers. The cross-sectional descriptive tables were brought to the statistical analysis environment. The spatial and temporal analysis of groundwater quality had revealed the process of decreasing water quality in the sub-watersheds and their aquifers. This study shows using point map data and various districts and times for statistical analysis. In addition, the results are a warning to decision-makers who try to reform the policies in proper groundwater management.

* Corresponding Author's Email: jamaliaa@maybodiau.ac.ir.

Keywords: map, spatial, temporal, groundwater, water quality, statistic

BEHAVIORAL GOALS

After reading this section, the reader is expected to:

1. Understand the use of geospatial data on the quality and quantity of water in large watersheds, along with topographic data in water management and geospatial analysis.
2. How to intersect the vector layers and extract the table data in the GIS for water quality management and geospatial techniques.
3. Learn how to extract and categorize the tabular data for statistical analysis.
4. Considering the conditions for the studied watersheds in applying statistical analysis and trends.
5. Familiarize yourself with AQ.Qa specialized software with a variety of water quality charts.

1. INTRODUCTION

Underground water in the watershed has played a crucial role in providing human demands from the ancient eras. Underground waters form about 4% of the water collection in the world's aquatic ecosystem (Abell et al., 2008; Besbes et al., 2019). Underground water is one of the main resources of water supply in all countries of the world and its irrigation, agriculture and industrial use is increasing. In Iran, due to its low rainfall and high evapotranspiration, more than 50% of the required water is supplied through groundwater. Therefore, it is necessary to protect this water resource in the aquifers with water management programs against pollution. Particularly, groundwater contamination is more important than water surface contamination, since the interval between the onset of contamination of subsurface water and its effects sometimes takes several years. Therefore, it is very costly to find the source of pollution and eliminate it (Elubid et al., 2019; Tavakoly et al., 2019; Rashid et al., 2019). For this reason, it is necessary to maintain them through the proper management of these resources in the basin, watershed, and catchment areas. Currently, many water pollutants, without drainage systems, release toxic and dangerous substances such as heavy metals into groundwater aquifers in the watershed areas (Baun and Christensen, 2004; Zhai et al., 2019). Therefore, studying the pollution and quality of groundwater by GIS is one of the most important environmental management and geospatial techniques in hydrogeochemistry issues.

Geospatial hydrogeochemistry assessment, like any other engineering project, requires the basic information and accurate planning (Kamali et al., 2015). The selection of multiple factors leads to the abundance of information layers and attempts to find a suitable solution for analyzing many information layers and obtaining the correct result. Decision makers unconsciously lead to the use of systems that, in addition to the high accuracy in terms of the speed of operation and the ease of doing operations is at a high level. One of the best solutions to solve the problem is the use of geographic information systems (GIS) (Jamali et al., 2018; Jamali and Ghorbani Kalkhajeh 2019; Ghorbani Kalkhajeh and Jamali 2019). GIS can be widely used in environmental planning and engineering issues. On the other hand, one of the most important GIS capabilities is the ability to intersect and overlay data for geospatial hydrogeochemistry modeling. On the other hand, the use of descriptive information extracted from these layers in statistical analyzes after the intersection of the layers and after producing multiple descriptive cross-sectional tables, has the special ability to geospatial hydrogeochemistry and groundwater resources assessments.

2. Materials and Methods

In general, considering the nature of the research on the geospatial analysis of hydrogeochemistry with intersectional layers and statistical analysis (Ghorbani Kalkhajeh and Jamali 2019), library studies, the preparation of maps using GIS, field and laboratory studies, the use of databases, computer software, and statistical analysis is required by statistical software based on the following steps and the flow chart presented in Figure 1.

Step 1: Select the area or aquifer.

Step 2: Collecting required groundwater quality data in the form of information or attribution tables on water quality and pollutants from relevant organizations.

Step 3: Transform the attribute tables into the map layers, plotting the polygonal layers related to the study area, obtaining point maps from the obtained data, intersecting the maps with the polygon layers, extracting the intersection tables, and finally transferring the tables in the statistical software.

Step 4: Normalize the data and analyze their nonparametric results in Kruskal-Wallis test and Mann-Whitney U test (upon on data groups' number) and analyze the results.

Step 5: Identify and evaluate the reasons for changing or reducing the quality of groundwater in different parts of the area or aquifer and comparing them with each other.

Step 6: Final, report preparation.

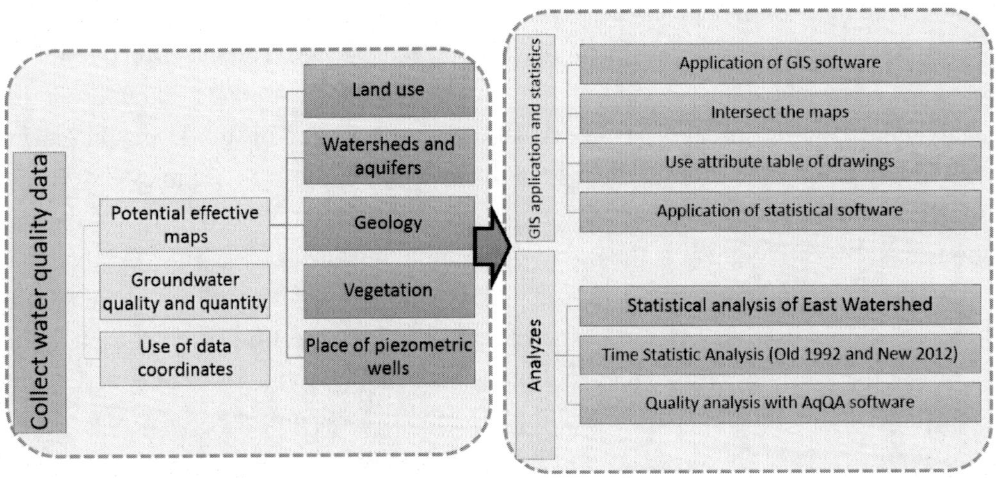

Figure 1. Flow chart for intersection of layers and statistical analysis.

2.1. Geographic Information System

The GIS is used for recording, storing, controlling, overlay, deploying, analyzing, and displaying data to make optimal geospatial decision-making. In addition to providing basic maps, this information system is nowadays used in different issues for urban water planning (Czajkowski et al., 2018), the quantity and quality of aquifer (Shooshtarian et al., 2018), traffic management, land use management and environmental studies are also used. On the one hand, GIS integrates systematic, database and computer technologies with satellite information and remote sensing technologies and global information networks, and on the other hand, it be included all human activities and natural phenomena that have spatial and temporal dimensions in a layered layout of digital map overlays. Then the capabilities and benefits of this system will be identified in quick access to information, aggregation of different types of information, comprehensive analysis of information, updating, accuracy, and speed of operation.

2.2. Statistical Software

Statistical software such as Statistical Package for Social Science (SPSS) is a comprehensive and flexible statistical analyzer and an appropriate data management system (Badie-Neige et al., 2014) for the watershed, geospatial analysis and hydrogeochemistry studies. In statistical assessment, all types of files can be imported and generated in tabular reports, maps, distribution maps, distribution and trends, development, descriptive statistics, and statistical analysis of complex flow and combined data behavior in mass management of data. In the aquifer; this statistical and geospatial assessing is an application to the geospatial analysis of hydrogeochemistry, which makes statistical analyzes available to users at all levels more and more without the need to compile grammatical rules (Singh et al., 2004).

2.3. Kruskal-Wallis Test

Kruskal-Wallis test uses as nonparametric analysis of variance for three and more than three independent groups (three sub-watersheds, three zones, three quality, etc.) in the geospatial hydrogeochemistry analysis. The Kruskal-Wallis method tests the hypothesis K of a sample group from a common statistical society or a similar statistical society similar in averages. Kruskal-Wallis's one-way analysis of variance is a useful tool for deciding whether to depend on a K group of independent samples of different statistical societies. Obviously, specimens have the differences without exception, but it is necessary to determine the source of the difference observed in the samples due to differences in societies or randomness. The null hypothesis in this test, contrary to its hypothesis, emphasizes the no difference between the groups. This hypothesis, based on the meanings, takes the basis on the similarity of the K model to common society, namely, two hypotheses, null and one, were shown in the Equation (1)

$$\begin{cases} H_0: U_1 = U_2 = \cdots = U_k \\ H_1: U_i \neq U_j \cdots \quad i \neq j \end{cases} \tag{1}$$

In this test, all K samples are integrated to obtain N view. Then each N view is ranked as a rating. The smallest value is ranked one and the highest is the last rank (rank N). Now for each K group, the total rank is so different that it cannot be said that they are out of a common statistical society. The Kruskal-Wallis statistical index is shown in Equation (2)

$$H = \frac{12}{N(N+)} \sum_{j=1}^{k} \frac{R_j^2}{n_j} - 3(N+1) \tag{2}$$

where K is the number of groups, n_j is the number of the samples in each group, N is the total number of samples among all the groups and R_j is the sum of ratings in each group. The high index for large specimens (the number of samples in each group is greater than 5) has almost x2 distribution with a degree of freedom of K-1, which can be used for the x2 distribution table based on the significant level determined for rejection or acceptance of the null hypothesis. If the number of samples in each group be less than five and the number of groups be three, then the Kruskal-Wallis test method and table should be used (Gaddis and Gaddis 1990)

2.4. Mann-Whitney U test

Mann-Whitney U test is a nonparametric test for comparing the status of two independent groups. When the data of a study is qualitatively sequential, it is better to use this test as the equivalent of testing two independent samples t-test. Because the average of the variables measured on a sequential scale, due to the unequal spacing of the units, does not mean the true meaning of the mean. For example, when the water quality of the two sub-watersheds is compared, it is better to sort the water samples by salinity and determine their rank to compare the mean salinity. Suppose a comparison of two aquifers is considered in terms of water quality. For this purpose, N wells were selected randomly and the sample was randomly assigned to each of the n1 samples in the first aquifer and n2 in the second aquifer (n1 + n2 = N), and after the end of the test period, samples of their quality number were written and they are ranked. Then, the total rank of each group is computed and assigned as R1 and R2 respectively in Equation (3)

$$w_0 = n_1 \times n_2 + \frac{n_1(n_1+1)}{2} - R_1$$
$$w_1 = n_1 \times n_2 + \frac{n_2(n_2+1)}{2} - R_2 \quad (3)$$

If the size of the two groups is not equal, n1 is the smaller group size and n2 is the larger group size. Of course, the calculation of one of the two w upper ones is enough, because having one of them is different from the Equation (4)

$$w_0 + w_1 = n_1 \times n_2 \quad (4)$$

This test is one of the strongest nonparametric and useful substitute tests for independent t-tests.

If the sample size is less than 20, then the Mann-Whitney U method and table should be used to deny the null hypothesis. You should consider the smaller w in Equation (5) and compare the value of the table. The smaller the w of the table value, however, suppresses the hypothesis of null, which means that the values of R1 and R2 are smaller, and therefore the difference between the two groups is high and the null hypothesis is rejected. If the sample size is greater than 20, according to the mean and variance, w will use the U = Z index (Equation 5) with normal distribution and its value will be compared at a level of 0.05 with a value of 1.96 in the standard normal table (Bergmann et al., 2000).

$$U = Z = \frac{\left|W_1 - \frac{n_1 n_2}{2}\right|}{\sqrt{\frac{n_1 n_2 (n_1 n_2 + 1)}{12}}} \quad (5)$$

2.5. Analysis of Aquifer Water Quality in Watershed Management with Aq.QA Software

Aq.QA software is software for analyzing and adjusting the parameters of chemical water quality in the watersheds. This software is used to determine the chemical group of water, chemical analysis of the data obtained from the test and plotting the chemical charts of water.

2.6. Schoeller Chart

The Schoeller diagram is a graphical method for classifying drinking water quality. In this chart, the waters understudy in the watershed are divided into six groups of good, acceptable, moderate, inappropriate, unpleasant, and unpredictable. The most important qualitative criteria for drinking water classification using the Schoeller chart are the amount of soluble salts (anions and cations), total dry residue and total water hardness, total soluble solids, or dry residues. TDS is a very effective measure for creating a taste of drinking water. Water with a TDS of fewer than 500 milligrams per liter is very good in terms of drinking water quality standards. TDS between 500 and 1000 is desirable and in the range of 1,000 to 1,500 is allowed for drinking, but water with TDS does not have more than 1500 drinkable properties. (Hosseinifard and Aminiyan 2015). The concentrations of Na +, -Cl, -SO4 ions as well as TDS and TH are the water quality assessment criteria in Schoeller diagrams, which are classified according to Table 1.

To investigate the capability of drinking groundwater in each area, due to salinity changes, the Schoeller chart is used for zoning and classification of the waters by considering the salinity level of the aquifer, and a sample is selected in each area and its qualitative status is plotted and evaluated.

Table 1. Quality criteria for drinking water according to Schoeller (mg/L)

Quality	Na^+	Cl^-	SO_4^-	TDS	TH
Good	<115	<175	<145	500>	250>
Acceptable	115-230	175-350	145-280	1000-500	500-250
Medium	230-460	350-700	280-580	2000-1000	1000-500
Inappropriate	460-920	700-1400	580-1150	4000-2000	2000-1000
Totally inappropriate	920-1840	1400-2800	1150-2240	8000-4000	4000-2000
Non-drinkable	>1840	>2800	>2240	8000<	4000<

Water resources in regions with ionic abundances in this order rCa < rMg < rNa and rSo4 <rCl < rHco3 (prefix r means concentration in eq/L, One equivalent per liter is equal to one thousand milligram-equivalents per one thousand milliliters, meq/mL) due to the fine texture of alluvium and the role of saline formations will place in the acceptable class. One of the areas where underground water is in a good group is the area where sodium ion and bicarbonate are affected and are low by the recharging of the river. This district is the most suitable as drinking water. The electrical conductivity of these waters varies from 185 to 382 micromhos per cm. Water quality in areas affected by surface currents of the river is also rank in the good group. The electrical conductivity of this group varies from 312 to 775 micromhos per centimeter and TDS varies from 197 to 488 mg/L. The quality of water in the areas after the fans and due to river floods affection, their drinking water is limited and rank in the poor group. The electrical conductivity of these waters is more than 4000 micrometers per centimeter and TDS is more than 2500 mg/L. Sodium and chloride ions are predominant in this sector and their ionic abundance is rCa < rMg < rNa and rHco3 < rCl < rSo4. In addition, the quality of groundwater along the rivers towards the outlet of the plain is different in terms of drinking water is ranked in the acceptable and bad category. The dominant ion in this area is sodium and sulfate. The order of the ionic frequency of these waters is rCa < rMg < rNa and rHco3 < rCl < rSo4.

3. Results

Statistical analysis with data obtained from the intersection of layers was done. To investigate the relationship between water quality and distribution of ions, Kruskal-Wallis test was used for two years in three sub-basins of Mashhad Plain as a case study.

3.1. Analysis of the Difference between the Three Sub-Watersheds in Terms of Groundwater Quality in 1992

Based on the results of this test, water quality variation was significant at 95% level (0.05), but potassium ion (K) and water content of sodium content in 1992 in three sub-watersheds (west, middle, and east) of Mashhad had no significant difference (Table 2). Figure 2 shows the position of these three sub-watersheds.

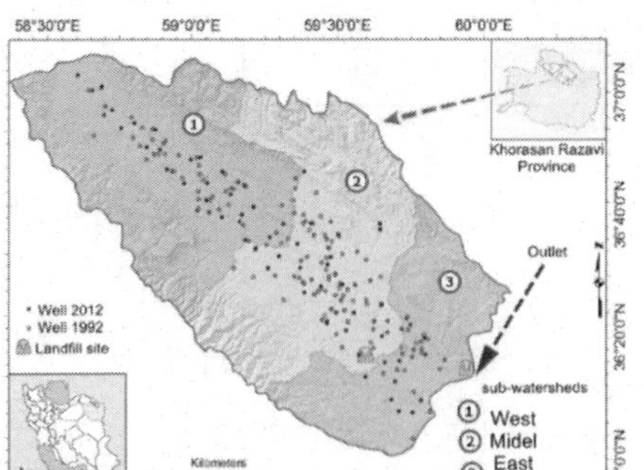

Figure 2. Study of groundwater associated with groundwater quality in Mashhad Plain.

Table 2. Kruskal-Wallis test results for independent samples in the study of the difference between three sub-watersheds for groundwater quality in 1992

Null	Sig.	Null hypothesis description
Reject	0.01	The distribution of calcium ion (Ca) in the study groups is the same.
Reject	0.002	The distribution of magnesium ion (Mg) in the study groups is the same.
Reject	0.003	The distribution of sodium ion (Na) in the study groups is the same.
Accept	1.000	The distribution of potassium ion (K) in the study groups is the same.
Reject	0.002	The distribution of cation in the study groups is the same.
Accept	0.273	The distribution of sodium percent (%Na) in the study groups is the same.
Reject	0.01	The distribution of sodium absorption (SAR) in the study groups is the same.
Reject	0.003	The distribution of TH in the study groups is the same.
Reject	0	The distribution of EC in the study groups is the same.
Reject	0.001	The distribution of TDS in the study groups is the same.
Reject	0.02	The distribution of PH in the study groups is the same.
Reject	0.009	The distribution of CO3 in the study groups is the same.
Reject	0.003	The distribution of HCO3 in the study groups is the same.
Reject	0	The distribution of Cl in the study groups is the same.
Reject	0	The distribution of SO4 in the study groups is the same.
Reject	0.001	The distribution of anions in the study groups is the same.

3.2. Analysis of the Difference between Three Sub-Watersheds in Terms of Groundwater Quality in 2012

Similarly, in 1992, the Kruskal-Wallis test was used for 2012 at 95% (0.05) to analyze the water quality difference in three sub-watersheds. The results obtained (Table 2) showed that only potassium and bicarbonate null hypothesis in three sub-watersheds had no significant difference in terms of these ions. For the rest of this study, for example, the distribution of calcium ion (Ca) in studied sub-watersheds groups in 1992 was shown and identified an increase trend toward inlet in both years (Figures 3 and 4). The detection of water flows are toward east and this can concentrate contaminant and ions in the east.

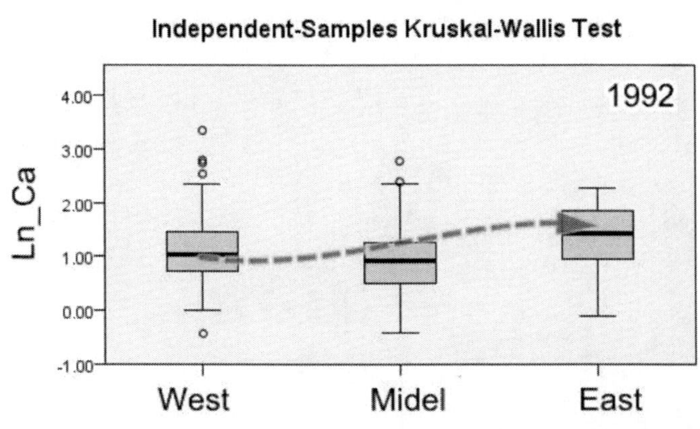

Figure 3. Calcium ion (Ca) distribution in the sub-watersheds groups of 1992 in Mashhad Plain.

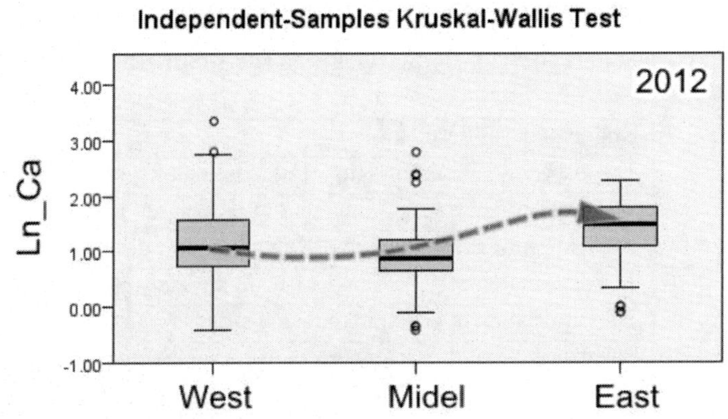

Figure 4. Calcium ion (Ca) distribution in the sub-watershed in 2012 in Mashhad Plain.

3.3. Analysis of Cations and Anions Based on the Stiff Diagram

This diagram is used to assess water quality changes in a place like a watershed and a period (Erdoğan et al., 2018). The Stiff diagram consists of four parallel horizontal axes that cut off a zero-valued vertical axis. The values of each of the major anions and cations are expressed in terms of milli-equivalent per liter (meq/L) on one of the horizontal axes, and the level of these graphs represents the amount of total dissolved solids (TDS) (Hounslow, 2018). In each review, based on the Stiff diagram, one sample can be tested. In this study, analysis of cations and anions was carried out separately for the years 1992, 2002, and 2012. According to the results of this chart in 1992, high anion chloride can be mentioned, while cation levels are almost equal in this year (Figure 5). The results of the Stiff Chart in 2002 showed that the amount of chlorine was lower than that of 1992, but the amount of SO_4, carbonate, bicarbonate and cations increased in comparison with the year 1992 (Figure 6). In addition, the results of the review of the Stiff chart in 2012 showed that the amounts of anions was approximately equal, and the cations were higher in comparison with the anions and magnesium ions were the highest in cations (Figure 7).

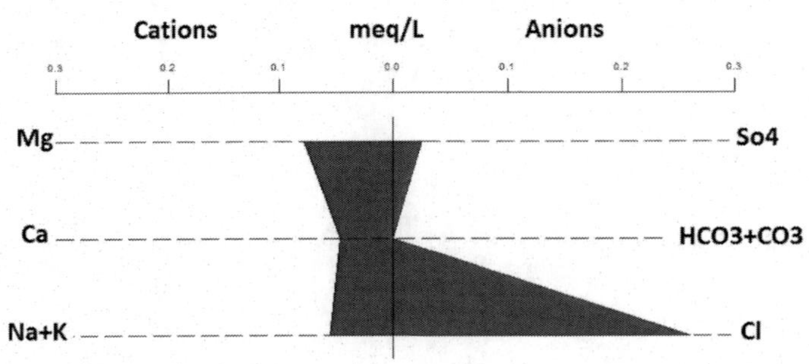

Figure 5. Distribution of the amounts of anions and cations in studied sub-basin groups in Mashhad plain, based on the Stiff diagram of 1992.

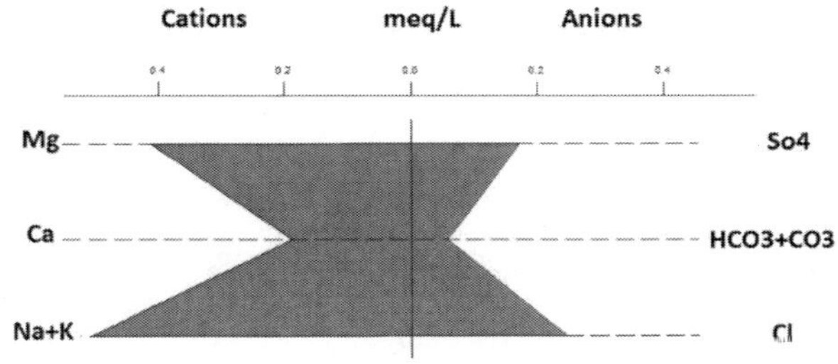

Figure 6. Distribution of the amounts of anions and cations in studied sub-aquatic groups in Mashhad Plain, based on the Stiff diagram of 2002.

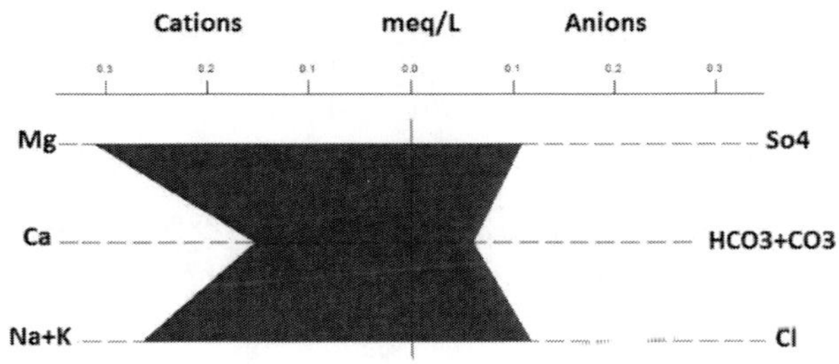

Figure 7. Distribution of the amounts of anions and cations in studied sub-watersheds in Mashhad plain based on the Stiff chart of 2012.

3.4. Analysis of Cations and Anions Based on Ionic Balance Scale

The ionic balance diagram of cation and anion shows the complex samples in terms of electrical equivalence. According to the ionic equilibrium chart drawn for the year 1992, it is deduced that the amounts of anions is higher than cations (Figure 8).

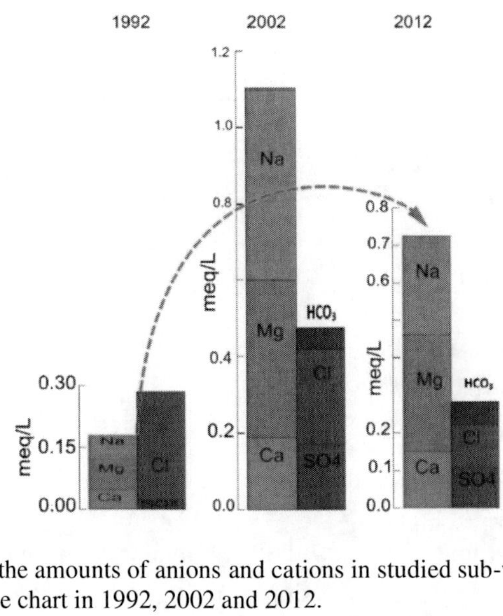

Figure 8. Distribution of the amounts of anions and cations in studied sub-watersheds in Mashhad plain based on the ionic balance chart in 1992, 2002 and 2012.

However, this chart for the year 2002 indicates an increase in cations against anions. The cations amount to more than meq/L, while the amounts of anions are slightly more than 0.4 meq/L (Figure 8). The ionic balance chart for 2012 shows that among cations, magnesium ion has the highest and calcium ion have the lowest amount. Among the anions, chlorine ion has the highest and bicarbonate has the lowest amount (Figure 8).

3.5. Analysis of Cations and Anions Based on Piper Diagram

This diagram is used to classify samples and determine the chemical group of water. The total amount of anions and cations is 100%, and the percentage of ions is marked on the side triangles and the corresponding points are depicted on the diamond on the middle ridge. Finally, the judgments about the water quality group are performed by the Piper chart according to the focus area of the points. The Piper chart shows the major ionic compounds of any number of samples. Figure 9 shows cations and anions in the years 1992, 2002, and 2012. According to the results of this chart, the highest amount of chlorine is in the year 1992, and the total amount of calcium and magnesium ion of this year is also more than the two years of 2002 and 2012. However, the total amount of sodium and potassium in 2002 shows the highest amount than the other two years. In addition, the chart shows the highest carbonate and bicarbonate content in 2012 compared to these three years.

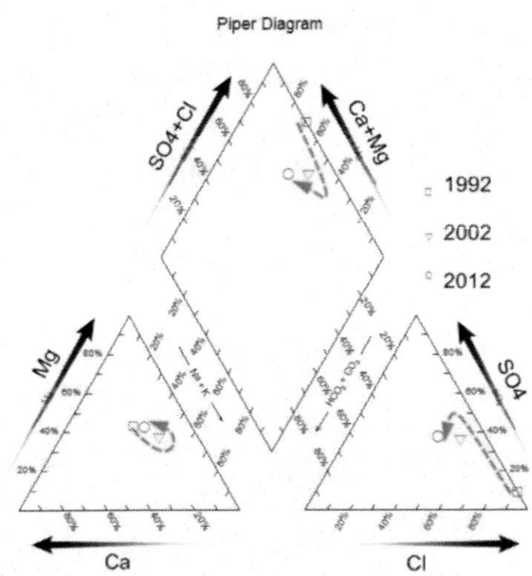

Figure 9. Distribution of the amounts of anions and cations in studied sub-watershed groups in Mashhad Plain, based on Piper chart in three years.

3.6. Analysis of Cations and Anions Based on Schoeller Charts

Schoeller's semi-logarithm graph is used to display the main ions in meq/kg or meq/L, as well as the chemical difference of the samples in a graph. In a hydrological report, for drinking water classification, Schoeller charts are usually used. In this study, to evaluate the amounts of anions and cations of groundwater in Mashhad Plain for years 1992, 2002, and 2012, this diagram is used as described in Figure 10.

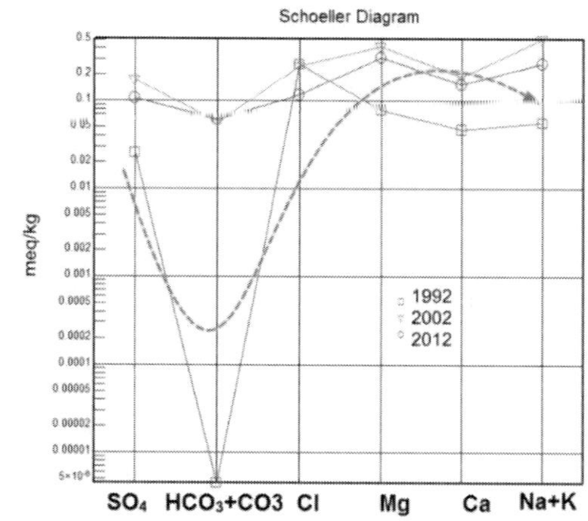

Figure 10. Distribution of the amounts of anions and cations in study sub-watersheds in Mashhad plain based on Schoeller diagrams over three years.

3.7. Analysis of Cations and Anions Based on Triangular Diagrams

In triangular diagrams, three variables are compared each other. In this study, the amount of sodium, calcium and magnesium contained in groundwater of Mashhad plain is compared with each other in years pf the 1992, 2002 and 2012 and is shown in Figure 11. The results show that sodium content in 2002 was more than two other years, as well as calcium and magnesium amount in 1992 was more than two other years.

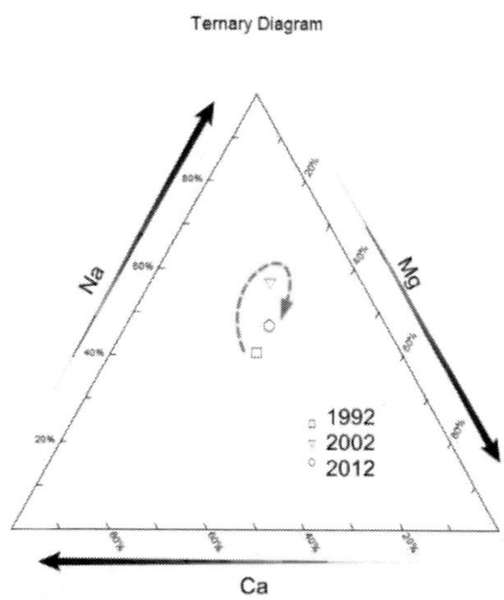

Figure 11. Distribution of the amounts of anions and cations in studied sub-watersheds in Mashhad plain based on triangular diagrams in three years.

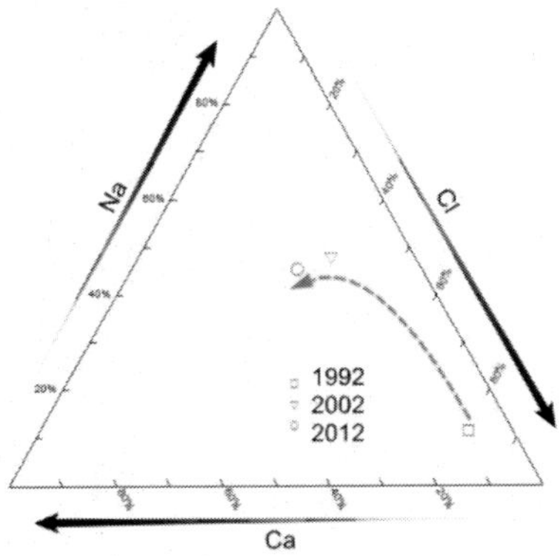

Figure 12. Distribution of the amounts of anions and cations in study sub-watersheds in Mashhad plain based on triangular diagrams in three years.

In another study, sodium, calcium, and chlorine ions were studied for these three years. Figure 12 shows the triangular diagram of these three variables. In the results of the study, chloride amount was highest in 1992 and in 2012, was the lowest in these three years.

3.8. Analysis of Cations and Anions According to Durov Chart

Durov's diagram is like the Piper graph in that the results of the analyzes of anions and cations are on separate triangles (Matiatos et al., 2018). This diagram is in fact an expanded form of the Piper graph (Hounslow, 2018). The Durov chart was also used to further analyze of the probable controller processes of water chemistry. The Durov chart shows the original ionic composition, pH, and TDS content of each number of samples. In this sample, the Durov diagram shows that the pH value was eight in the years of 2002 and 2012. This is while this amount is less than two in 1992. Also, the TDS value in 2002 is more than other two years and shows a number greater than 1200 milligrams per liter (Figure 13).

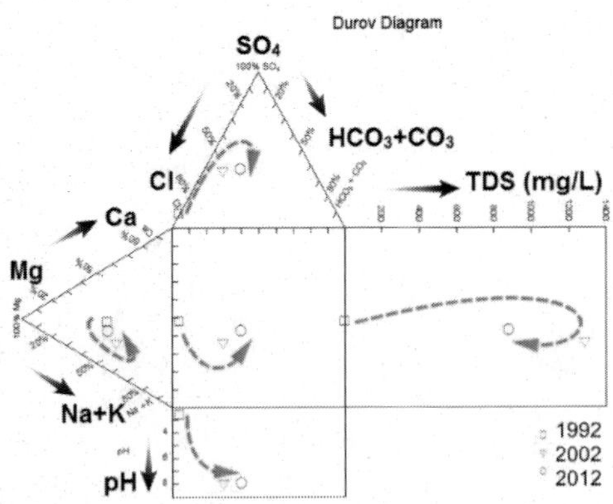

Figure 13. Distribution of the amounts of anions and cations in the studied sub-watersheds in Mashhad plain based on Durov chart in three years.

3.9. Analysis of Cations and Anions Based on Quality Drawing of the Specimen Series

The drawing of the qualitative series of 1992, 2002 and 2012 in Figure 14 shows that the sodium, sulfate, magnesium and calcium content increased significantly in 2002 compared with two years, but the amount of chloride ion in years of 1992 and 2002 are near each other and is higher than 2012. The amount of bicarbonate in 2012 is more than two other years.

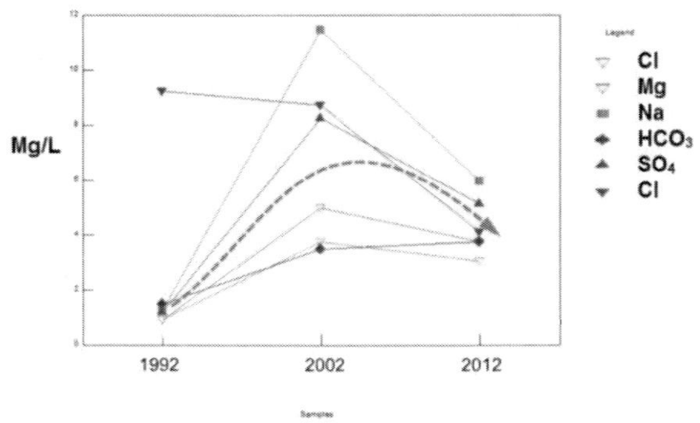

Figure 14. Distribution of the amounts of anions and cations in studied sub-basin groups in Mashhad Plain based on qualitative series drawing in three years.

3.10. Analysis of Cations and Anions Based on Intersection Drawings

The analysis of sodium and anion chloride cation in the intersection plot showed that sodium content was less than 1 mg/kg in 1992, and the chloride content in this year was more than 9 mg/kg while in 2012, sodium content was higher than chloride. However, in 2002, the amounts of both ions were high, indicating a sodium cation more than 11 mg/kg and the anion of chloride was greater than 8 mg/kg (Figure 15).

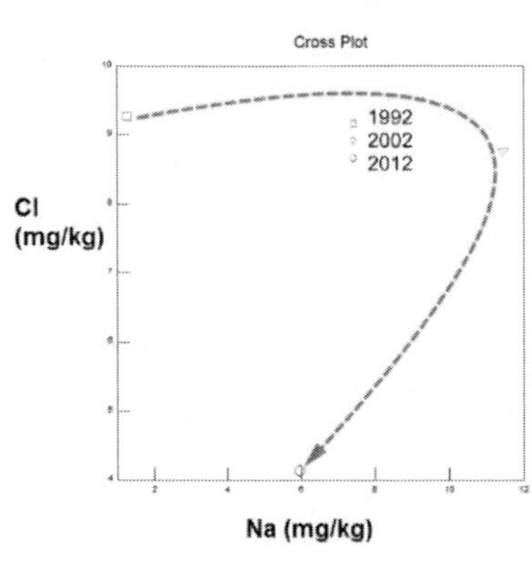

Figure 15. Distribution of the amounts of anions and cations in studied sub-watersheds in Mashhad plain based on an intersection plot of three years.

Conclusion

In geospatial techniques for hydrogeochemistry studies, there is a need to prioritize the aquifers according to the target and budget. Dividing the district to some aquifer and watersheds for initial identification is a valuable geospatial technique. Appropriate method for these planning is the use of modern geospatial techniques that utilize spatial and descriptive data using multiple layer intersections and extracting useful information from these data using both geospatial and statistical software and techniques. This information can be used to prioritize the various water resources in terms of hydrogeochemistry. On the other hand, the aquifer hydrogeochemistry categorization can determine the process of qualitative and quantitative spatial and temporal changes in groundwater resources. If timely data is available over many years, the manager of the watershed and the aquifer can, based on the changes made over the locations and the times, take appropriate planning to prevent, rehabilitate, or stopping exploitation.

Practice

1. How using coordinated data, water quality and quantity changes in large watersheds and topographic data analyzed and managed?
2. How intersect attributes and geospatial data identify as a modern way for statistical analysis of spatial factors in aquifers and watersheds management and studies?
3. How a non-parametric statistical analysis and trend process can be conducted in districts with more than three sub-watersheds or aquifers?
4. For which purpose non-parametric Mann Whitney-U test is used in aquifers and watersheds?

References

Abell, R., Thieme, M. L., Revenga, C., Bryer, M., Kottelat, M., Bogutskaya, N. & Stiassny, M. L. (2008). Freshwater ecoregions of the world: a new map of biogeographic units for freshwater biodiversity conservation. *BioScience*, *58*(5), 403-414. https://doi.org/10.1641/B580507.

Baun, D. L & Christensen, T. H (2004), "Speciation of heavy metals in landfill leachate: a review," *Waste Management & Research,* 22(1), 3-23. https://doi.org/10.1177/0734242X04042146.

Bergmann, R., Ludbrook, J., & Spooren, W. P. (2000). Different outcomes of the Wilcoxon-Mann-Whitney test from different statistics packages. *The American Statistician*, 54(1), 72-77. https://doi.org/10.1080/00031305.2000.10474513.

Besbes, M., Chahed, J., & Hamdane, A. (2019). Fifty Years of Water Policies, 1960-2010. In National Water Security (pp. 57-92). Springer, *Cham*. https://doi.org/10.1007/978-3-319-75499-4_3.

Czajkowski, J., Engel, V., Martinez, C., Mirchi, A., Watkins, D., Sukop, M. C & Hughes, J. D (2018), Economic impacts of urban flooding in south Florida: Potential consequences of managing groundwater to prevent salt water intrusion, *Science of The Total Environment*, 621, 465-478. https://doi.org/10.1016/j.scitotenv.2017.10.251.

Elubid A, B., Huag, T., H Ahmed, E., Zhao, J., Elhag, M., Abbass, W., & M Babiker, M. (2019). Geospatial Distributions of Groundwater Quality in Gedaref State Using Geographic Information System (GIS) and Drinking Water Quality Index (DWQI). *International journal of environmental research and public health*, 16(5), 731. https://doi.org/10.3390/ijerph16050731.

Erdoğan, B. U., Gökbulak, F., Serengil, Y., Yurtseven, İ & Özçelik, M. S (2018), Changes in selected physical water quality characteristics after thinning in a forested watershed, *Catena,* 166, 220-228. https://doi.org/10.1016/j.catena.2018.04.010.

Gaddis, G. M., & Gaddis, M. L. (1990). Introduction to biostatistics: Part 5, Statistical inference techniques for hypothesis testing with nonparametric data. *Annals of emergency medicine*, 19(9), 1054-1059. https://doi.org/10.1016/S0196-0644(05) 82571-5

Ghorbani Kalkhajeh, R., & Jamali, A. A. (2019). Analysis and Predicting the Trend of Land Use/Cover Changes Using Neural Network and Systematic Points Statistical Analysis (SPSA). *Journal of the Indian Society of Remote Sensing,* 1-15. https://doi.org/10.1007/s12524-019-00995-7.

Hosseinifard, S. J., & Aminiyan, M. M. (2015). Hydrochemical characterization of groundwater quality for drinking and agricultural purposes: a case study in Rafsanjan plain, Iran. Water Quality, *Exposure and Health*, 7(4), 531-544. https://doi.org/ 10.1007/s12403-015-0169-3.

Hounslow, A (2018), Water quality data: analysis and interpretation. *CRC press*. 397 p. https://doi.org/10.1201/9780203734117.

Jamali, A. A., & Ghorbani Kalkhajeh, R. (2019). Urban environmental and land cover change analysis using the scatter plot, kernel, and neural network methods. *Arabian Journal of Geosciences,* 12(3), 100. https://doi.org/10.1007/s12517-019-4258-7.

Jamali, A. A., Randhir, T. O., & Nosrati, J. (2018). Site suitability analysis for subsurface dams using boolean and fuzzy logic in arid watersheds. *Journal of Water Resources Planning and Management,* 144(8), 04018047. https://doi.org/10.1061/(ASCE)WR. 1943-5452.0000947.

Kamali, M. I., Nazari, R., Fridhosseini, A., Ansari, H., Eslamian, S., 2015, The Determination of Reference Evapotranspiration for Spatial Distribution Mapping Using Geostatistics, *Water Resources Management,* Vol. 29 (11), 3929–3940.

Matiatos, I., Paraskevopoulou, V., Lazogiannis, K., Botsou, F., Dassenakis, M., Ghionis, G & Poulos, S. E (2018), Surface-ground water interactions and hydrogeochemical evolution in a fluvio-deltaic setting: The case study of the Pinios River delta, *Journal of Hydrology*, 561, 236-249. https://doi.org/10.1016/j.jhydrol.2018.03.067.

Rashid, A., Khattak, S. A., Ali, L., Zaib, M., Jehan, S., Ayub, M., & Ullah, S. (2019). Geochemical profile and source identification of surface and groundwater pollution of District Chitral, Northern Pakistan. *Microchemical Journal*, 145, 1058-1065. https://doi.org/10.1016/j.microc.2018.12.025.

Shooshtarian, M. R., Dehghani, M., Margherita, F., Gea, O. C & Mortezazadeh, S (2018), Land use change and conversion effects on ground water quality trends: An integration of land change modeler in GIS and a new Ground Water Quality Index developed by fuzzy multi-criteria group decision-making models, *Food and Chemical Toxicology*, 114, 204-214. https://doi.org/10.1016/j.fct.2018.02.025.

Singh, K. P., Malik, A., Mohan, D., & Sinha, S (2004), Multivariate statistical techniques for the evaluation of spatial and temporal variations in water quality of Gomti River (India)-a case study, *Water Research*, 38(18), 3980-3992. https://doi.org/10.1016/j.watres.2004.06.011.

Tavakoly, A. A., Habets, F., Saleh, F., Yang, Z. L., Bourgeois, C., & Maidment, D. R. (2019). An integrated framework to model nitrate contaminants with interactions of agriculture, groundwater, and surface water at regional scales: The STICS-EauDyssée coupled models applied over the Seine River Basin. *Journal of Hydrology*, 568, 943-958. https://doi.org/10.1016/j.jhydrol.2018.11.061.

Zhai, Y., Ma, T., Zhou, J., Li, X., Liu, D., Wang, Z., & Du, Q. (2019). Impacts of leachate of landfill on the groundwater hydrochemistry and size distributions and heavy metal components of colloids: a case study in NE China. *Environmental Science and Pollution Research,* 26(6), 5713-5723. https://doi.org/10.1007/s11356-018-4053-0.

In: Advances in Hydrogeochemistry Research
Editors: S. Saeid Eslamian and Faezeh Eslamian

ISBN: 978-1-53617-829-6

Chapter 8

A COMPARATIVE HYDROCHEMICAL STUDY OF BOTTLED MINERAL WATERS

Daniel Marcos Bonotto[1,*]***, Gabrielle Roveratti***[2] ***and Saeid Eslamian***[3,4]

[1]Departamento de Petrologia e Metalogenia, Universidade Estadual Paulista (UNESP), Câmpus de Rio Claro, Rio Claro, São Paulo, Brasil
[2]Claretiano Faculdade, Rio Claro, São Paulo, Brasil
[3]Department of Water Engineering, Collage of Agriculture, Isfahan University of Technology, 8415683111, Isfahan, Iran
[4]Center of Excellence in Risk Management and Natural Hazards, Isfahan University of Technology, 8415683111, Isfahan, Iran

ABSTRACT

In the last few decades, the consumption of natural drinking water, either spring or mineral (bottled or not), increased in several countries. Despite drinking water being mostly used as the tap water accessible in every household, many people believe that naturally occurring waters are healthy and/or can be utilized for health remedies, thus, exhibiting better quality than the tap water. Additionally, economic reasons have also favored their use as bottled waters therefore widely increasing the commercialization of mineral waters. Thermal and mineral waters use in Brazil is not recent due to arrival of European immigrants, mainly from Portugal. The construction of thermal and non-thermal spas for therapeutic and leisure purposes reached a maximum number in the 1930s and 1950s, mainly at São Paulo (SP) and Minas Gerais (MG) states. The Brazilian Code of Mineral Waters (BCMW) was established in this time, under French influence, by Register 7841 published on 8 August 1945. The BCMW classifies the mineral and potable waters for uses in spas and bottling purposes. There are several parameters considered by the BCMW, for instance, the radioactivity due to dissolved ^{222}Rn and ^{220}Rn. EuroGeoSurveys

[*] Corresponding Author's Email: danielbonotto@yahoo.com.br.

(The Geochemistry Group of the European Geological Surveys) managed a common European sampling campaign of bottled mineral and spring waters (analysis of 884 samples for more than 70 chemical parameters in one laboratory) whose results were published in 2010. The hydrogeochemical study held within the framework of the project involved different approaches like analytical techniques, major constituents, trace elements, radionuclides, stable isotopes, mapping, waters classification, statistical treatment of hydrochemical data, human health, etc. This chapter reports a comparative hydrochemical study of mineral waters bottled in different countries based on Brazilian and international guidelines with the aim of increasing the information generated by EuroGeoSurveys.

Keywords: mineral waters, hydrochemical parameters, hydrogeochemical diagrams, water quality, Brasilia

INTRODUCTION

Hippocrates has been recognized as the author of several texts, among them, "On Airs, Waters, and Places," in which there is reference to the quality of waters ingested by people, according to some physical characterisitics, i.e., light and clear fresh waters or hard and salty water.

The roman architect and engineer *Marcus Vitruvius Pollio* wrote the book "*De Architectura"* in ~year 40 BC that describes the construction of thermae, bathhouses, aqueducts, wells and cisterns, as well the properties of different types of waters and devices used for finding and transporting water. *Thermae* was the name given by the romans for the large imperal sites dedicated to public baths, whereas *Balnea* corresponded to smaller size facilities for public or private baths.

Gaius Plinius Secundus (Pliny the Elder) was born in Como (Italy) in 23 AC and wrote "Historia Naturalis," *an* encyclopedia describing, among other subjects, the unusual properties of waters, the beneficial and maleficent types of waters, techniques for prospecting waters, constructing wells and transporting waters, thermal springs, medicinal properties of the waters, sea water desalination, and springs able to heal wounds, fractures, gout, foot, sciatic, fever, psoriasis, eyes disease, ears, headache, insanity, and female sterility.

Vitruvius and Pliny the Elder recognized several types of thermal spring waters and medicinal waters very similar to the classification adopted in the present days in several countries. For instance, sulphurous springs (for relieving the muscles and tendons weakness), aluminous springs (for the paralysis treatment), bituminous springs (for purifying and healing of "inner diseases"), alkaline springs (for purification and tumors reduction), and acid springs (for dissolving bladder stones).

The romans developed a net of springs of mineral and thermal waters in their conquered territories, however, after the Roman Empire decline, it was verified an interest decrease by baths and their therapeutic use. The Christianity contributed to such depletion due to its opposition to some practices occurring in the roman baths like food and wine abundance, and occasional sexual activity. In the Middle Ages, the baths were restrict to rich families as the accentuated poverty was not able to provide appropriate places for their practice.

The king Charles IV of Bohemia, Czech Republic, discovered in 1350 the springs of *Karlovy Vary*, whose waters have been used for health treatment since that time. The interest by medicinal springs came back in Europe with the beginning of the Renaissance in 14th century. In the same period, a blacksmith in Spa city situated in a valley of the mountain chain of Ardennes, Liège Province, southeast of Belgium, utilized the waters of a spring to cure his rheumatism. After that, he founded a bathhouse for health treatment and the word *spa* has been used since then in order to describe premises constructed for the same purpose.

The water use for ritual baths, hygiene or health treatment also happened in other cultures. For instance, the *Mikvah* has been a bath used since ~2000 BC for the purpose of ritual immersion in Judaism to achieve ritual purity. In China, between 1100 and 220 BC (Zhou dynasty), there are records of baths with herbs. The Japanese public bathhouses (Sentō*) have been used since* 710 AC, initially as religious practices in temples. The ancient Finnish word *Sauna* has been adopted to describe a traditional finnish bath (dry or wet between 70 and 100°C) and the bathhouse. The Turkish baths, *hamam*, spread during the Otoman Empire, between 1300 and 1923 AC, taking an important role in the Middle East cultures.

In Brazil, the thermal and mineral waters use is not recent due to arrival of European immigrants, mainly from Portugal. The discovery of the thermal waters in Caldas Velhas city (Goiás State) occurred in 1722 that was followed by the identification of the therapeutic properties of the waters of Araxá city (Minas Gerais State). The construction of thermal and non-thermal spas for therapeutic and leisure purposes reached a maximum number in the 1930s and 1950s, mainly at São Paulo (SP) and Minas Gerais (MG) states (Mourão, 1992). The Brazilian Code of Mineral Waters (BCMW) was established in this time, under French influence, by Register 7841 published on 8 August 1945 (DFPM, 1966). The commercialization of mineral waters in the country has widely increased, reaching circa 20 million consumers (SEBRAE, 2012) as many people believe that they are healthy and/or can be utilized for health cures, thus, exhibiting better quality than the tap water. DNPM-National Department of Mineral Production manages in Brazil the production and commercialization of mineral waters. The amount of bottled mineral waters at SP in 2007 was higher than 1.5 billion liters (34% of the total in the country) (CPRM, 2012a).

According to temperature, the BCMW establishes the following waters classification (DFPM, 1966): cold- values lower than 25°C; hypothermal- values ranging from 25 to 33°C; mesothermal- values ranging from 33 to 36°C; isothermal- values between 36 and 38°C; hyperthermal- values higher than 38°C.

The BCMW also establishes different waters classification based on chemical aspects (DFPM, 1966). Thus, they are: *radiferous* if contain dissolved radioactive substances that sustain a permanent radioactivity; *bicarbonate-alkaline* if contain dissolved alkaline compounds corresponding to a minimum of 0.2 g per liter of sodium bicarbonate; *earth-alkaline* if contain dissolved earth-alkaline compounds corresponding to a minimum of 0.12 g per liter of sodium carbonate; *sulfated* if contain dissolved sulfate corresponding to a minimum of 0.1 g per liter combined with sodium, potassium and magnesium; *sulfured* if contain dissolved sulfur corresponding to a minimum of 0.001 g per liter; *nitrated* if contain dissolved nitrate of mineral origin corresponding to a minimum of 0.1 g per liter; *chlorinated* if contain dissolved chloride corresponding to a minimum of 0.5 g per liter of sodium chloride; *ferruginous* if contain dissolved iron corresponding to a minimum of 0.005 g per liter; *radioactive* if contain dissolved radon corresponding to a minimum of 5 Mache Unit per liter; *thoriferous* if contain dissolved thoron corresponding to a minimum of 2 Mache Unit per liter; and *carbogaseous* if contain dissolved carbon dioxide gas corresponding to a minimum of 200 mL per liter at 20 °C temperature and 760 mmHg pressure.

EuroGeoSurveys (The Geochemistry Group of the European Geological Surveys) managed a common European sampling campaign of bottled mineral and spring waters (analysis of 884 samples for more than 70 chemical parameters in one laboratory) whose results were published in 2010. The water sources were from Croatia (Peh et al., 2010), Estonia (Bityukova and Petersell, 2010), Germany (Birke et al., 2010a, 2010b), Greece (Demetriades, 2010; Dotsika et al., 2010), Hungary (Fugedi et al., 2010), Italy (Dinelli et al., 2010; Cicchella et al., 2010), Norway/Sweden/Finland/Iceland (Frengstad et al., 2010), Portugal (Lourenço et al., 2010), Serbia (Petrović et al., 2010), Slovakia (Dušan et al., 2010), and Slovenia (Brenčič and Vreča, 2010; Brenčič et al., 2010). Eslamian et al. (2009, 2018) described the main characteristics and parameters affecting waters from thermal springs in Iran. In Brazil, Bertolo et al. (2007) and Bonotto (2016) reported different hydrogeochemical approaches focusing natural spring and bottled mineral waters.

This chapter reports a comparative hydrochemical study from a novel database of bottled mineral waters consisting of diverse brands of those focused in EuroGeoSurveys and in past reports developed in Brazil. Such investigation contributes to the increase of the information related to this important resource worldwide, taking into account guidelines specified by the Brazilian and international legislation.

MATERIALS AND METHODS

The database of this study consisted on 139 bottled mineral waters providing from Brazil (94), Argentina (2), Paraguay (1), Australia (1), Japan (1), United Arab Emirates

(1), Austria (3), England (2), France (5), Germany (2), Greece (8), Hungary (3), Italy (6), Norway (1), Poland (2), Scotland (2), Slovenia (1), and Turkey (4) (Figure 1).

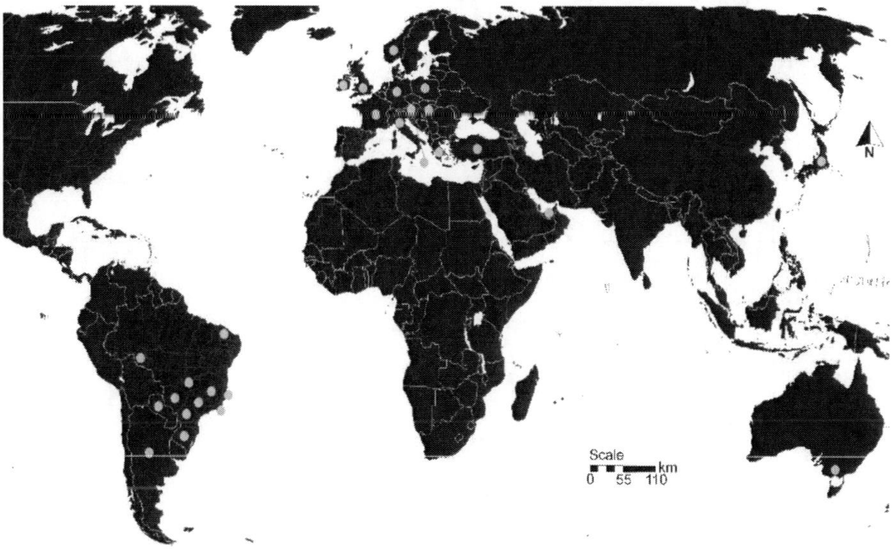

Figure 1. Location of the cities where the exploitation of the bottled mineral waters focused in this study takes place. The circles represent the sites, most of them situated in South America and Europe.

The hydrochemical data were available in the labels of the water bottles, and, in some cases, more than one dataset was recorded, representing bottling in different occasions (year following the brand name, or capital letter if the year is not available). Table 1 shows the countries, corresponding brands and lithologies/ages associated to the water sources studied.

Table 1. The bottled mineral waters focused in this study

Country	No.	Brand	City/Province[1]	Remark[2]	Lithology/Age	Reference
Brazil	1	Levity (2005)	Potirendaba (SP)	Fluoridated, vanadic	Paraná basin	Milani et al. (2007)
	2	Levity (2006)	Potirendaba (SP)	Fluoridated, vanadic	Paraná basin	Milani et al. (2007)
	3	Vittalev (A)	Bauru (SP)	Fluoridated, vanadic	Serra Geral Formation/Cretaceous	Milani et al. (2007)
	4	Vittalev (B)	Bauru (SP)	Fluoridated, hypothermal	Serra Geral Formation/Cretaceous	Milani et al. (2007)
	5	Lindoya (A)	Lindóia (SP)	Fluoridated	Socorro massif granite/Precambrian	Bizzi et al. (2003)
	6	Lindoya (B)	Lindóia (SP)	Fluoridated, radioactive	Socorro massif granite/Precambrian	Bizzi et al. (2003)
	7	Lindoya (C)	Lindóia (SP)	Fluoridated, radioactive	Socorro massif granite/Precambrian	Bizzi et al. (2003)
	8	Lindoya (D)	Lindóia (SP)	Radioactive	Socorro massif granite/Precambrian	Bizzi et al. (2003)

Table 1. (Continued)

Country	No.	Brand	City/Province[1]	Remark[2]	Lithology/Age	Reference
	9	Villa Fonte	Boa Esperança do Sul (SP)	Fluoridated	Paraná basin	Milani et al. (2007)
	10	H7	Analândia (SP)	Fluoridated	Serra Geral Formation/Cretaceous	Milani et al. (2007)
	11	Kiki (2004)	Bragança Paulista (SP)	Fluoridated	Bragança Paulista Suite/Proterozoic	Bizzi et al. (2003)
	12	Kiki (A)	Bragança Paulista (SP)	Fluoridated	Bragança Paulista Suite/Proterozoic	Bizzi et al. (2003)
	13	Santa Rita	Campo Limpo Paulista (SP)	Fluoridated	Crystalline basement/ Proterozoic	Bizzi et al. (2003)
	14	Prata (2002)	Águas da Prata (SP)	Alkaline-bicarbonate	Poços de Caldas alkaline massif	Schorscher and Shea (1992)
	15	Prata (2007)	Águas da Prata (SP)	Alkaline, fluoridated	Poços de Caldas alkaline massif	Schorscher and Shea (1992)
	16	Prata (A)	Águas da Prata (SP)	Fluoridated, radioactive	Poços de Caldas alkaline massif	Schorscher and Shea (1992)
	17	Cantagalo	Itirapina (SP)	Fluoridated	Paraná basin	Milani et al. (2007)
	18	Daflora (A)	Mococa (SP)	Fluoridated	Paraná basin	Milani et al. (2007)
	19	Daflora (B)	Mococa (SP)	Fluoridated	Paraná basin	Milani et al. (2007)
	20	Minalice (A)	São Simão (SP)	Fluoridated, vanadic	Paraná basin	Milani et al. (2007)
	21	Minalice (B)	São Simão (SP)	Fluoridated, vanadic	Paraná basin	Milani et al. (2007)
	22	Minalice (C)	São Simão (SP)	Fluoridated, hypothermal	Paraná basin	Milani et al. (2007)
	23	Indaiá	Campos do Jordão (SP)	Fluoridated, radioactive	Crystalline rocks/ Precambrian	Bizzi et al. (2003)
	24	Genuína	Águas de Lindóia (SP)	Fluoridated, radioactive	Socorro massif granite/Precambrian	Bizzi et al. (2003)
	25	Genuína	Águas de Lindóia (SP)	Alkaline-bicarbonate, Fluoridated, radioactive	Socorro massif granite/Precambrian	Bizzi et al. (2003)
	26	CrisÁgua	Brotas (SP)	Fluoridated	Paraná basin	Milani et al. (2007)
	27	Attiva	Bananal (SP)	Fluoridated	Colluvial deposits/ Pleistocene	Bizzi et al. (2003)
	28	Acquamin	Serra Negra (SP)	Fluoridated, radioactive	Bambuí Group	Bizzi et al. (2003)
	29	MariaÁgua	Rio Claro (SP)	Fluoridated	Rio Claro Formation/ Permian	Gimenez et al. (2016)
	30	Jorabel	Ipeúna (SP)	Fluoridated	Itaqueri Formation	Gimenez et al. (2016)
	31	Bioleve	Lindóia (SP)	Fluoridated, radioactive	Socorro massif granite/Precambrian	Bizzi et al. (2003)
	32	Crystal	Mogi das Cruzes (SP)	Fluoridated	Açungui Group/ Proterozoic	Bizzi et al. (2003)

Country	No.	Brand	City/Province[1]	Remark[2]	Lithology/Age	Reference
	33	Embaúna	Rio Claro (SP)	Fluoridated	Rio Claro Formation/ Permian	Gimenez et al. (2016)
	34	Fratelli Vita	Jaguariúna (SP)	Fluoridated, radioactive	Crystalline basement/ Precambrian	Bizzi et al. (2003)
	35	Acqua Vita	Araçatuba (SP)	Fluoridated, hypothermal	Bauru Group/ Cretaceous	Milani et al. (2007)
	36	Petra	Potirendaba (SP)	Fluoridated, vanadic	Bauru Group/ Cretaceous	Milani et al. (2007)
	37	Santa Inês	Presidente Prudente (SP)	Fluoridated, hypothermal	Paraná basin	Milani et al. (2007)
	38	Clara	Indaiatuba (SP)	Fluoridated, radioactive	Tubarão Group/ crystalline basement	Milani et al. (2007)
	39	Schincariol	Itu (SP)	Fluoridated, radioactive	Tubarão Group/ crystalline basement	Milani et al. (2007)
	40	Quilombo	Iacanga (SP)	Fluoridated, alkaline, hypothermal	Bauru Group/ Cretaceous	Milani et al. (2007)
	41	Santa Bárbara	Águas de Santa Bárbara (SP)	Hypothermal	Paraná basin	Milani et al. (2007)
	42	Pompeia	Socorro (SP)	Fluoridated	Socorro-Guaxupe Nappe/ Cretaceous-Tertiary	CPRM (2008)
	43	Varginha	Varginha (MG)	Fluoridated	Socorro-Guaxupe Nappe/ Cretaceous-Tertiary	CPRM (2008)
	44	Jacutinga (A)	Jacutinga (MG)	Fluoridated, radioactive	Amparo Complex	Petolchckny et al. (2009)
	45	Jacutinga (B)	Jacutinga (MG)	Fluoridated, radioactive	Amparo Complex	Petolchckny et al. (2009)
	46	Tropical	Igarapé (MG)	Radioactive	Belo Horizonte Complex/ Transamazonic	Noce et al. (1997)
	47	Igarapé	Igarapé (MG)	Fluoridated, radioactive	Belo Horizonte Complex/ Transamazonic	Noce et al. (1997)
	48	São Lourenço (2004)	São Lourenço (MG)	Carbogaseous, fluoridated	Mantiqueira Mountain Range	Santos (2007)
	49	São Lourenço (A)	São Lourenço (MG)	Carbogaseous, alkaline-bicarbonate, fluoridated	Mantiqueira Mountain Range	Santos (2007)
	50	São Lourenço (2014)	São Lourenço (MG)	Carbogaseous, fluoridated	Mantiqueira Mountain Range	Santos (2007)
	51	Viva (A)	Itaúna (MG)	Fluoridated, radioactive	Iron Quadrangle/ Precambrian	Santos (2007)
	52	Viva (B)	Itaúna (MG)	Radioactive	Minas Gerais Iron Quadrangle/ Precambrian	Santos (2007)
	53	Santa Rita (A)	Santa Rita do Sapucaí (MG)	Fluoridated	Monte Sião and Camanducaia faults	Massoli (1980)
	54	Santa Rita (B)	Santa Rita do Sapucaí (MG)	Fluoridated	Monte Sião and Camanducaia faults	Massoli (1980)

Table 1. (Continued)

Country	No.	Brand	City/Province[1]	Remark[2]	Lithology/Age	Reference
	55	Krenak (2011)	Resplendor (MG)	Fluoridated, hypothermal	São Tomé Formation	Bizzi et al. (2003)
	56	Krenak (2014)	Resplendor (MG)	Hypothermal	São Tomé Formation	Bizzi et al. (2003)
	57	Hélios	Dona Euzébia (MG)	Fluoridated	São Tomé Formation	Bizzi et al. (2003)
	58	Bella	Mário Campos (MG)	Fluoridated, radioactive	Três Irmãos Mountain Range	Moraes et al. (2017)
	59	Natureza (1993)	Campo Belo (MG)	Fluoridated, radioactive	Campo Belo Metamorphic Complex/ Archean	Carneiro and Barbosa (2008)
	60	Natureza (2009)	Campo Belo (MG)	Fluoridated	Campo Belo Metamorphic Complex/ Archean	Carneiro and Barbosa (2008)
	61	Pouso Alto	Pouso Alto (MG)	Fluoridated	São Francisco Craton	Kuster (2014)
	62	Aqualeve	Ponte Nova (MG)	-	Mantiqueira Complex	Bizzi et al. (2003)
	63	Passa Quatro (2003)	Passa Quatro (MG)	Fluoridated, radioactive	Passa Quatro Alkaline Complex	Silva (2014)
	64	Passa Quatro (2005)	Passa Quatro (MG)	Fluoridated, radioactive	Passa Quatro Alkaline Complex	Silva (2014)
	65	Passa Quatro (2012)	Passa Quatro (MG)	Fluoridated, radioactive	Passa Quatro Alkaline Complex	Silva (2014)
	66	Ingá	Brumadinho (MG)	Fluoridated, radioactive	Granite-Gneiss Complex	Silva (2014)
	67	Caxambu	Caxambu (MG)	Carbogaseous, radioactive, hypothermal, fluoridated	Brasília and Ribeira belts	Trouw (2007)
	68	Aguaí	São Joaquim de Bicas (MG)	Fluoridated, radioactive	Iron Quadrangle/ Precambrian	Noce et al. (1997)
	69	Roda d'Água	Juatuba (MG)	Fluoridated, radioactive	Rio das Velhas greenstone belt	Barbosa (1988)
	70	Divina Pureza	Matipó (MG)	Fluoridated	Mantiqueira Province/ Precambrian	Liccardo and Jordt-Evangelista (2013)
	71	Ouro de Minas	Barbacena (MG)	Fluoridated, radioactive	Barbacena- South Paraíba Association	Queiroga et al. (2012)
	72	Cristalina	Cássia (MG)	Fluoridated	Araxá Group	Azzi (2009)
	73	Santa Elizabeth	Sabará (MG)	Fluoridated, radioactive	Iron Quadrangle/ Precambrian	Noce et al. (1997)
	74	Bonaqua	Itabirito (MG)	Fluoridated, hypothermal	Metamorphic rocks	Oliveira (2013)
	75	Caldas Novas	Marzagão (GO)	Fluoridated, hypothermal	Amazonic Craton	Nogueira and Riccomini (2006)
	76	Água Cristalina	Anápolis (GO)	-	Anápolis Complex/ Neoproterozoic	Silva and Martins (2016)

Country	No.	Brand	City/Province[1]	Remark[2]	Lithology/Age	Reference
	77	Lia	Anápolis (GO)	Fluoridated, hypothermal	Anápolis Complex/ Neoproterozoic	Silva and Martins (2016)
	78	Água Iza	Goiânia (GO)	Fluoridated	Goiano Massif/ Proterozoic	Silva and Martins (2016)
	79	Minalinda	Porto Velho (RO)	Fluoridated, hypothermal	Jaciparaná and Solimões formations	Adamy and Romanini (1990)
	80	Kaiary (A)	Porto Velho (RO)	Fluoridated, hypothermal	Jaciparaná and Solimões formations	Adamy and Romanini (1990)
	81	Kaiary (2002)	Porto Velho (RO)	Radioactive, hypothermal	Jaciparaná and Solimões formations	Adamy and Romanini (1990)
	82	Lind'Água	Candeias do Jamary (RO)	Fluoridated, hypothermal	Ariquemes Domain	CPRM (2018)
	83	Água Clara	Aquiraz (CE)	-	Crystalline rocks, sedimentary formations and alluvial deposits	CPRM (1998)
	84	Indaiá	Horizonte (CE)	Fluoridated, radioactive, hypothermal	Crystalline rocks, sedimentary formations and alluvial deposits	CPRM (1998)
	85	Naturágua	Fortaleza (CE)	Hypothermal	Messejana Migmatites	Vidal et al. (2005)
	86	Dupote	Rio de Janeiro (RJ)	Fluoridated, hypothermal	Pedra Branca and Tijuca massifs	CPRM (2012b)
	87	Aquarel	Petrópolis (RJ)	Fluoridated	Migmatites and granitoids/Precambrian	Guerra et al. (2007)
	88	Água Campinho	Domingos Martins (ES)	Fluoridated	South Paraíba Complex	Burgos et al. (2006)
	89	Flor do Pantanal	Jardim (MS)	Fluoridated, vanadic, hypothermal	Amazonic Craton	GEMS (2016)
	90	LonRita	Londrina (PR)	Hypothermal	Serra Geral Formation/Cretaceous	Milani et al. (2007)
	91	Ouro fino	Campo Largo (PR)	Fluoridated	Atuba Complex/Proterozoic	Rocha (2014)
	92	Fontana Oro	Cascavel (PR)	Fluoridated	Serra Geral Formation/Cretaceous	Rocha (2014)
	93	Sarandi	Barra Funda (RS)	Fluoridated	Paraná basin	Milani et al. (2007)
	94	Fonte Ijuí	Ijuí (RS)	Fluoridated, alkaline-bicarbonate	Serra Geral Formation/Cretaceous	Milani et al. (2007)
Argentina	95	Villavicencio	Mendoza	Fluoridated	Barreal-Uspallata and Bermejo-Mendoza basins	Peralta (2003)
	96	Eco	Mendoza	-	Barreal-Uspallata and Bermejo-Mendoza basins	Peralta (2003)
Paraguay	97	Dasani	San Lorenzo	-	Areguá sandstones/ Cenozoic	Orué (1996)

Table 1. (Continued)

Country	No.	Brand	City/Province[1]	Remark[2]	Lithology/Age	Reference
Australia	98	Summit	Melbourne	-	Neoproterozoic-Carboniferous-Mesozoic	VSG (2015)
United Arab Emirates	99	Alain	Abu Dhabi	-	Oman Mountains/ Eocene-Oligocene	Eltokhi et al. (2012)
Japan	100	Water Forest	Kamikochi	-	Volcanic (greenstone)-granitoid belts	Taira (2001)
Turkey	101	Hamildiye	Istanbul	-	Triassic and Cretaceous sedimentary rocks	Ündül and Tuğrul (2006)
	102	Akmina	Istanbul	-	Triassic and Cretaceous sedimentary rocks	Ündül and Tuğrul (2006)
	103	Yasam Pinarim	Sakarya	-	Eocene Series	Görür and Elbek (2013)
	104	Damla	Sapanca	-	Eocene Series	Görür and Elbek (2013)
Scotland	105	Highland	Perth and Kinross	-	Igneous intrusions/ Caledonian	Banks (2007)
	106	Gleneagles	Blackford	-	Grampian Group	Banks (2007)
England	107	Harrogate	Harrogate	-	Carboniferous rocks, Permian limestone escarpments	Hudson (1938)
	108	Thirsty Planet	Harrogate	-	Carboniferous rocks, Permian limestone escarpments	Hudson (1938)
Italy	109	San Benedetto (1993)	Scorzè – Veneto	-	Italian Alps (Dolomites) - Miocene	Goldsmith and Goldsmith (1989)
	110	San Benedetto (2000)	Scorzè - Veneto	-	Italian Alps (Dolomites) - Miocene	Goldsmith and Goldsmith (1989)
	111	San Benedetto (2004)	Scorzè - Veneto	-	Italian Alps (Dolomites) - Miocene	Goldsmith and Goldsmith (1989)
	112	Levissima	Valdisotto	-	Alps belt/ Paleozoic	Guidottia et al. (2015)
	113	Tione	Tione di Trento	-	Sedimenatry rocks	Calamita and Deiana (1988)
	114	Acqua Panna	Florence	-	Northern Apennines/ Miocene to Quaternary	Carmignani et al. (2013)
France	115	Evian	Évian-les-Bains	-	French Alps/Cenozoic	Gavillot et al. (2019)
	116	Roches Fraîches	Chorges	-	French Alps/Cenozoic	Gavillot et al. (2019)
	117	Canyon	Alet-les-Bains	-	Alet-les-Bains block/ Ordovician-Carboniferous sediments	Gavillot et al. (2019)

Country	No.	Brand	City/Province[1]	Remark[2]	Lithology/Age	Reference
	118	Edena	La Réunion	-	Edena Mountain quadrangle/ Paleozoic	Edena (2019)
	119	Volvic	Volvic	-	Volcanic complex	Rouquet et al. (2012)
Greece	120	Vikos	Zagorohoria	-	Ionian formations	van Hinsbergen (2006)
	121	Loutraki (1994)	Loutraki		Alpine mountain/ Mesozoic	Taylor et al. (2011)
	122	Loutraki (1995)	Loutraki	-	Alpine mountain/ Mesozoic	Taylor et al. (2011)
	123	Loutraki (1998)	Loutraki	-	Alpine mountain/ Mesozoic	Taylor et al. (2011)
	124	Avra	Achaea	-	Achaia, Pindos zone, Gavrovo zone formations	Taylor et al. (2011)
	125	Korpi	Aetolia-Acarnania	-	Ionian formations	van Hinsbergen (2006)
	126	Apollonio	Agia Varvara	-	Volcanic rocks	van Hinsbergen (2006)
	127	Dirfis	Steni Dirfyos	-	Karya Formation/ Pliocene	van Hinsbergen (2006)
Slovenia	128	Bonaqua	-	-	Igneous and metamorphic rocks	UOL (2019)
Norway	129	Bonaqua	Fyresdal	-	Granite and granite gneiss	Stout (1972)
Germany	130	Bonaqua	-	-	-	-
	131	Sawell	Emsdetten	-	South German Keuper basin	Banning (2012)
Austria	132	Römerquelle	Edelstal	Bicarbonated, sulfated	Paleozoic outcrops similar to the Styrian Remschnig/Sausal areas	Hubmann et al. (2006)
	133	Vöslauer	Bad Vöslau	-	Calcareous Alpine at Vienna Basin	Wessely et al. (2007)
	134	Gasteiner	Bad Gastein	-	Northern Calcareous Alps	Wessely et al. (2007)
Hungary	135	Fonyódi	Fonyód	-	Volcanic rocks	Balderer et al. (2014)
	136	Guizza	Budaörs	-	Dolomites/ Triassic	Balderer et al. (2014)
	137	Theodora	Budapest	Fluoridated, Bicarbonated	Dolomites/ Triassic	Balderer et al. (2014)
Poland	138	Aretuza	Opacz	-	Sedimentary rocks/ Proterozoic	Hakenberg and Ska (2001)
	139	Nałęczowianka	Naleczow	-	Neocomian strata/ Jurassic	Hakenberg and Ska (2001)

[1]SP=São Paulo State, MG=Minas Gerais State, GO=Goiás State, RO=Rondônia State, CE=Ceará State, RJ=Rio de Janeiro State, ES=Espírito Santo State, MS=Mato Grosso do Sul State, PR=Paraná State, RS=Rio Grande do Sul State; [2]as highlighted in the label.

RESULTS AND DISCUSSION

Table 2 reports the physicochemical data of the bottled mineral waters. The water temperature range was 5.4-32.2°C (mean=22.3°C) and, according to the BCMW guidelines for temperature (DFPM, 1966), they are cold (<25°C, 85 water sources) and hypothermal (25-33°C, 24 water sources). The pH of the water sources was quite variable, ranging from acid (4.12) to alkaline (9.55) conditions, with the mean value (6.58) tending to reach neutrality. One of the most important compositional parameter used in the classification of drinking water is the amount of total dissolved solids (TDS). In the labels of the bottled waters, this information is reflected by two parameters, total solids residue (dry residue, DR) and electrical conductivity (EC). In general, both are directly linked to the different types of waters (Hem, 1985) and such relationship was also obbeyed in this study (Figure 2).

Table 2. Physicochemical parameters of the bottled mineral waters focused in this study

No.	Temperature (°C)	pH	Conductivity (µS/cm)	Dry Residue (mg/L)
1	25.2	8.84	163.0	117.7
2	24.4	9.33	133.0	124.4
3	28.2	8.30	121.9	74.2
4	28.2	8.60	108.0	98.2
5	24.3	7.12	209.0	153.0
6	21.0	6.30	80.0	92.0
7	22.0	6.30	142.0	114.2
8	21.0	6.00	520.0	52.0
9	23.8	4.12	24.6	20.2
10	23.4	6.54	49.8	53.9
11	22.0	6.60	134.8	110.6
12	21.9	6.42	98.1	92.8
13	22.5	7.69	143.0	107.6
14	22.8	7.01	584.0	270.3
15	22.7	7.10	322.0	205.8
16	21.9	5.87	101.4	57.2
17	21.0	5.20	840.0	57.0
18	23.5	6.49	73.6	94.0
19	23.1	-	99.2	104.2
20	26.5	8.75	120.2	84.3
21	26.4	8.79	131.3	75.1
22	26.0	7.30	225.0	89.2
23	17.9	7.63	166.0	100.2
24	20.9	6.24	897.0	66.0
25	24.0	7.10	393.0	224.9
26	23.4	5.01	17.5	20.3

No.	Temperature (°C)	pH	Conductivity (μS/cm)	Dry Residue (mg/L)
27	24.2	6.95	210.0	144.4
28	19.8	6.04	42.7	56.2
29	22.6	4.62	13.7	7.6
30	23.7	4.73	10.9	8.6
31	21.2	7.00	104.0	83.9
32	23.4	6.99	200.0	126.4
33	22.8	4.56	13.2	13.7
34	22.7	6.02	77.9	80.0
35	25.4	5.80	16.0	27.0
36	24.4	9.33	133.0	124.4
37	25.4	4.81	16.0	27.0
38	21.0	7.90	210.0	-
39	24.1	7.25	374.0	239.4
40	26.0	9.55	306.0	194.8
41	27.0	7.90	125.0	104.0
42	22.5	6.56	173.0	650.0
43	19.9	5.83	25.0	27.7
44	21.8	6.55	96.5	102.1
45	21.8	6.34	105.2	108.1
46	22.4	5.70	14.6	17.3
47	22.5	6.02	15.7	15.6
48	22.2	5.20	-	21.6
49	21.4	5.67	563.0	295.9
50	22.0	5.37	464.0	277.9
51	20.1	5.16	8.1	5.9
52	21.6	5.07	790.0	4.6
53	21.8	6.66	102.0	77.0
54	21.7	6.65	88.0	85.5
55	25.3	4.20	83.2	44.3
56	25.2	4.23	97.3	48.0
57	24.3	6.27	32.5	32.4
58	22.8	7.18	180.0	101.6
59	19.0	5.60	514.0	42.0
60	19.1	6.30	34.0	48.0
61	20.8	6.30	71.6	64.0
62	-	6.14	21.7	29.1
63	20.5	6.00	107.0	67.6
64	20.5	6.08	60.7	59.3
65	20,1	6.05	87.5	76.1
66	22.1	5.40	10.0	12.4
67	25.5	5.21	146.0	102.8
68	22.2	6.18	94.2	97.2
69	22.2	5.52	20.9	15.6
70	23.0	6.04	60.6	81.4
71	24.4	5.41	133.0	124.4
72	22.6	6.32	61.5	76.8

Table 2. (Continued)

No.	Temperature (°C)	pH	Conductivity (μS/cm)	Dry Residue (mg/L)
73	21.6	6.46	192.2	129.8
74	28.1	7.87	380.0	245.7
75	25.7	8.19	78.4	72.0
76	23.6	5.23	6.6	4.0
77	26.2	6.75	62.5	46.2
78	24.5	5.00	560.0	12.0
79	22.5	4.60	134.8	107.6
80	26.0	4.79	12.3	6.3
81	26.5	4.54	17.1	7.9
82	26.2	4.47	13.3	4.6
83	-	-	-	-
84	30.5	7.24	352.0	253.8
85	32.2	5.60	7.2	33.0
86	-		-	-
87	17.6	5.53	18.0	18.08
88	22.6	5.40	69.4	47.3
89	26.3	6.71	108.7	90.3
90	23.2	6.60	13.4	42.0
91	18.5	7.35	251.0	148.5
92	20.3	5.22	6.5	5.4
93	21.0	9.35	335.0	223.0
94	21.4	8.00	320.0	217.0
95	-	-	-	-
96	-	-	-	-
97	-	5.97	-	24.0
98	20.0	6.40	110.0	-
99	-	7.30	-	110.0
100	-	7.10	-	27.9
101	10.3	6.96	153.5	0.5
102	-	6.25	1974.0	3920.3
103	-	-	-	111.4
104	-	7.73	110.3	-
105	-	-	-	170.0
106	-	7.78	-	235.0
107	-	7.00	-	-
108	-	7.00	-	-
109	16.0	7.72	354.0	233.0
110	16.7	7.68	400.0	250.0
111	16.2	7.21	435.0	274.8
112	5.4	7.80	114.0	76.3
113	-	7.20	213.0	184.0
114	10.3	8.00	222.0	141.0
115	-	7.20	-	309.0

No.	Temperature (°C)	pH	Conductivity (μS/cm)	Dry Residue (mg/L)
116	-	7.60	-	240.0
117	-	-	-	-
118	-	-	-	-
119	-	7.00	-	109.0
120	25.0	7.30	485.5	290.0
121	18.0	7.90	665.0	342.0
122	18.0	7.90	710.0	37.6
123	18.0	8.05	645.0	332.0
124	-	7.10	641.2	317.5
125	-	-	-	-
126	18.0	7.85	993.0	-
127	18.0	7.60	500.0	290.0
128	-	-	-	-
129	-	-	-	-
130	-	-	-	-
131	-	-	-	-
132	-	-	-	-
133	-	-	-	-
134	13.6	6.71	263.0	157.0
135	-	-	-	761.0
136	-	-	-	424.0
137	-	-	-	1600.0
138	-	6.80	-	631.0
139	-	-	-	650.0

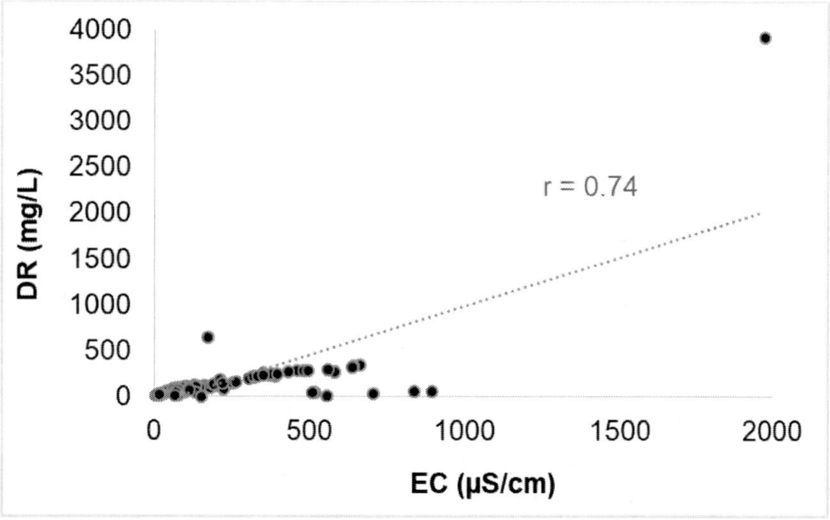

Figure 2. The relationship between the total solids residue (dry residue, DR) and electrical conductivity (EC) in the bottled mineral waters of this study.

Dinelli et al. (2010) classified the Italian bottled mineral waters according to EC in very low mineral concentration (<77 μS/cm), low mineral concentration (77-769 μS/cm), intermediate mineral concentration (769-2308 μS/cm), and high mineral concentration (>2308 μS/cm). According to these classes, the bottled mineral waters studied are of very low mineral concentration (32), low mineral concentration (77), and intermediate mineral concentration (4).

The EU mineral water directive proposed four TDS classes (van der Aa, 2003): very low mineral concentration (TDS < 50 mg/L); low mineral concentration (TDS 50-500 mg/L); intermediate mineral concentration (TDS 500-1500 mg/L); high mineral concentration (TDS > 1500 mg/L). According to such criterion, the bottled mineral waters studied are of very low mineral concentration (36), low mineral concentration (82), intermediate mineral concentration (4), and high mineral concentration (2). The EC and TDS criteria indicated predominance of the water sources with low mineral concentration. However, there is no total coincidence between them as the parameter EC pointed out none brand of high mineral concentration, whereas the TDS allowed to group 2 brands in this class.

Tables 3 and 4 report all chemical data of the bottled mineral waters, whereas Table 5 shows the standard error (SE) for all parameters considered in this chapter. SE was estimated from the equation $SE = SD/(n)^{1/2}$, where SD is the standard deviation of each parameter in the dataset and n represents the number of data available for the parameter. SE increased according to the SD increase and decreased for the higher n values.

The Regulation on the Hygienic Acceptability of Potable Water given by the World Health Organization (WHO) defines the Maximum Acceptable Concentration of chemical substances in water for public water supply. Despite the mineral waters are not used for such purpose, this guideline is helpful for the data comparison. For nitrate, the reference level is 50 mg/L (WHO, 2011), which was not surpassed in any case as the concentration range is 0.01-14.1 mg/L. For barium, the reference level is 0.7 mg/L (WHO, 2011), which also was not exceeded in none brand (the concentration range corresponded to 0.002-0.38 mg/L). For fluoride, the reference level is 1.5 mg/L (WHO, 2011), which was slightly surpassed in one analysis (year 2002) of the Brazilian brand Prata (concentration=1.59 mg/L, Table 4), whose water source occur in the Poços de Caldas alkaline massif. High-fluoride groundwaters are a recognized feature of a number of aquifers across the world (see, for instance, Smedley et al., 2002; Reddy et al., 2010; Edmunds and Smedley, 2013), inclusive in the Poços de Caldas Plateau (Bonotto, 1993, 2005). They have required effective actions from the national health agencies as dental fluorosis may develop from successive exposures to high F^- concentrations during the water ingestion. This disease is a developmental disturbance leading to dental enamel with lower mineral content and increased porosity (Alvarez et al., 2009; Edmunds and Smedley, 2013). The criterion for fluoride in mineral waters according to the EU mineral water directive is $F^- > 1$ mg/L and this was found in 6 water sources.

Table 3. Chemical parameters (in mg/L) of the bottled mineral waters focused in this study

No.	Na	K	Ca	Mg	Sr	Ba	Li	V
1	27.51	0.84	7.85	0.64	-	-	-	0.070
2	30.40	0.48	3.19	0.40	0.25	0.010	-	0.34
3	17.00	1.74	7.65	0.34	0.35	-		-
4	16.80	1.50	3.24	1.71	0.032	0.027	-	-
5	18.10	4.38	16.70	4.69	0.21	0.061	0.050	-
6	11.71	5.63	26.73	19.56	0.17	0.38	-	-
7	3.78	4.86	-	-	0.09	0.030	0.060	-
8	-	4.61	-	3.73	-	-	-	-
9	1.02	0.66	0.30	0.27	-	0.024	-	-
10	0.35	4.32	6.62	0.70	0.023	-	0.056	-
11	7.94	2.68	13.70	2.93	0.16	0.038	-	-
12	8.80	3.14	7.20	1.18	0.12	-	0.073	-
13	16.20	2.00	14.32	1.07	0.36	0.011	-	-
14	60.60	4.38	26.39	8.19	-	-	-	-
15	16.04	3.99	33.97	11.91	-		-	-
16	2.95	3.09	7.02	2.84	-	-	-	-
17	-	4.61	17.59	18.23	-	-	-	-
18	4.09	2.80	5.60	2.95	0.038	0.059	-	-
19	4.90	4.40	8.15	2.88	0.056	0.11	-	-
20	20.70	0.65	5.44	1.26	0.21	-	-	0.013
21	18.86	0.50	4.73	1.08	0.18	-	-	-
22	-	5.73	-	10.23	-	-	-	-
23	1.10	1.20	16.77	8.85	0.013	0.017	-	-
24	8.20	2.50	5.09	2.24	0.042	0.042	-	-
25	15.87	3.30	38.50	12.96	0.10	-	0.040	-
26	0.67	0.48	0.65	0.48	-	0.049	-	-
27	6.93	2.59	25.37	5.30	0.083	-	-	-
28	4.78	1.14	2.50	0.60	0.032	0.019	-	-
29	0.58	0.24	0.19	0.07	-	-	-	-
30	0.37	0.25	0.03	0.03	-	-	-	-
31	2.20	2.60	9.10	5.50	0.026	0.018	-	-
32	35.54	2.70	3.70	1.05	0.045	0.119	-	-
33	0.39	0.14	2.20	0.042	-	-	-	-
34	5.76	4.27	6.93	1.28	0.065	0.092	-	-
35	1.95	1.41	2.79	2.53	0.020	0.030	-	-
36	30.40	0.48	3.19	0.40	0.25	0.010	-	0.34
37	1.09	4.64	0.88	0.63	0.015	0.062	-	-
38	38.01	2.82	12.41	11.97	0.25	0.30	0.030	-
39	34.48	2.08	26.40	10.30	-	-	-	-
40	-	3.71	9.79	3.25	0.07	-	-	-
41	-	6.66	-	-	0.05	-	-	-
42	12.60	3.34	20.80	3.44	0.08	-	-	-
43	1.74	0.47	1.16	0.82	-	0.032	-	-
44	7.41	1.54	8.20	4.06	-	-	-	-
45	8.27	1.74	8.98	4.45	2.33	-	-	-
46	0.24	0.10	1.29	0.63	-	-	-	-

Table 3. (Continued)

No.	Na	K	Ca	Mg	Sr	Ba	Li	V
47	0.71	0.19	1.34	0.68	-	-	-	-
48	24.70	25.50	24.44	9.48	-	-	0.25	-
49	37.44	37.52	29.36	13.45	-	-	0.036	-
50	33.49	32.93	28.41	12.38	0.049	-	0.028	-
51	0.50	0.40	0.12	0.02	-	-	-	-
52	0.59	0.40	0.09	-	0.001	0.005	-	-
53	4.28	2.23	9.19	3.73	0.062	0.045	-	-
54	4.79	2.21	8.28	3.19	-	-	-	-
55	7.49	1.48	0.51	0.80	-	0.11	-	-
56	9.72	1.37	<0.50	0.78	-	0.12	-	-
57	2.27	0.71	2.40	0.83	0.018	0.033	-	-
58	0.71	0.86	18.19	10.33	0.016	0.080	-	-
59	3.80	1.00	1.00	0.36	0.010	0.020	-	-
60	3.54	0.86	1.75	0.83	0.026	0.030	0.020	-
61	3.33	1.36	4.40	3.07	-	-	-	-
62	1.94	1.71	1.51	0.60	-	-	-	-
63	7.05	3.62	7.71	2.29	-	-	-	-
64	7.10	2.50	4.35	1.44	-	-	-	-
65	10.34	2.76	5.34	1.72	-	-	-	-
66	0.23	0.24	0.59	0.29	-	-	-	-
67	7.46	10.10	13.87	2.25	-	-	0.012	-
68	6.99	0.76	6.71	4.82	0.068	-	-	-
69	0.098	0.25	0.91	0.72	-	0.009	-	-
70	7.15	1.47	3.72	0.89	0.035	0.026	0.003	-
71	1.13	0.71	0.85	0.39	-	-	-	-
72	3.20	3.05	9.06	3.87	-	-	-	-
73	4.81	1.53	22.67	6.96	-	-	0.013	-
74	0.82	1.78	56.11	-	-	-	-	-
75	6.38	1.90	8.15	1.21	0.061	-	-	-
76	0.25	-	0.39	0.05	0.003	0.004	-	-
77	1.77	0.85	4.97	3.48	0.031	0.052	-	-
78	-	0.51	-	0.54	-	-	-	-
79	0.50	0.45	0.23	0.08	0.002	0.002	-	-
80	0.86	0.23	0.12	0.10	-	0.006	-	-
81	0.25	0.69	0.26	0.03	-	-	-	-
82	0.19	0.19	0.14	-	-	0.030	-	-
83	6.00	0.80	14.00	0.70	-	-	-	-
84	33.45	7.83	17.18	13.04	0.17	0.25	0.014	-
85	8.80	2.60	0.80	0.48	-	-	-	-
86	2.80	1.10	1.40	1.21		-	-	-
87	1.57	0.59	0.64	0.28	0.006	-	-	-
88	5.09	1.42	2.59	1.72	0.022	0.028	-	-
89	3.98	1.39	8.89	3.75	0.06	0.033	-	0.02
90	-	0.15	-	-	-	-	-	-
91	0.88	0.53	30.48	15.67	0.017	-	-	-
92	0.20	-	0.27	0.12	0.003	0.007	-	-
93	71.00	6.00	2.00	-	-	0.010	-	-
94	-	0.70	-	-	0.040	-	-	-
95	128.00	5.20	39.20	40.80	-	-	-	-

No.	Na	K	Ca	Mg	Sr	Ba	Li	V
96	10.00	4.00	30.00	3.00	-	-	-	-
97	9.05	3.09	-	3.40	-	-	-	-
98	-	-	8.00	1.60	-	-	-	-
99	8.00	2.00	8.00	13.00	-	-	-	-
100	2.02	0.16	0.67	0.27	-	-	-	-
101	18.11	-	9.69	3.08	-	-	-	-
102	22.20	5.36	393.20	28.80	-	-	-	
103	3.86	-	14.90	1.24	-	-	-	-
104	2.06	-	-	0.0002	-	-	-	-
105	5.60	0.70	40.50	10.10	-	-	-	-
106	11.60	0.70	46.00	7.30	-	-	-	-
107	35.00	-	45.00	14.00	-	-	-	-
108	42.00	-	46.00	13.00	-	-	-	-
109	7.40	1.00	44.00	25.00	-	-	-	-
110	6.80	1.10	46.00	30.00	-	-	-	-
111	6.90	1.00	48.20	29.40	-	-	-	-
112	1.80	1.70	20.60	1.70	-	-	-	-
113	12.50	18.00	15.20	5.30	-	-	-	-
114	6.70	0.90	32.00	6.40	-	-		-
115	1.00	1.00	78.00	24.00	-	-	-	-
116	1.40	0.40	63.00	10.20	-	-	-	-
117	11.90	1.80	55.34	24.80	-	-	-	-
118	8.20	1.30	11.20	8.10	-	-	-	-
119	9.40	5.70	9.90	6.10	-	-	-	-
120	2.40	0.60	102.00	1.10	-	-	-	-
121	16.10	0.80	9.60	77.30	-	-	-	-
122	16.80	0.80	4.00	88.60	-	-	-	-
123	17.90	0.80	12.80	70.50	-	-	-	-
124	8.40	0.70	110.80	9.90	-	-	-	-
125	4.70	0.80	101.20	2.10	-	-	-	-
126	64.80	6.00	31.60	64.70	-	-	-	-
127	9.40	1.60	59.30	20.90	-	-	-	-
128	2.50	0.60	62.10	41.00	-	-	-	-
129	1.40	0.40	2.70	0.40	-	-	-	-
130	180.00	-	68.00	28.00	-	-	-	-
131	16.50	-	61.60	3.36	-	-	-	-
132	14.00	1.90	144.20	65.80	-	-	-	-
133	13.30	1.40	112.30	40.00	-	-	-	-
134	17.70	2.80	30.00	1.77	-	-	-	-
135	141.10	3.00	32.20	14.50	-	-	-	-
136	6.40	1.10	49.50	29.50	-	-	-	-
137	37.00	12.90	280.00	57.00	-	-	-	-
138	89.60	11.30	43.30	15.60	-	-	-	-
139	11.00	2.80	110.20	23.10	-	-	-	-

Table 4. Chemical parameters (in mg/L) of the bottled mineral waters focused in this study

No.	Bicarbonate	Carbonate	Sulfate	Chloride	Fluoride	Bromide	Nitrate	Phosphate
1	71.14	8.80	7.50	1.31	0.10	-	-	-
2	87.26	-	3.59	0.70	0.12	-	0.10	-
3	65.97	3.28	0.30	0.09	0.060	-	-	-
4	58.06	3.11	0.36	0.11	0.072	-	0.12	0.060
5	124.20	-	2.26	0.85	0.30	0.010	0.49	0.060
6	-	-	-	-	-	-	3.62	-
7	29.84	-	-	0.81	-	-		-
8	-	-	-	-	-	-	4.93	-
9	1.23	-	-	0.74	0.030	-	5.20	-
10	29.34	-	0.41	0.06	0.020	-	1.05	0.29
11	72.60	-	1.30	0.86	0.28	-	5.50	-
12	46.68	-	0.50	0.95	0.10	-	7.20	-
13	84.99	3.99	1.72	0.076	0.056	-	-	-
14	250.06	-	20.60	4.81	1.59	-	-	-
15	192.69	-	11.50	1.36	0.84	-	-	-
16	39.63	-	4.20	-	0.23	-	1.10	-
17	7.90	-	6.37	2.85	0.22	-	-	-
18	38.00	-	0.38	0.90	0.19	0.010	3.86	0.45
19	43.23	-	0.30	0.86	0.12	-	14.10	-
20	73.88	2.31	0.17	0.080	0.060	-	0.060	-
21	70.45	-	0.22	0.040	-	-	-	-
22	-	-	-	-	-	-	0.030	-
23	99.77	-	0.10	0.090	0.030	-	0.20	-
24	43.95	-	0.54	1.54	0.25	0.040	4.21	0.18
25	231.91	-	4.41	2.17	0.18	0.030	2.08	-
26	1.89	-	-	0.25	0.040	-	4.63	-
27	118.20	-	5.37	0.81	0.16	0.030	0.040	-
28	23.08	-	0.33	0.30	0.030	0.020	1.08	-
29	0.70	-	-	0.36	0.020	-	1.50	-
30	0.090	-	-	0.23	0.020	-	1.10	-
31	62.90	-	0.10	0.17	0.010	0.0040	1.62	-
32	107.16	-	1.40	1.27	0.93	-	-	-
33	6.20	-	0.050	0.27	0.010	-	1.44	-
34	38.40	-	0.10	1.02	0.050	-	9.40	-
35	-	-	-	0.35	-	-	4.25	-
36	87.26	-	3.59	0.70	0.12	-	0.10	-
37	5.70	-	0.30	1.20	0.010	-	8.00	-
38	98.08	-	-	4.39	1.48	-	0.27	-
39	151.89	-	13.85	28.19	0.14	-	9.65	-
40	-	67.57	-	-	0.38	-	0.60	0.050
41	-	-	-	-	-	-	4.32	-
42	110.98	-	5.15	0.84	0.19	0.010	0.36	0.080
43	12.03	-	0.05	0.61	0.020	-	0.80	-
44	58.77	-	0.11	2.14	0.10	-	4.62	0.12
45	68.61	-	0.17	1.85	0.090	-	-	-
46	6.70	-	0.15	0.17	-	-	0.74	-
47	7.76	-	0.10	0.54	0.010	-	0.80	-

No.	Bicarbonate	Carbonate	Sulfate	Chloride	Fluoride	Bromide	Nitrate	Phosphate
48	218.77	-	2.80	2.01	0.11	-	1.90	-
49	308.00	-	3.20	1.08	0.11	-	0.90	-
50	282.80	-	1.79	1.53	0.12	-	1.83	0.46
51	1.89	-	-	0.070	0.010	-	0.40	-
52	2.10	-	-	0.040	-	-	-	0.020
53	60.78	-	0.35	0.050	-	-	0.13	0.070
54	56.00	-	0.62	-	0.050	-	0.12	-
55	-	-	0.90	11.50	0.020	-	10.60	-
56	-	-	0.88	14.64	-	0.030	8.43	-
57	12.61	-	1.75	1.45	-	0.020	1.32	-
58	103.56	-	5.05	0.080	0.060	-	0.32	-
59	16.50	-	-	-	-	-	-	-
60	19.88	-	0.060	0.020	0.060	-	0.14	-
61	34.19	-	2.35	0.18	0.020	-	2.28	-
62	12.61	-	0.21	0.43	0.010	-	0.28	-
63	47.58	-	4.80	2.88	0.12	-	-	-
64	33.73	-	5.50	0.90	0.14	-	0.70	-
65	42.96	-	7.78	1.52	0.20	-	0.84	-
66	3.88	-	-	0.050	0.020	-	0.20	-
67	88.16	-	0.90	-	0.19	-	-	-
68	60.57	-	0.28	1.00	0.060	-	1.73	-
69	6.34	-	0.26	0.11	0.020	-	0.18	-
70	34.62	-	0.20	0.23	0.030	-	2.10	-
71	7.41	-	0.060	0.37	0.020	-	0.49	-
72	57.78	-	0.30	-	0.080	-	1.10	-
73	83.92	-	23.20	2.12	0.11	-	2.00	-
74	99.20	-	112.50	0.16	0.13	-	-	-
75	50.19	-	-	0.090	0.060	-	-	0.20
76	2.09	-	-	0.070	-	-	-	-
77	37.84	-	0.40	0.090	0.040	-	0.010	-
78	-	2.09	-	-	0.070	-	-	-
79	-	-	0.30	0.31	-	-	2.20	-
80	2.22	-	0.10	0.23	0.10	-	0.30	-
81	1.28	-	-	0.66	-	-	3.00	-
82	0.41	-	0.10	0.16	0.020	-	0.40	-
83	-	-		-	-	-	-	-
84	151.43	-	6.13	33.88	0.28	0.070	0.53	-
85	11.74	-	0.85	10.93	-	-	0.56	-
86	9.26	-	1.81	3.90	0.13	-	-	-
87	5.20	-	0.40	1.45	0.20	-	0.20	-
88	8.48	-	0.90	6.23	0.020	0.080	12.00	-
89	57.60	-	-	0.19	0.15	-	0.10	0.20
90	-	-	-	-	-	-	-	-
91	168.65	-	-	0.97	-	-	-	-
92	2.09	-	-	-	0.010	-	-	-
93	89.01	21.54	51.86	0.83	1.19	-	-	-
94	-	9.39	-	2.43	1.16	-	5.77	-
95	384.00	-	148.00	27.60	1.18	-	-	-
96	79.00	-	44.00	3.00	-	-	3.00	-
97	-	-	4.70	14.00	-	-	7.00	-
98	22.00	-	-	14.00	-	-	0.68	-

Table 4. (Continued)

No.	Bicarbonate	Carbonate	Sulfate	Chloride	Fluoride	Bromide	Nitrate	Phosphate
99	-	1.00	5.00	40.00	0.10	-	<0.30	-
100	-	-	-	-	-	-	-	-
101	-	-	14.43	20.20	-	-	-	-
102	1383.00	-	9.52	5.05	0.54	-	1.24	0.060
103	80.00	-	8.40	1.20	0.030	-	2.10	-
104	-	-	8.60	-	-	-	-	-
105	150.00	-	5.30	6.10	-	-	3.10	-
106	136.00	-	10.70	42.70	0.10	-	9.00	-
107	211.00	-	32.00	36.00	-	-	1.00	-
108	185.00	-	26.00	58.00	-	-	1.00	-
109	262.30	-	6.50	2.10	-	-	-	-
110	293.00	-	4.90	2.80	<0.10	-	6.80	-
111	306.00	-	3.80	1.90	0.060	-	8.20	-
112	58.50	-	14.40	-	0.20	-	1.60	-
113	90.00	-	7.00	10.40	-	-	19.20	-
114	103.00	-	22.70	9.00	-	-	2.90	-
115	357.00	-	10.00	4.50	-	-	3.80	-
116	173.20	-	51.30	1.00	-	-	2.00	-
117	286.70	-	15.80	9.10	-	-	2.70	-
118	79.30	-	2.20	6.50	<0.10	-	5.70	-
119	65.30	-	6.90	8.40	-	-	-	-
120	291.00	-	10.50	7.80	-	-	<0.10	-
121	402.70	-	4.80	30.10	-	-	-	-
122	390.00	-	-	37.30	-	-	0.050	-
123	373.40	-	12.00	30.10	-	-	6.20	-
124	308.80	-	39.70	8.00	-	-	35.60	-
125	305.50	-	4.30	10.20	-	-	4.30	-
126	451.00	-	29.10	82.00	-	-	0.42	-
127	285.60	-	9.60	19.50	-	-	0.50	-
128	369.40	-	14.10	2.80	0.050	-	8.90	-
129	9.70	-	2.40	1.50	0.20	-	-	-
130	381.00	-	-	259.00	-	-	-	-
131	132.00	-	37.10	38.20	-	-	-	-
132	410.40	-	292.80	8.10	0.40	-	-	-
133	254.50	-	223.30	16.70	-	-	-	-
134	79.90	-	36.00	11.10	0.78	-	-	-
135	-	-	-	6.20	-	-	0.50	-
136	308.00	-	4.00	2.40	0.15	-	8.10	-
137	1110.00	-	-	-	1.02	-	-	-
138	343.20	-	-	68.00	-	-	-	-
139	453.70	-	-	9.20	0.30	-	-	-

Table 5. Standard error (SE) for the parameters focused in this study

Parameter	SE	Parameter	SE	Parameter	SE
Temperature (°C)	0.36	Magnesium (mg/L)	1.46	Sulfate (mg/L)	3.87
pH	0.11	Strontium (mg/L)	0.047	Chloride (mg/L)	2.38
Conductivity (μS/cm)	25.20	Barium (mg/L)	0.012	Fluoride (mg/L)	0.037
Dry residue (mg/L)	36.30	Lithium (mg/L)	0.016	Bromide (mg/L)	0.069
Sodium (mg/L)	2.24	Vanadium (mg/L)	0.076	Nitrate (mg/L)	0.48
Potassium (mg/L)	0.45	Bicarbonate (mg/L)	17.10	Phosphate (mg/L)	0.039
Calcium (mg/L)	4.23	Carbonate (mg/L)	6.38		

The chemical criteria for mineral waters classification according to the EU mineral water directive are of easier use than the BCMW, what justifies their use here. The following classes were defined: containing bicarbonate ($HCO_3^- > 600$ mg/L), containing sulfate ($SO_4^{2-} > 200$ mg/L), containing chloride ($Cl^- > 200$ mg/L), containing sodium ($Na^+ > 200$ mg/L), containing calcium ($Ca^{2+} > 150$ mg/L), containing magnesium ($Mg^{2+} > 50$ mg/L), and suitable for low sodium diets ($Na^+ < 20$ mg/L).

Thus, based on the EU mineral water directive, the waters Akmina (Turkey) and Theodora (Hungary) exhibited $Ca^{2+} > 150$ mg/L (Table 3) and $HCO_3^- > 600$ mg/L (Table 4). The waters Römerquelle and Vöslauer from Austria exhibited $SO_4^{2-} > 200$ mg/L (Table 4). The water Bonaqua (Germany) possessed $Cl^- > 200$ mg/L (Table 4). The waters Loutraki and Apollonio from Greece, Theodora (Hungary) and Römerquelle (Austria) exhibited $Mg^{2+} > 50$ mg/L (Table 3). 112 bottled waters contain $Na^+ < 20$ mg/L and are suitable for low sodium diets (Table 3).

A wide range of mixing models have been developed in the literature for investigating influences among three components of the hydrological cycle (sea water, rain water, and groundwater) like the ternary diagram proposed by van Wirdum (1991). The modeling takes into account *lithotrophic* water (a calcium-bicarbonate type of water, usually owing its characteristic composition to a contact with soil), *atmotrophic* water (a type of water with low concentrations of most constituents, usually owing its characteristic composition to atmospheric precipitation), and *thalassotrophic* water (a saline sodium-chloride type of water as found in the oceans). Waters forming a series between two of these types, or a wide cluster obviously derived from one type are denoted as *clines*, e.g., an *atmo-lithocline* or a *litho-thalassocline* (van Wirdum, 1991). In the van Wirdum diagram, the analyses are classified based on the EC and Ion Ratio (IR) calculated as follows:

$$IR\ (\%) = 100 \times [1/2\ Ca^{2+}] / \{[1/2\ Ca^{2+}] + [Cl^-]\}$$

Where: Ca and Cl are given in mEq/L.

Figure 3 shows the results of the analysis of the bottled mineral waters in this study plotted in the van Wirdum diagram. It is possible verify that the mineral waters are dominantly distributed in the lithocline (calcium-rich, fresh) field, despite there also are mineral waters that are characterized as atmocline (rain water). Moreover, there is no suggestion of sea water influence on the mineral waters composition.

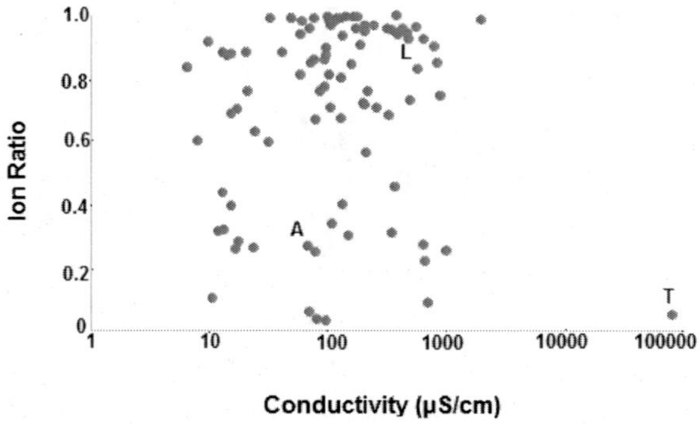

Figure 3. Data of the bottled mineral waters in this study plotted in the van Wirdum diagram. A = Atmocline, L = Lithocline, and T = Thalassocline.

The Piper (1944) diagram is another useful tool for grouping waters exhibiting similar chemical characteristics. It has been constructed using the Aquachem 4.0 software (Waterloo Hydrogeologic, 2003), considering three major geographical areas (South America, Europe and Middle East/Asia/Oceania), as shown in Figures 4-6. In terms of dissolved anions, they are HCO_3^--dominated, except in the Middle East/Asia/Oceania where occur predominance of sulfate and chloride in some brands. In terms of dissolved cations, the trend of the mineral waters in South America is mixed>Ca>Na+K>Mg, whereas in Europe is Ca>mixed>Mg. In the Middle East/Asia/Oceania, the limited available database indicates that practically there is no well-defined preponderance of any hydrochemical facies based on the dissolved cations as they are equally distributed.

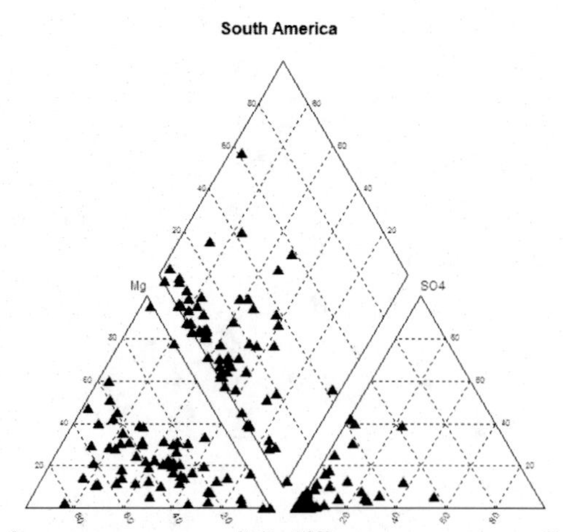

Figure 4. Data of the bottled mineral waters from South America plotted in the Piper (1944) diagram.

Another aspect highlighted from the dataset is the hydrochemical facies changes according to the bottling period. The Aquachem 4.0 software (Waterloo Hydrogeologic, 2003) allowed identify the dominant cations/anions present in the water sources bottled in different occasions, as shown in Table 6. It is possible verify that bicarbonate was always the dominant anion, except in Krenak (Nos. 55 and 56) and Kaiary (No. 81), whose hydrochemical facies changed to mixed. In terms of the dissolved cations, the changes to the mixed character were more frequent, possibly reflecting climate, seasonal variations, rainfall regime, flow rate, and/or other unidentified factors. The hydrochemical facies also changes for the same brand as illustrated in Table 6 for Bonaqua (Nos. 74, 128, 129, and 130) bottled in different countries (Brazil, Slovenia, Norway, and Germany). In such cases, the diverse lithologies of each site may take a major role for explaining the variability on the hydrochemistry.

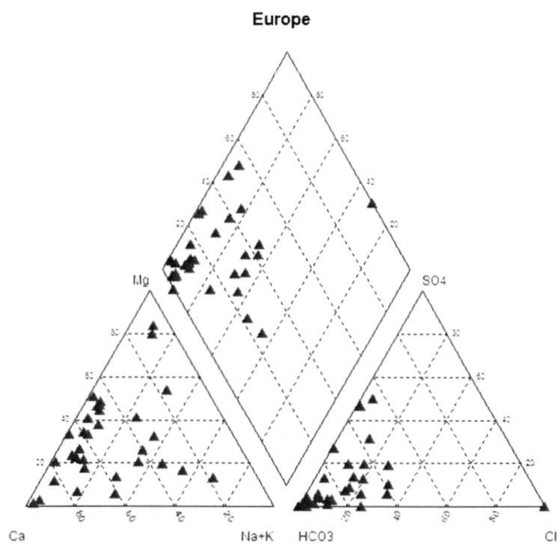

Figure 5. Data of the bottled mineral waters from Europe plotted in the Piper (1944) diagram.

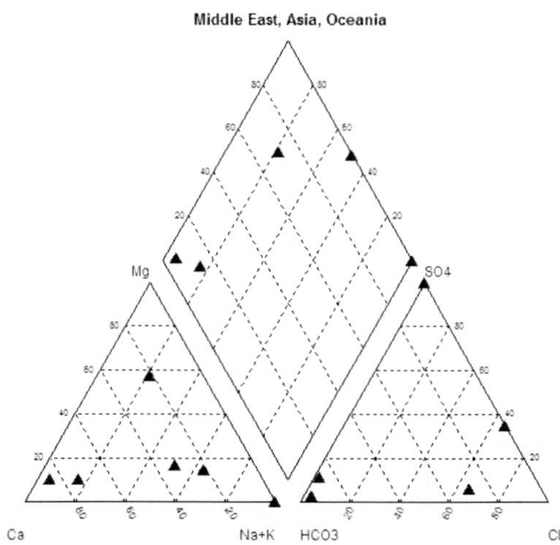

Figure 6. Data of the bottled mineral waters from Middle East, Asia and Oceania plotted in the Piper (1944) diagram.

Table 6. Water type of the bottled mineral waters as determined by the Aquachem 4.0 software

No.	Brand	Water type	No.	Brand	Water type
1	Levity (2005)	Na-Ca-HCO_3	48	São Lourenço (2004)	Ca-Na-Mg-HCO_3
2	Levity (2006)	Na-HCO_3	49	São Lourenço (A)	Na-Ca-Mg-HCO_3
3	Vittalev (A)	Na-Ca-HCO_3	50	São Lourenço (2014)	Na-Ca-Mg-HCO_3
4	Vittalev (B)	Na-HCO_3	53	Santa Rita (A)	Ca-Mg-HCO_3
5	Lindoya (A)	Ca-Na-HCO_3	54	Santa Rita (B)	Ca-Mg-Na-HCO_3
6	Lindoya (B)	Mg-Ca-Na	55	Krenak (2011)	Na-Cl-NO_3
7	Lindoya (C)	Na-K-HCO_3	56	Krenak (2014)	Na-Cl-NO_3
8	Lindoya (D)	Mg-K-HCO_3	59	Natureza (1993)	Na-Ca-Mg-HCO_3
11	Kiki (2004)	Na-Ca-HCO_3	60	Natureza (2009)	Na-Mg-HCO_3
12	Kiki (A)	Ca-Mg-Na-HCO_3	63	Passa Quatro (2003)	Ca-Na-HCO_3
14	Prata (2002)	Ca-Mg-HCO_3	64	Passa Quatro (2005)	Na-Ca-HCO_3
15	Prata (2007)	Na-Ca-HCO_3	65	Passa Quatro (2012)	Na-Ca-HCO_3
16	Prata (A)	Na-Ca-HCO_3	80	Kaiary (A)	Na-HCO_3
18	Daflora (A)	Ca-Mg-Na-HCO_3	81	Kaiary (2002)	K-Ca-NO_3-HCO_3-Cl
19	Daflora (B)	Ca-Mg-Na-HCO_3	109	San Benedetto (1993)	Ca-Mg-HCO_3
20	Minalice (A)	Ca-Na-Mg-HCO_3	110	San Benedetto (2000)	Mg-Ca-HCO_3
21	Minalice (B)	Na-Ca-Mg-HCO_3	111	San Benedetto (2004)	Mg-Ca-HCO_3
22	Minalice (C)	Mg-HCO_3	121	Loutraki (1994)	Mg-HCO_3
24	Genuína (A)	Na-Ca-Mg-HCO_3	122	Loutraki (1995)	Mg-HCO_3
25	Genuína (B)	Ca-Mg-HCO_3	123	Loutraki (1998)	Mg-HCO_3
44	Jacutinga (A)	Ca-Mg-Na-HCO_3	74	Bonaqua	Ca-SO_4-HCO_3
45	Jacutinga (B)	Ca-Mg-Na-HCO_3	128	Bonaqua	Mg-Ca-HCO_3
51	Viva (A)	Na-K-HCO_3	129	Bonaqua	Ca-Na-HCO_3
52	Viva (B)	Na-K-HCO_3	130	Bonaqua	Na-Ca-Cl-HCO_3

Statistical correlation tests among the selected parameters for the mineral waters yielded several significant values of Pearson correlation coeficiente, reflecting the major hydrochemical trends. Ca and HCO_3^- are directly related (r=0.90) as a consequence of the direct relationships of the total solids residue (DR) with Ca (r=0.92) and HCO_3^- (r=0.88) (Figure 7). The DR-EC relationship (Figure 2) also implied on direct relationships of the EC with Ca (r=0.76) and HCO_3^- (r=0.88) (Figure 7).

Other significant relationships can be also highlighted from the available dataset. For instance, like Ca, Mg as well correlates with HCO_3^- (r=0.68) (Figure 8), whilst the other earth alkaline element barium correlates with Ca (r=0.60) and Mg (r=0.81) (Figure 8). However, despite Ca and Mg correlates significantly with bicarbonate, Ba preferentially exhibits relationship with chloride (r=0.66, Figure 8). This perhaps happens because $BaCl_2$ is one of the most common water-soluble salt of barium as confirmed by its high solubility value in water at 20°C (35.8 g/100 mL) (Brackett et al., 1963). Such hypothesis is supported by additional statistical tests. The lowest determination coefficient was found between Ca and Ba (R^2=0.3577) (Figure 8). The two-tailed P value estimated by GraphPad software from the Pearson correlation coefficient of this dataset (n=42) equals 0.0192 by conventional criteria, suggesting this difference is statistically significant. The remaining higher determination coefficients yielded P values indicating that the relationships are very statistically significant and extremely statistically significant. In terms of outlier data, the most striking one was found for sample No. 102 (Akmina, Turkey) that exhibits quite large values for electrical conductivity (1974 µS/cm), dry residue (3920 mg/L), Ca (393 mg/L), and bicarbonate (1383 mg/L). The carbonates dissolution from the rock matrices is certainly favored at that site, but these data were not excluded during the statistical treatment as they did not affect the acquisition of significant relationships.

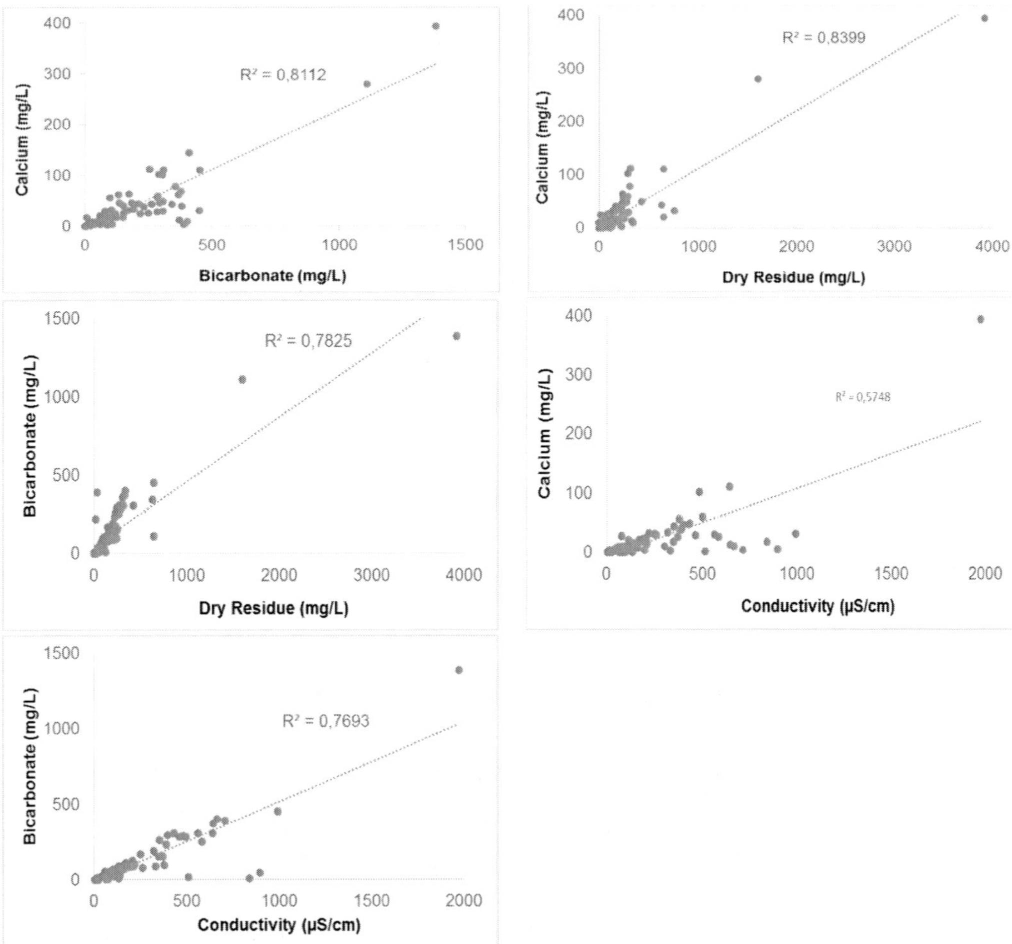

Figure 7. The relationships among calcium, bicarbonate, dry residue, and conductivity in the bottled mineral waters of this study.

Sodium tends to be transported associated with the halogens anions Cl^- and F^-, whose monovalent negative charges are essential for neutralizing the positive ones of the cations Na^+. Such situation has been described in several aquifers worldwide as pointed out by Freeze and Cherry (1979), Appelo and Postma (2005), among other. It has been evidenced in this study by the significant correlations of Na with Cl^- (r=0.68) and F^- (r=0.73) (Figure 8).

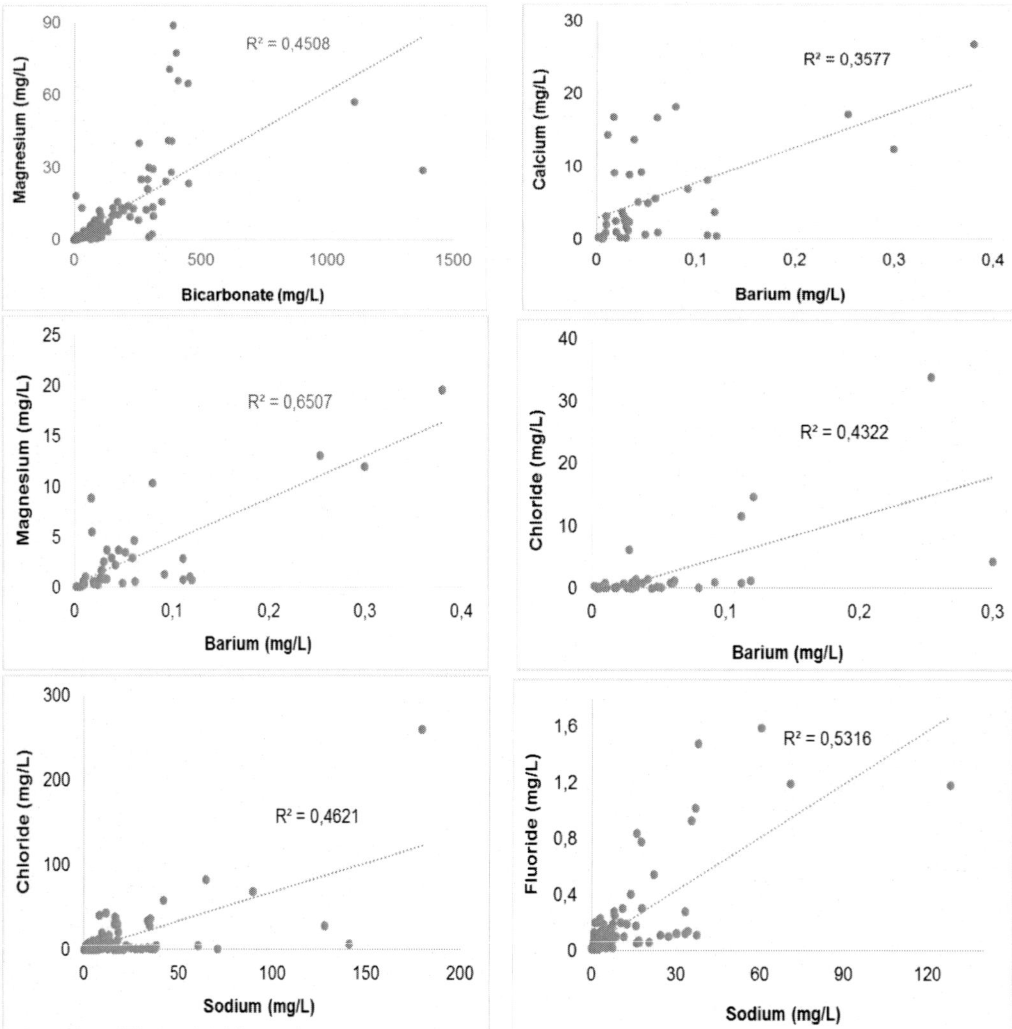

Figure 8. Significant relationships among dissolved cations and anions in the bottled mineral waters of this study.

Water-rock interactions are certainly the processes responsible by the dominant mineral waters distribution in the lithocline field, as shown in Figure 3. The Aquachem 4.0 software (Waterloo Hydrogeologic, 2003) allows realize Saturation Index (SI) calculations from the input parameters, suggesting possible mineral phases dissolution (SI<0) responsible by the hydrochemical composition. For instance, high negative SI values can be modelled for the following minerals from the composition of the water exhibiting the highest EC value (Akmina, Turkey, 1974 μS/cm): fluorite (-1.17), dolomite (-1.27), gypsum (-2.17), anhydrite (-2.40), and halite (-8.36). Another example is for the Brazilian brand Prata (bottling year 2002), that is neutral (pH=7) and whose fluoride concentration slightly exceeded the guideline reference level of 1.5 mg/L (WHO, 2011). The modelled minerals phases and corresponding higher negative SI values were fluorite (-0.96), dolomite (-1.47), gypsum (-2.61), anhydrite (-2.84), and halite (-8.09). In both cases, the minerals assemblage is compatible with the lithologies occurring in the sites of their occurrence (Table 1).

CONCLUSION

Natural spring or mineral waters are an important resource worldwide. Their consumption (bottled or not) as drinking water has drastically increased in several countries in the last few decades, partially due to the poor quality and scarcity of the ingested water in some regions or due to beliefs of many people on their healing properties. For instance, the Brazilian National Department of Mineral Production (DNPM) that is responsible for managing in the country the production and commercialization of mineral waters estimated, in São Paulo State, an amount of about 1.5 billion liters of bottled mineral waters in 2007, which corresponded to 34% of the total in the country in that year. Important initiatives have been realized for focusing these resources under different approaches. For example, we had the creation of the Commission on Mineral and Thermal Waters in 1968 (that is a branch of the IAH-International Association of Hydrogeologists), as well the extensive study of European bottled mineral and spring waters as managed by EuroGeoSurveys (The Geochemistry Group of the European Geological Surveys) and published in 2010. This chapter described a comparative study of bottled mineral waters, extending beyond Europe the brands evaluated according to their major physicochemical and chemical aspects. Thus, it was also included South America, and some water sources of Middle East, Asia and Oceania. Despite the lack of the inclusion of North American waters in the comparison, which is a subject of further investigation in the near future, the available database allowed apply the same methodological approach, enhancing the knowledge about other non-European types of bottled mineral waters. The waters analyzed are cold (<25°C, 85 water sources) and hypothermal (25-33°C, 24 water sources) and their pH ranged from acid (4.1) to alkaline (9.6) conditions, with the mean value (6.6) tending to reach neutrality. According to the electrical conductivity (EC) they are of very low mineral concentration (<77 μS/cm, 32 water sources), low mineral concentration (77-769 μS/cm, 77 water sources), and intermediate mineral concentration (769-2308 μS/cm, 4 water sources). The well-known relationship between EC and total solids residue (dry residue, DR) was also recognized in this study. A number of aquifers across the world exhibit concentration of dissolved fluoride exceeding the WHO guideline reference level of 1.5 mg/L and such situation occurred in one analysis of the Brazilian brand Prata (concentration=1.59 mg/L) that is exploited from rocks belonging to the Poços de Caldas alkaline massif. The EU mineral water directive allowed suggest the following water classes: Akmina (Turkey) and Theodora (Hungary) contain calcium (Ca^{2+} > 150 mg/L) and bicarbonate (HCO_3^- > 600 mg/L); Römerquelle and Vöslauer (Austria) contain sulfate (SO_4^{2-} > 200 mg/L); Bonaqua (Germany) contains chloride (Cl^- > 200 mg/L); Loutraki and

Apollonio (Greece), Theodora (Hungary), and Römerquelle (Austria) contain magnesium (Mg^{2+} > 50 mg/L). Statistical correlation tests allowed determine significant Pearson correlation coefficients among several hydrochemical parameters: Ca and HCO_3^- (r=0.90), DR and Ca (r=0.92), DR and HCO_3^- (r=0.88), EC and Ca (r=0.76), EC and HCO_3^- (r=0.88), Mg and HCO_3^- (r=0.68), Ba and Ca (r=0.60), Ba and Mg (r=0.81), Ba and Cl^- (r=0.66), Na and Cl^- (r=0.68), Na and F^- (r=0.73). The Piper diagram evidenced that the waters are mainly HCO_3^--dominated, whereas, in terms of dissolved cations, they are chiefly mixed in South America and Ca-dominated in Europe. The Ion Ratio (IR) *vs.* EC van Wirdum diagram permitted verify that the mineral waters are dominantly distributed in the lithocline (calcium-rich, fresh) field, despite there also are mineral waters that are characterized as atmocline (rain water). Moreover, there is no suggestion of sea water influence on the mineral waters composition.

ACKNOWLEDGMENTS

D.M.B. thanks CNPq (National Council for Scientific and Technologic Development) and FAPESP (Foundation for Supporting Research in São Paulo State), Brazil, for financial support of several studies of mineral waters in Brazilian spas.

REFERENCES

Adamy, A., and S. J. Romanini. 1990. *Geology of the Porto Velho – Abunã region. Sheets Porto Velho (SC.20-V-B-V), Mutumparaná (SC.20-V-C-VI), Jaciparaná (SC.20-V-D-I) and Abunã (SC.20-V-C-V), states of Rondônia and Amazonas*. Brasília, Brazil: DNPM/CPRM.

Alvarez, J. A., Rezende, K. M. P. C., Marocho, S. M. S., Alves, F. B. T., Celiberti, P., and A. L. Ciamponi. 2009. "Dental fluorosis: Exposure, prevention and management." *Journal of Clinical and Experimental Dentistry* 1(1): e14-18.

Appelo, C., and D. Postma. 2005. *Geochemistry, Groundwater and Pollution*. Rotterdam: Balkema.

Azzi, A. de A. 2009. "*Geology of the central east portion of the Cássia Sheet (1:50,000), MG*." Monograph, UNESP-São Paulo State University, Brazil.

Balderer, W., Arno Synal, H., Deák, J., Fórizs, I., and F. Leuenberger. 2014. "Origin of Thermal Waters in Budapest Based on Chemical and Isotope Investigations Including Chlorine-36." *Environmental Earth Sciences*: 49:1-60.

Banks, C. J. 2007. "Exceptional preservation of sedimentary structures in metamorphic rocks: an example from the upper Grampian Group, Creag Stalcair, Perthshire." *Scottish Journal of Geology* 43: 9-14.

Banning, A. 2012. "*Natural arsenic and uranium accumulation and remobilization in different geological environments*." PhD Thesis, RWTH Aachen University, Germany.

Barbosa, M. P. 1988. "*Study of the genetic relationships of the geologic features of the southern Espinhaço region and surroundings (MG)*." PhD Thesis, USP-São Paulo University, Brazil.

Bertolo, R., Hirata, R., and A. Fernandes. 2007. "Hydrogeochemistry of Brazilian bottled mineral waters." *Brazilian Review of Geosciences* 37:2-15.

Birke, M., Rauch, U., Lorenz, H., and R. Kringel. 2010a. "Distribution of uranium in German bottled and tap water." *Journal of Geochemical Exploration* 107:272-282.

Birke, M., Rauch, U., Harazim, B., Lorenz, H., and W. Glatte. 2010b. "Major and trace elements in German bottled water, their regional distribution, and accordance with national and international standards." *Journal of Geochemical Exploration* 107:245-271.

Bityukova, L., and V. Petersell. 2010. "Chemical composition of bottled mineral waters in Estonia." *Journal of Geochemical Exploration* 107:238-244.

Bizzi, L. A., *Schobbenhaus*, C., Vidotti, R. M., and J. H. Gonçalves. 2003. *Geology, Tectonics and Mineral Resources in Brazil*. Brasília, Brazil: CPRM.

Brenčič, M., and P. Vreča. 2010. "The use of a finite mixture distribution model in bottled water characterization and authentication with stable hydrogen, oxygen and carbon isotopes – Case study from Slovenia." *Journal of Geochemical Exploration* 107:391-399.

Brenčič, M., Ferjan, T., and M. Gosar. 2010. "Geochemical survey of Slovenian bottled waters." *Journal of Geochemical Exploration* 107:400-409.

Bonotto, D. M. 1993. "Enhancement of Uranium 234 in spring waters of Águas da Prata (SP), Brazil." *Water Resources Research* 29:2041-2048.

Bonotto, D. M. 2005. "The U-isotopes modeling in aquifers from Poços de Caldas plateau, Brazil." *Environmental Geology* 48:507-523.

Bonotto, D. M. 2016. "Hydrogeochemical study of spas groundwaters from southeast Brazil." *Journal of Geochemical Exploration* 169:60-72.

Brackett, E. B., Brackett, T. E., and R. L. Sass. 1963. "The Crystal Structures of Barium Chloride, Barium Bromide, and Barium Iodide." *Journal of Physical Chemistry* 67(10):2132.

Burgos, C. D., Fornaciari, A. F., and C. T. Wanderley. 2006. "Characterization of the staggered heights of south Capixaba with focus in the rocks exfoliation processes (Espirito Santo - Brazil)." *Proceedings of the VI National Symposium of Geomorphology* 1:492.

Calamita, F. and G. Deiana. 1988. "The arcuate shape of the Umbria-Marche-Sabina Apennines (Central Italy)." *Tectonophysics* 146(1):139-147. doi: 10.1016/0040-1951(88)90087-X.

Carmignani, L., Conti, P., Cornamusini, G., and A. Pirro. 2013. "Geological map of Tuscany (Italy)." *Journal of Maps* 9(4):487-497. doi: 10.1080/17445647.2013.820154.

Carneiro, M. A., and M. S. C. Barbosa. 2008. "Geologic and tectonic implications of the magnetometric interpretation from Oliveira region, Minas Gerais." *Brazilian Review of Geophysics* 26(1):87-98.

Cicchella, D., Albanese, S., De Vivo, B., Dinelli, E., Giaccio, L., Lima, A., and P. Valera. 2010. "Trace elements and ions in Italian bottled mineral waters: Identification of anomalous values and human health related effects." *Journal of Geochemical Exploration* 107:336-349.

CPRM (Brazilian Geological Survey). 1998. "*Diagnosis of Aquiraz municipality*." http://rigeo.cprm.gov.br/xmlui/bitstream/handle/doc/15782/Rel_Aquiraz.pdf?sequence=1.

CPRM (Brazilian Geological Survey). 2008. *Geology of Brazil Program – Sheets Varginha and Itajubá*. Rio de Janeiro, Brazil: CPRM-UFRJ.

CPRM (Brazilian Geological Survey). 2012a. "*The Brazilian industry of mineral waters*." <http://www.cprm.gov.br/>.

CPRM (Brazilian Geological Survey). 2012b. *Geology and Mineral Resources of the Guanabara Bay Sheet (SF-23-Z-B-IV)*. Belo Horizonte, Brazil: CPRM.

CPRM (Brazilian Geological Survey). 2018. "*CPRM reports geologic and mineral resources products in Rondônia*." https://www.cprm.gov.br/publique/Noticias/CPRM-lanca-produtos-de-geologia-e-recursos-minerais-em-Rondonia-5280.html.

Demetriades, A. 2010. "General ground water geochemistry of Hellas using bottled water samples." *Journal of Geochemical Exploration* 107:283-298.

DFPM (Division for Supporting the Mineral Production). 1966. *The mining code, the mineral waters code and how applying research in a mineral deposit*. Rio de Janeiro, Brazil: DFPM.

Dinelli, E., Lima, A., De Vivo, B., Albanese, S., Cicchella, D., and P. Valera. 2010. "Hydrogeochemical analysis of Italian bottled mineral waters: Effects of geology." *Journal of Geochemical Exploration* 107:317-335.

Dotsika, E., Poutoukis, D., Raco, B., and D. Psomiadis. 2010. "Stable isotope composition of Hellenic bottled waters." *Journal of Geochemical Exploration* 107:299-304.

Dušan, B., Jozef, K., Igor, S., Peter, M., Pavel., L., Daniel, P., Jarmila, B., and M. Daniel. 2010. "Mineral Waters in Slovakia – Evaluation of chemical composition stability using both historical records and the most recent data." *Journal of Geochemical Exploration* 107:382-390.

Edena. 2019. "*Welfare natural sources from Reunion Island*." www.edena.re/Decouvrir-123.html.

Edmunds, W. M. and P. L. Smedley. 2013. "Fluoride in natural waters." In *Essentials of Medical Geology*, edited by Olle Selinus, Brian Alloway, Jose A. Centeno, Robert B.

Finkelman, Ron Fuge, Ulf Lindh, and Pauline L. Smedley, 311-336. Heidelberg, Germany: Springer.

Eltokhi, M., Arman, H., Abdelghany, O., Hashem, W., and A. El-Saiy. 2012. "Isotope Stratigraphy of Oligocene Limestone in Al-Ain City, United Arab Emirates." *Arabian Journal for Science and Engineering* 37(5). Accessed March 10, 2019. doi: 10.1007/s13369-012-0253-y.

Eslamian, S., Amiri, M. J., and W. Balderer. 2009. "A review on thermal springs in Iran." In *Groundwater, Thermal and Mineral Water in Areas of Arid Conditions: Consequences for the Current Situation of Climate Change and the Increasing Population of Egypt*, Proceedings of the IAH-CMTWWorkshop. Cairo, Egypt: IAH.

Eslamian, S., Sayahi, M., Ostad-Ali-Askari, K., Basirat, S., Ghane, M., and M. Matouq. 2018. "Saturation." In *Encyclopedia of Engineering Geology - Encyclopedia of Earth Sciences Series*, edited by P. Bobrowsky and B. Marker. Heidelberg, Germany: Springer.

Freeze, R. A. and Cherry, J. A. 1979. *Groundwater*. Englewood Cliffs: Prentice-Hall Inc.

Frengstad, B. S., Lax, K., Tarvainen, T., Jæger, Ø., and B. J. Wigum. 2010. "The chemistry of bottled mineral and spring waters from Norway, Sweden, Finland and Iceland." *Journal of Geochemical Exploration* 107:350-361.

Fugedi, U., Kuti, L., Jordan, G., and B. Kerek. 2010. "Investigation of the hydrogeochemistry of some bottled mineral waters in Hungary." *Journal of Geochemical Exploration* 107:305-316.

Gavillot, Y., Uzunlar, N., and J. Nômade. J. 2019. "*Geology/Alpine Tectonics Field Camp - French Alps*." http://geologyfieldcamp.sdsmt.edu/France%20Field%20Camp.htm.

GEMS (Mato Grosso do Sul State Government). 2016. "*Geoenvironments of the border strip GTNF/MS*." http://www.imasul.ms.gov.br/wp-content/uploads/2016/02/Geoambientes-da-Faixa-de-Fronteira-Versao-2016.pdf.

Gimenez, V. B., Morales, D., and G. L. Luvizotto. 2016. "*Characterization of the structural pattern of Irati Formation (Permian) applied to the migration of hydrocarbons from Paraná basin at Rio Claro and Ipeúna regions (SP)*." Monograph, UNESP-São Paulo State University, Brazil.

Goldsmith, J., and A. Goldsmith. 1989. *The Dolomites of Italy*. Buchanan: Hunter Publishing Inc.

Görür, N., and Ş. Elbek. 2013. "Tectonic events responsible for shaping the Sea of Marmara and its surrounding region." *Journal Geodinamica Acta* 26(1-2):1-11.

Guerra, A. J. T., Lopes, P. B. M., and R. D. dos S. Filho. 2007. "Geographic and geomorphologic characteristics of APA Petrópolis, RJ." *Brazilian Review of Geomorphology* 8(1):77-86.

Guidottia, L., Carini, F., Rossi, R., Gattia, M., Cenci, R. M., and G. M. Beonea. 2015. "Gamma-spectrometric measurement of radioactivity in agricultural soils of the Lombardia region, northern Italy." *Journal of Environmental Radioactivity* 142:36-44.

Hakenberg, M., and S. I. J. Ska, S. I. J. 2001. "Cretaceous Basin Evolution in the Lublin Area along the Teisseyre-Tornquist Zone (SE Poland)." *Proceedings of the Polish Geological Society* 71:1-20.

Hubmann, B., Suttner, T. J., and F. Messner, F. 2006. "Geologic Frame of Paleozoic Reefs in Austria with Special Emphasis on Devonian Reef-Architecture of the Graz Palaeozoic." *Joannea Geology and Paleontology* 8:47-72.

Hudson, R. G. S. 1938. "The General Geology and Topography of the Harrogate District." *Proceedings of the Geologists' Association* 3:295-304.

Kuster, K. 2014. "*Geology of Andrelândia Group – south of Minas Gerais*." Monograph, UFRJ-Federal University of Rio de Janeiro, Brazil.

Liccardo, A., and H. Jordt-Evangelista. 2013. "Occurrences of corundum in the east portion of Minas Gerais." *Geonomos* 8(1):39-45.

Lourenço, C., Ribeiro, L., and J. Cruz. 2010. "Classification of natural mineral and spring bottled waters of Portugal using Principal Component Analysis." *Journal of Geochemical Exploration* 107:362-372.

Marras, S. 2004. *About the virtuous waters - formation and occurrence at a Brazilian spa*. Belo Horizonte, Brazil: UFMG.

Massoli, M. 1980. "Geology of the Santa Rita do Passa Quatro Sheet." *Geological Institute Review* 1(1):7-14.

Milani, E. J., Melo, J. H. G., Souza, P. A., Fernandes, L. A., and A. B. França. 2007. "The Paraná basin." *Petrobras Geosciences Bulletin* 15(2):265-287.

Moraes, M. F., Costa e Silva, D. L., Bastos, M. F., and P. M. Souza. 2017. "Multitemporal analysis of the soils use and occupation at Mário Campos-MG." *Proceedings of the XVIII Brazilian Symposium of Remote Sensing* 1:554-561.

Mourão, B. M. 1992. *Hydrological medicine – modern therapy of the mineral waters and healing spas*. Poços de Caldas, Brazil: Municipal Secretary of Education.

Noce, C. M., Teixeira, W., and N. Machado. 1997. "Geochemistry of TTGs gneisses and Neoarchean granitoids from Belo Horizonte Complex, Iron Quadrilateral, Minas Gerais." *Brazilian Review of Geosciences* 27(1):25-32.

Nogueira, A. C. R., and C. Riccomini. 2006. "The Araras Group (Proteozoic) in the north portion of Paraguai Strip and south of the Amazon Craton, Brazil." *Brazilian Review of Geosciences* 36(4):576-587.

Oliveira, S. A. 2013. "*Geoenvironments and soils from RPPN Córrego Seco, Itabirito, MG*." MS diss., Federal University of Viçosa, Brazil.

Orué, D. 1996. "*Synthesis of the geology of eastern Paraguai Oriental with emphasis on the associated alkaline magmatism*." MS diss., USP-São Paulo University, Brazil.

Peh, Z., Šorša, A., and J. Halamić. 2010. "Composition and variation of major and trace elements in Croatian bottled waters." *Journal of Geochemical Exploration* 107:227-237.

Peralta, S. H. 2003. "*An introduction to the geology of the Precordillera, western Argentina*." http://www.insugeo.org.ar/publicaciones/docs/misc-10-03.pdf.

Petolchckny, A. F., Artur, A. C., and L. S. A. Simões. 2009. "*Geologic-structural mapping and economic potentiality from Jacutinga surroundings – MG*." Monograph, UNESP-São Paulo State University, Brazil.

Petrović, T., Zlokolica-Mandić, M., Veljković, N., and D. Vidojević. 2010. "Hydrogeological conditions for the forming and quality of mineral waters in Serbia." *Journal of Geochemical Exploration* 107:373-381.

Piper, A. M. 1944. "A graphic procedure in the geochemical interpretation of water analyses." *Transactions of the American Geophysical Union* 25:914-928.

Queiroga, G., Novo, T., and A. C. Pedrosa-Soares. 2012. "Geologic mapping of the Turvo mountain range, Caratinga (MG), south sector of the Araçuaí orogeny." *Geonomos* 20(1):23-31.

Reddy, D. V., Nagabhushanam, P., Sukhija, B. S., Reddy, A. G. S., and P. Smedley. 2010. "Fluoride dynamics in the granitic aquifer of the Wailapally watershed, Nalgonda District, India." *Chemical Geology* 269:278-289.

Rocha, D. da L. 2014. "Geographical aspects and geological outline of Serra Geral's aquifer, in Paraná's state." *Geographia Opportuno Tempore* 1:140-152.

Rouquet, S., Boivin, P., Lachassagne, P., and E. Ledoux. 2012. "A 3-D genetic approach to high-resolution geological modelling of the volcanic infill of a paleo-valley system. Application to the Volvic catchment (Chaîne des Puys, France)." *Bulletin of the French Geological Society* 183 (5): 395-407, 2012.

Santos, S. B. dos. 2007. "*Analysis of the soil use and occupation from areas of permanent preservation utilizing SIG tools in the management of hydrographic basins: the case of São Lourenço stream hydrographic basin*." MS diss., Federal University of Itajubá, Brazil.

Schorscher, J. H. D., and M. E. Shea. 1992. "The regional geology of the Poços de Caldas alkaline complex: mineralogy and geochemistry of selected nepheline syenites and phonolites." *Journal of Geochemical Exploration* 45:25-51.

SEBRAE (Service for Supporting the Small Businesses in São Paulo State). 2012. "*Comércio de água mineral [Mineral water trade]*." <http://www.sebrae-sc.com.br/ideais/default.asp?vcdtexto= 31586&%5E%5E>.

Silva, J. C. L. da. 2014. "*Geologic recognition of the Passa Quatro Alkaline Complex*." Monograph, UFRRJ-Federal Rural University do Rio de Janeiro, Brazil.

Silva, W. F., and E. R. Martins. 2016. "Use of Debris Program of Civil Construction Waste Management in Containment Residential Erosion Geovanni Braga in Anapolis in the State of Goiás." *American Journal of Civil Engineering* 3:207-216.

Smedley, P. L., Nicolli, H. B., Macdonald, D. M. J., Barros, A. J., and J. O. Tullio. 2002. "Hydrogeochemistry of arsenic and other inorganic constituents in groundwaters from La Pampa, Argentina." *Applied Geochemistry* 17:259-284.

Stout, J. H. 1972. "The Precambrian rocks of the Telemark area in south central Norway." *Norwegian Journal of Geology* 52:23-41.

Taira, A. 2001. "Tectonic Evolution of the Japanese Island Arc System." *Annual Review of Earth Sciences* 29:109–34.

Taylor, B., Weiss, J. R., Goodliffe, A. M., Sachpazi, M., Laigle, M., and A. Hirn. 2011. "The structures, stratigraphy and evolution of the Gulf of Corinth rift, Greece." *Geophysical Journal International* 185(3):1189–1219.

Trouw, C. C., Medeiros, F. F. F. de, and R. A. J. Trouw. 2007. "Tectonic evolution of the Caxambu shear zone, MG." *Brazilian Review of Geosciences* 37(4):767-776.

Ündül, Ö., and A. Tuğrul. 2006. "The engineering geology of Istanbul, Turkey." *Proceedings 10th IAEG International Congress* 1: 392.

UOL (University of Ljubljana). 2019. "*Geology of Slovenia*." https://www.ntf.uni-lj.si/og/en/corse/geology-of-slovenia/.

van der Aa, M. 2003. "Classification of mineral water types and comparison with drinking water standards." *Environmental Geology* 44:554-563.

van Hinsbergen, D. J. J., van der Meer, D. G., Zachariasse, W. J., and J. E. Meulenkamp. 2006. "Deformation of western Greece during Neogene clockwise rotation and collision with Apulia." *International Journal of Earth Sciences (Geol. Rundsch)* 95:463–490.

van Wirdum, G. 1991. "*Vegetation and hydrology of floating rich-fens*." PhD Thesis, University of Amsterdam, The Netherlands.

Vidal, F. W. H., Sales, F. A. C. B., Roberto, F. A. da C., Sousa, J. F. de, and Mattos, I. C. 2005. *Industrial rocks and minerals from Ceará State*. Fortaleza, Brazil: Funcap.

VSG (Victoria State Government). 2015. "*Geology of Victoria*." http://earthresources.vic.gov.au/earth-resources/geology-of-victoria.

Waterloo Hydrogeologic. 2003. *AquaChem User´s Manual: Water Quality Data Analysis, Plotting & Modeling*. Waterloo, Canada: Waterloo Hydrogeologic.

Wessely, G., Ćorić, S., Rögl, F., and I. Zorn. 2007. "Stratigraphy of the Thermal Water Area (Baden - Bad Vöslau) at the Western Border of the Southern Vienna Basin." *Scripta Proceedings of the Natural Sciences Faculty, Masaryk University, Brno, Czech Republic, Geology* 36: 39-44.

WHO (World Health Organization). 2011. *Guidelines for drinking water quality*. Geneva, Switzerland: WHO Press.

In: Advances in Hydrogeochemistry Research
Editors: S. Saeid Eslamian and Faezeh Eslamian
ISBN: 978-1-53617-829-6

Chapter 9

PHYSICOCHEMICAL ASSESSMENT OF GROUNDWATER QUALITY IN A SEMIARID PART OF NORTH EAST NIGERIA

***Gabriel Ike Obiefuna*[1,*] *and Saeid Eslamian*[2,3]**
[1]Department of Geology, Modibbo Adama University of Technology, Yola, Adamawa State, Nigeria
[2]Department of Water Engineering, Collage of Agriculture, Isfahan University of Technology, 8415683111, Isfahan, Iran
[3]Center of Excellence in Risk Management and Natural Hazards, Isfahan University of Technology, 8415683111, Isfahan, Iran

ABSTRACT

Physicochemical assessment of groundwater quality in a semiarid part of Northeastern Nigeria was determined using geochemical model in order to assess its suitability for domestic and irrigation purposes. One hundred and seven water samples collected from surface water and wells tapping shallow to deep aquifers were used. The water samples were analyzed for major cations: Na^+, Ca^{2+}, K^+ and anions: Cl^-, HCO_3^{-1}, SO_4^{2-} and NO_3^-. The important constituents that influence the water quality for irrigation such as Electrical Conductivity (EC), Total Dissolved Solids (TDS), Sodium Adsorption Ratio (SAR), Magnesium Adsorption Ratio (MAR), Permeability Index (PI), Kelly's Ratio (KR), and Residual Sodium Bicarbonate (RSBC) were assessed and compared with standard limits. The values of total dissolved solids (<250mg/l), electrical conductivity (<3ds/m), soluble sodium percentage (2.22% to 85%), permeability index (28.7% to 131.36%), residual sodium bicarbonate (<1.0) and sodium adsorption ratio (<9) were found to be within the safe limits and thus largely suitable for irrigation purposes. An attempt was also made to develop linear regression equations to predict the concentration of water quality having significant correlation coefficients with electrical conductivity (EC). The TDS and EC have

* Corresponding Author's Email: gobiefuna2000@yahoo.com.

perfect correlation coefficients whereas Nay and Cl^-, are highly correlated in all the water sources. Furthermore while Ca^{2+} and HCO_3^- are highly correlated in both the shallow and deep groundwater they have relatively lower correlation coefficients in the surface water samples. It was equally observed that Mg^{2+}, Ca^{2+}, NO_3, Cl^- and Fe are highly correlated with EC in surface water samples. The data also indicated that apart from surface water bodies Ca^{2+}, NO_3, CI and HCO_3 are poorly related with electrical conductivity at 5% level of significance. The study findings based on model calculations indicated that Na^+, Ca^{2+}, Mg^{2+} and K^+ are highly soluble and consequently will be bioavailable at the pH ranges of the tested wells. Iron is also expected to be bioavailable as 46.53% to 79.37% of the total iron is presented as Fe^{2+}. The lower values of Fe^{2+} are related to the organic complexation of the ions. Identifying ionic distributions is of particular importance as pollutants affect the groundwater environment by the chemical behavior of the ionic species and transformation of species than by total concentrations. The adverse effects of highly soluble free Fe^{2+} are important to groundwater chemistry because their inherent toxicity is related to their bioavailability. The result suggests that Na^+, Ca^{2+}, Mg^{2+} and K^+ at such level would be available for uptake by plants, animals and humans. It is also highly possible that such conditions may constitute a negative health impact. The model findings also suggest that Fe^{2+} levels represent a health risks in some water samples.

Keywords: physicochemical, assessment, groundwater, irrigation, bioavailable, complexation, semiarid, Yola Area NE Nigeria

INTRODUCTION

The quality must be considered in any assessment of groundwater quality from outcrop areas to those where the aquifer is confined as well as changes that occur vertically within an aquifer. The chemical and physical characteristics of groundwater determine whether or not it can be used for domestic, industrial or agricultural supply. However, the number of major dissolved constituents in groundwater is quite limited, and the natural variations are not as great as might be expected. Nevertheless, groundwater is a complex chemical substance that owes its composition mainly to the solution of soluble constituents in, and chemical reactions between, the water and the rock or soil masses through which it travels.

The possible pollutants in groundwater are virtually limitless. The sources and causes of groundwater pollution are closely associated with human use of water. A complex and interrelated series of modifications to natural water quality is created by the diversity of human activities impinging on the hydrologic cycle. Most pollution stems from disposal of wastes on or into the ground. In many groundwater assessment studies, evaluation of the quality of groundwater is as important as the quantity. Inasmuch as the useability of groundwater available is determined by its chemical, physical and bacteriological properties. A programme of study of the quality of groundwater envisages field observations regarding the source and environment of groundwater occurrence and sources and causes of pollution and other related aspects having a bearing on the quality of groundwater

Currently groundwater is intensively used for irrigation purposes and other human activities. Despite its importance, little is known about the natural phenomena that govern the chemical composition of groundwater or the anthropogenic factors that presently affect them (Garcia et al. 1998a).

This work therefore intends to assess the physicochemical assessment and groundwater quality of Yola area of Northeastern Nigeria

Location Extent and Accessibility of the Study Area

The study area fall within longitudes 12°20'E and 12°34'E and latitudes 9°11'N and 9°24'N and lies about 50 km south of the Hawal Massifs. It is bounded to the east by the Republic of Cameroun and to the west by Ngurore town. The northern boundary is demarcated by Gokra town and the southern boundary by the Mandarare town and occupies approximately 431 km^2 of the land surface. Location map of the study area is shown in Figures 1a, b

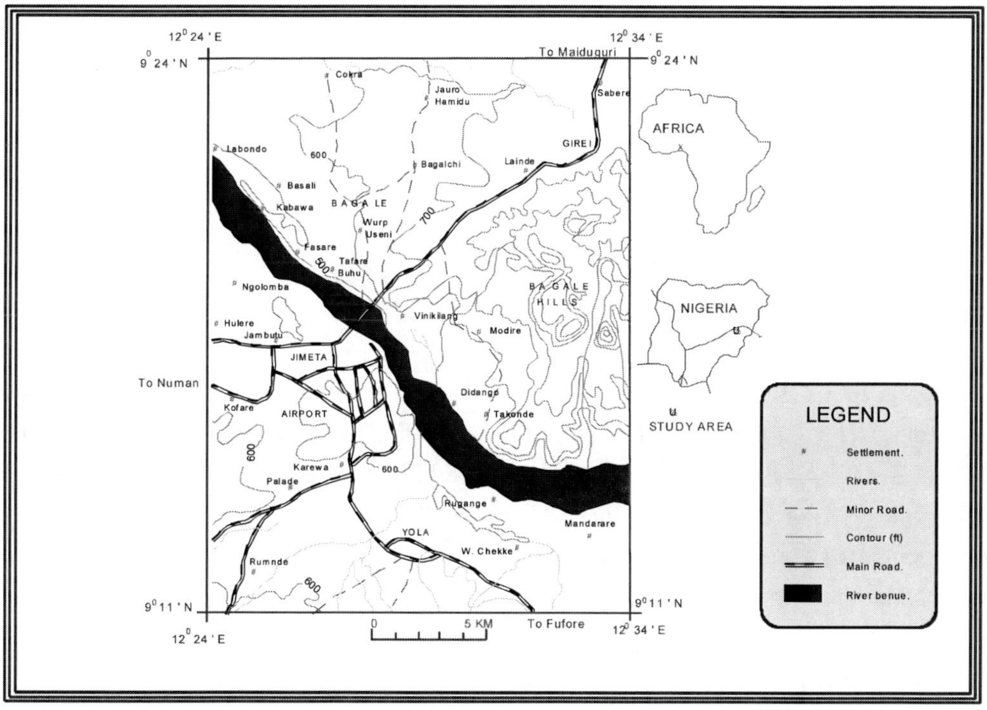

Figure 1a. Map of the study area showing access routes.

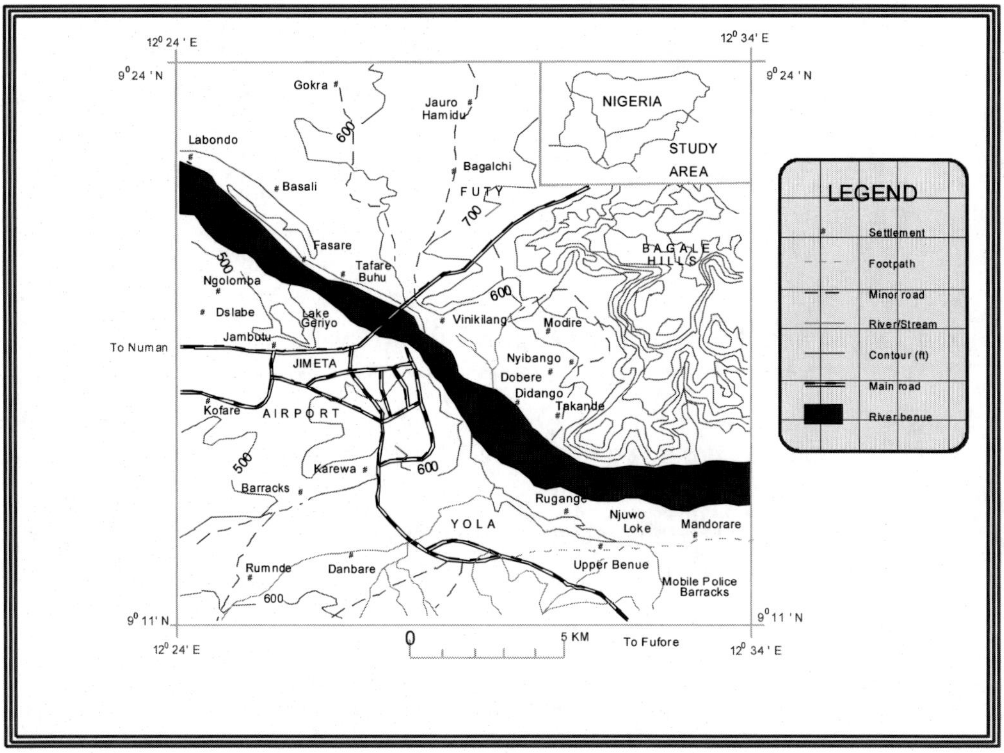

Figure 1b. Location map of study area.

Literature Review

An appreciable number of reports discussed the hydrogeochemical characterization and groundwater quality of groundwater system elsewhere (Stuyfzand 1986; Walraevens 1990; Garcia et al. 1998a; Garcia et al. 1998b; Carter and Sneed (2001); Garlindo et al. 2001; Mclean et al. 2000; Pannatier et al. 2000; Walraeven et al. 2003; Jezersky 2007); Saatsaz et al. (2013); Bano (2017); Bhunta et al. (2017); Seluakumar et al. (2017); Hamlat and Guidoum (2018); and Alfarrah and Walraeven (2018). However no detailed report on hydrogeochemistry and groundwater quality of the study area particularly as it relates to water-sediment interaction exists.

This is thus the first attempt towards detailed evaluation of the physicochemical characteristics and groundwater quality of the study area.

Aims and Objectives

The general objective of this research is to investigate among others the groundwater quality in Jimeta-Yola area and deduce a hydrochemical evaluation of the aquifer system based on the ionic constituents, water types, hydrochemical facies and factors controlling groundwater quality.

The investigation would specifically focus on the following;

(i) An attempt will be made to analyze the groundwater quality of the study area to determine the exact level of physicochemical parameters with special emphasis on its irrigation suitability for sustainable crop production.
(ii) To assess the physicochemical parameters of the groundwater of the study area in order to determine its quality and suitability for drinking purposes.
(iii) Estimate the activities of various ionic species of sodium, calcium, magnesium, potassium and iron in groundwater using geochemical models from the measured ion concentration.
(iv) Evaluation of the nature and extent of pollution of the groundwater as well as the hazards posed by the continual use of the water for domestic and agricultural purposes.

FLOW DIAGRAMS ILLUSTRATING DIGITIZATION PROCESSES AND DATA INTERPRETATION

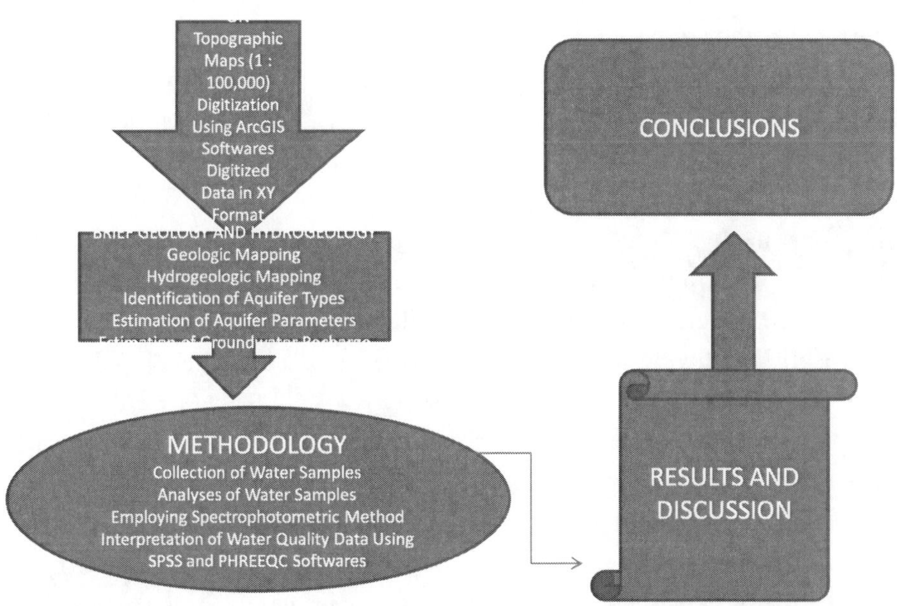

Data Source and Handling

The author acquired the topographic map used in this work as hard copies from the Lands andSurvey Department of the Federal Ministry of Works.

Map Digitization was done using GIS softwares such as ILWIS 3.3, ARC-GIS 9 (Arc Map) product of ESRI INC

The digitization produced a standard xy (position coordinate values) File

Methods

A total of one hundred and seven (107) water samples were collected from forty-eight (48)

Locations between the months of August and September 2008. Of this number, eighty-one (81) were collected from shallow groundwater, eleven (11) from deep groundwater, eleven (11) from surface water and five rainwater samples.

The water samples were filtered through a thin polycarbonate membrane with 0.45 m pore size

And collected in polyethylene bottles of four litre capacity with stopper. Each bottle was washed with 2% Nitric acid and then rinsed several times with distilled water and preserved in a coolplace in the laboratory prior to analyses.

Physicochemical parameters such as Temperature, pH, Turbidity, Dissolved oxygen (DO), Electrical conductivity (EC) and Total dissolved solids (TDS) were measured using water analyses kit model (Merk, DR Spectrophotometer 2400). All multi-probes of the kit were calibrated together using the same standards and procedure.

The total Alkalinity was measured by titrating the sample against N/50 solution of sulphuric acid using methy orange and phenalthalen indicator respectively. Chloride content was measured by titrating against N/50 solution of silver nitrate using potassium chromate as indicator. Flouride, sulphate, nitrate nitrogen and iron were determined spectrophotometrically employing standard procedures recommended by APHA, AWWA, WPCF (1998), USEPA (2003).

All the samples were assessed for charge balance and most of them fall within the acceptable range of ±5

In the interpretation of water quality data in this study we have subsequently applied the linear regression approach to develop a relationship between electrical conductivity and different water quality variables. For this purpose, regression equations were computed taking different water quality constituents as dependent variables and electrical conductivity as independent variable (Joarder et al. 2008).

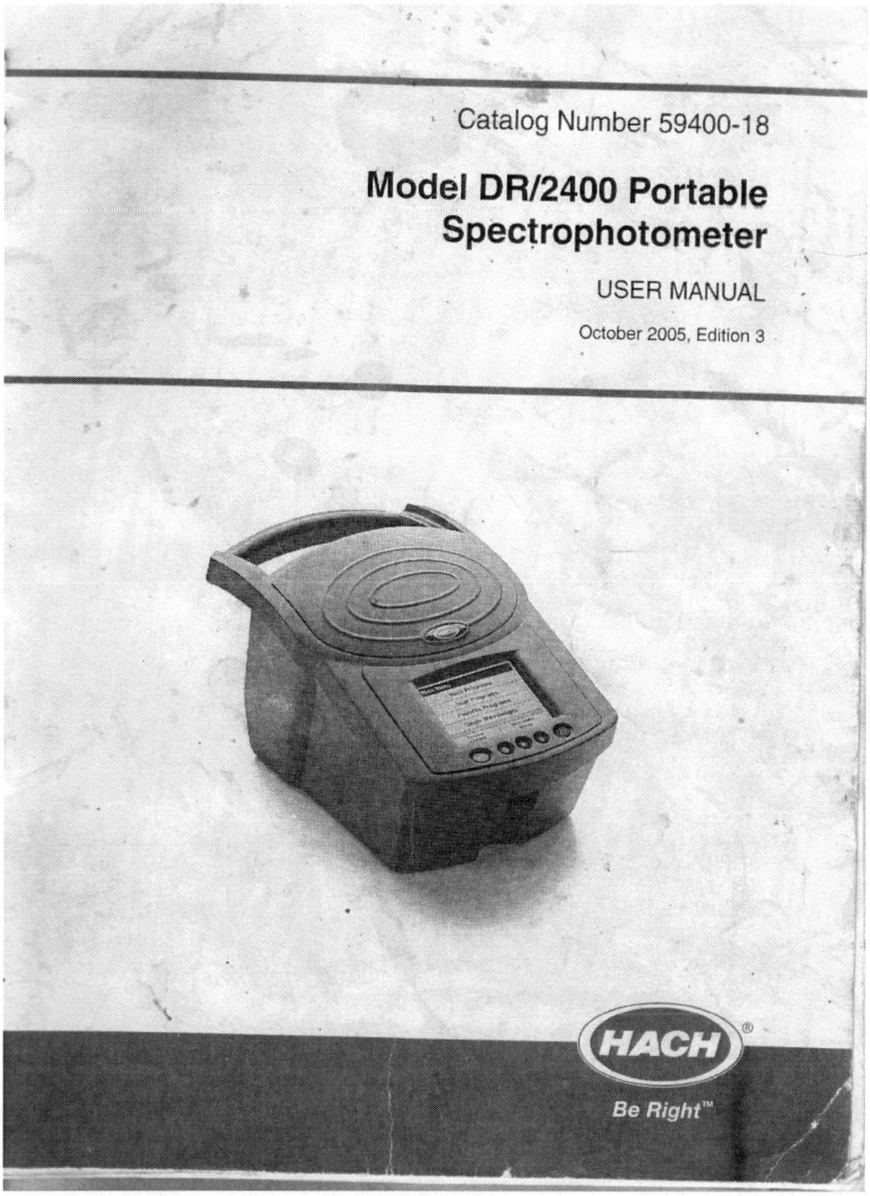

We have also used the Ordinary Least Square (OLS) regression with one regressor (IndependentVariable) in the form of $y_l=\hat{a}_0+\hat{a}_1x+e$. There are both practical and theoretical reasons to useOLS estimators of $\hat{a}_0$ and $\hat{a}_1$. Thus because OLS is the dominant method used in practice, it hasbecome the common language for regression analysis. The regression R^2 and adjusted R^2 tell uswhether the regressors are good at predicting or explaining the values of the dependent variable inthe sample of data at hand (Pindyck and Rubinfeld 1997). Adjusted R^2 considers the degree offreedom in estimating the parameters but both R^2 and Adjusted R^2 do not tell us whether;

1. an included variable is statistically significant
2. the regressors are a true cause of the movement in the dependent variable
3. there is an omitted variable bias and
4. we have chosen the most appropriate set of regressors (Stock and Watson 2003).

However in this study R^2 is used because it is the mostly used parameters in regression studies.

The Sodium Adsorption Ratio (SAR) was calculated by the following equation given byRichards (1954) as:

$$SAR = \frac{Na^+}{\sqrt{\left(\frac{Ca^{2+}+Mg^{2+}}{2}\right)}} \tag{1}$$

Where all the ions are expressed in meq/litre

The Soluble Sodium Percentage (SSP) was calculated by the following equation (Todd, 1995):

$$SSP = \frac{(Na^+ + K^+)\times 100}{Ca^{2+}+Mg^{2+}+Na^+ + K^+} \tag{2}$$

Where all the ions are expressed in meq/litre

The Residual Sodium Bicarbonate (RSBC) was calculated according to Gupta and Gupta (1987):

$$RSBC = HCO_3^{-} Ca \tag{3}$$

Where RSBC and the concentration of the constituents are expressed in meq/litre

The Permeability Index (PI) was calculated according to Doneen (1964) employing the following equation.

$$PI = \frac{Na^+ + \sqrt{HCO_3^-}\times 100}{Ca^{2+}+Mg^{2+}+Na^+} \tag{4}$$

Where all the ions are expressed in meq/litre

The Magnesium Adsorption Ratio (MAR) was calculated using the following equation (Raghunath 1987)

$$MAR = \frac{Mg^{2+}\times 100}{Ca^{2+}+Mg^{2+}} \tag{5}$$

Where all the ionic constituents are expressed in meq/litre

The Kellys Ratio was calculated employing the following equation (Kelly 1963) as:

$$KR = \frac{Na^{+}}{Ca^{2+}+Mg^{2+}} \tag{6}$$

Where all the ionic constituents are expressed in meq/litre

The Total Dissolved Solids (TDS) was calculated employing the following equation (Richards 1954):

$$\text{TDS= (0.64 x EC x 106) (Micro-mhos/cm)} \tag{7}$$

Where Electrical Conductivity (EC) and TDS are expresses in Micro-mhos/cm and mg/litre respectively.

Statistical/Computer Analysis

Statistical software called SPSS 15.0 for Windows Evaluation Version was used for thecalculation of correlation coefficients Among Water Quality Parameters whereas PHREEQC software was employed in the calculation of saturation indices and speciation modeling.

Geochemical modeling of the water composition was conducted with the PHREEQC modeldeveloped by the United States Geological Survey (Parkhurst and Appelo, 1999). The model can also be used to perform the calculations necessary to stimulate the contact of waste solutions with heterogeneous sediments or the interaction of groundwater with solid wastes (Jenne 1979; Felmy et al. 1984; Tawfic et al. 1997; Millero, 2001). It contains thermodynamic database which contains equilibrium constants for aqueous sample and complex species as well as solubility product and redox potential, and other equilibrium parameters. PHREEQC can calculate ion species/solubility, adsorption, oxidation-reduction, gas phase equilibrium and precipitation/dissolution of solid phases.

To formulate the input files for PHREEQC the following parameters such as equilibrium solution, pH values, ionic concentrations and temperature were used.

The hydrogeochemical computer model PHREEQC (Parkhurst and Appelo1999) performs Speciation calculations based on the law of mass action which states that for a reaction of the generalized type:

$$aA + bB \leftrightarrow cC + dD \tag{8}$$

the distribution at equilibrium of the species at the left and right side of the reaction is given by:

$$K = [C]c[D]d/[A]a[B]b \tag{9}$$

where;

K = equilibrium constant and the bracketed quantities denote activities or effective concentrations.

In the present investigations we equate activities with molal concentrations. PHREEQC (Parkhurst and Appelo 1999) is a computer model for performing hydrogeochemical calculations such as exchange and surface complexations, mineral reactions and equilibrations and mixing of water. PHREEQC operates with input files in which the calculation instructions are arranged according to key-words. These keywords are largely self-instructive as is illustrated by the example input file below where the keywords have been capitalized. Several databases are available with constants for the various reactions. The aqueous concentrations in a solution are defined with SOLUTION. The symbols for the elements are listed in the first column of SOLUTION_ MASTER_SPECIES in the database.

SOLUTION 1 Speciate and analyses, calculate saturation indices Temp 25 _ temperature in Degrees celcius, default = 25°C pH 7.0 _ default pH = 7 pe 4.0 _ pe = -log (electron activity), default = 4 -units mmol/kgw _ default units mmol/kg water

Ca 3.0 _ total dissolved Ca
Na 1.0
K 2.0
Mg 4.0
Alkalinity 3.8 as HCO3 _ mmol charge / kgw
S 1.0 _ total sulphur, mainly sulphate, S (6), at pe = 4
N (5) _ nitrogen in the form of nitrate, N (5) Cl1.0
Water 1 _ kg water default = 1 kg
Density 1 _ density, default = 1 kg/L
END

PHREEQC has been enhanced with several new features to increase applicability, such asOne dimensional transport with dispersion and diffusive exchange with stagnant zones, exchangers and surfaces that can be coupled to varying amounts of minerals or reactants, and generalized kinetics (Parkhurst and Appelo 1999).

GEOLOGY

Local Geologic Setting

The Bima Sandstone marks the base of the sedimentary succession in the Upper Benue Trough and it overlies uncomfortably on the undulating Precambrian Basement Complex. It varies in thickness from 100m to 300m and has its maximum thickness at the Lamurde anticline where it exceeds 3000m. The differing degree of sediment accumulation in the basin and the irregular relief of the crystalline basement on which the sediments accumulated are probably responsible for the variation in thickness. The Bima Sandstone rocks in the Upper Benue Basin have been sub-divided from base to top into three sandstone members. These include the Lower Aptian/Albian Lower Bima, the Middle Albian Bima Sandstone and the late Albian/Cenomanian Upper Bima Sandstone (Carter et al. 1963).

The study area is underlain by the upper member of the Bima Sandstone (B3) which is a cretaceous sedimentary unit of the Yola Arm of the Upper Benue Trough. The Upper Bima Sandstone (B3) was marked by the deposition, during the cenomanian? Of fluvio-deltaic sandstones and arkoses, which commenced in the south and extended progressively northwards? Several episodes of transgressions and regressions (often linked with sedimentary disturbances) are registered in the Bima Sandstone. The surface geologic units of the study area are the fine-medium grained sandstone to the north and south and the coarse grained sandstone to the northeast (Figures 2a, b and c). The depth to the bedrock varies from 30 meters to more than 45 meters. Stratigraphically, the Bima Sandstone consist of alternating layers of poorly to moderately consolidated fine to coarse grained sandstones, clay-shales, siltstone and mudstone with an average thickness of more than 250 meters as seen from their outcrops in the field. This geologic formation which reaches several hundred meters in thickness is of significant hydrogeologic interest. From field observations, exposures of Bima Sandstone in the study area is light brown to reddish brown in colour, feldspathic and fine to coarse grained in texture. It is highly crystalline and cemented in places especially north of Jimeta and Yola around Girei area.

The grain-sizes range from 2.2mm to 0.43mm indicating a fine to coarse grained sandstone that is poorly to moderately sorted. The mineralogical composition of the Bima Sandstone consist essentially of 50-60% quartz, 26-28% plagioclase feldspar, 4% microcline feldspar 8% clay matrix, 8% iron oxide and 3% calcite and are thus classified as arkosic sandstone.

In thin section, the quartz is sub-angular to sub-rounded and rimed by reddish brown colouration indicating iron-oxide. The feldspars are largely plagioclase and microcline parts of which have been altered to clay matrix. It is thus both texturally and mineralogically immature and hence competent. The predominance of quartz grains could be due to diagenetic effect of compaction and pressure solution at greater dense packing arrangements during the middle to later stages of diagenesis lending to reduction in porosity of the sandstone. The sandstone is thus highly indurated and has reduced porosity probably due increased siliceous cementation especially adjacent to lineaments. The Bima Sandstone has abundant soft-sediment deformation structures that include cusps, droplets, convolute bedding, deformed cross-bedding and sand volcanoes (Abubakar et al. 2006). The structures in most outcrops are sandwiched between undeformed cross-bedded strata that have no major textural differences with the sediments hosting them. The cusps are both simple internal cusps and interpenetrative cusps and are formed by post-depositional fluidization triggered by seismic shocks, where the interpenetrative cusps serve as conduits through which sands rose to the surface to form volcanoes (Abubakar et al. 2006). The droplets are the discrete type associated mostly with complex deformed cross-beddings while the convolute bedding forms concentric antiforms and synforms without any evidence of faulting and gradually die out vertically upward. The deformed cross-bedding is represented by both simple and complex recumbent folds of flood and seismically induced origin respectively. The source of the seismic shocks may be episodic syndepositional Mesozoic Volcanism of Jurassic to Albian times within the Upper Benue Trough (Abubakar et al. 2006). Overlying the Bima Sandstone in the study area is the river course alluvium which is composed of sands, silts, shales and clays and is confined mainly along the course of the River Benue and its tributaries. Field studies have shown that there is hydraulic connection between the river course alluvium and the underlying Bima Sandstone Formation (Obiefuna and Orazulike 2010 c).

HYDROGEOLOGY

Aquifer Systems in the Area

The main aquifer systems in the study area are defined based on data from borehole logs and field work. The study area is divided diagonally into two nearly equal halves by the River Benue and is limited by the Bagale hills to the east, Gongola Sub-basin to the West and rocks of the Hawal Massif and Precambrian Basement Complex to the north and south respectively.

The Bima Sandstone which underlies the area represents alluvial fan and braided river complexes deposited in an uneven topography created during the folding and lifting of the Upper Benue Trough along the west coast prior to the cretaceous? The lithology is characterized by mottled sublitharenite and quartzarenites with a 100m thick conglomerate at the base followed by fine to coarse grained sandstone and siltstones; altogether the maximum thickness of the Bima Formation in the study area exceed 240 meters based on borehole logs characterized by minor sandwiched iron-rich clay-shale, silty-clay and mudstones which affect the groundwater quality in terms of colour and smell and serve as semi-confining medium for the sandstone aquifers in the study area

The accurate definition of the limits and types of aquifers in the study area is not simple because of the heterogeneity of the Bima sedimentary sequence and the different criteria used in the lithological description of the well records. However, based on available borehole lithologic logs an unconfined to confined aquifer systems exist in areas underlain by the recent quaternary river coarse alluvial stratigraphic unit which laterally changes to upper unconfined aquifer system in areas underlain by the cretaceous sandstone. This is subsequently underlain by the lower semi-continued to confined aquifer systems (Obiefuna and Orazulike 2010 c).

Thus the recent quaternary sediments constitute the shallow upper alluvial aquifer whereas the older cretaceous sediments constitute the lower semi-confined to confined aquifer system.

Groundwater within the alluvial aquifer occur largely under water table conditions at outcrops and in places it is confined by interstitial clays especially in areas close to the banks of the River Benue which recharges both aquifer systems.

The Quaternary River Course Alluvial (Upper Unconfined) Aquifer System

In the study area the alluvial deposits, composed of recent sediments such as sands, silts, shales and clays that reaches more than 50 meters in thickness are fully saturated. This aquifer is probably recharged by direct exfiltration of the River Benue. The fine to coarse grained sand units form the aquifers which are separated by aquitards and aquicludes of poorly permeable silts and clays. They occur at depths of 20-40 and 45-80 meters respectively and were penetrated at shallow depths ranging from 5m to more than 50m below the surface and are thus subjected to little diagenetic alteration than the underlying Older Cretaceous sands with a long burial history. Diagonally along the southeast-northwest part, the alluvial aquifer decreases in thickness as it moves away from the River Benue interfingering with some saturated sand lenses in the blanketing clay-shales and siltstone of the Bima Sandstone Formation (at a depth of more than 30m) and within the deeper aquifers. The lateral extent of this aquifer system extends althrough the study area displaying semi-confining to confining conditions in places. It varies in thickness from 6meters in Bagalchi, Damare and Gokra in the Northern part to 48 meters in Sanda Primary School in Yola Town in the South with an overall average thickness of about 14.52 meters.

A confining condition was observed in Gokra and Jambutu areas whereas an unconfined situation was observed towards Yola town in the southern part of the study area. The blanketing clay-shales and siltstones is an aquiclude-aquitard system of local significance that acts as the natural relatively impermeable to semi-permeable barrier that limits the vertical infiltration of rainfall into the deeper aquifers within the study area. Field studies have shown that there is hydraulic connection between the upper unconfined (alluvial) aquifer and the underlying Bima Sandstone Formation. This aquifer present at a depth of 18 to 39 meters in the distal southern part of the alluvial deposits is under unconfined condition. The depth to static water levels in wells tapping these aquifers is generally relatively low whereas well yields are relatively high. The depth to static water level ranging from 1meter at Vinikilang to 56meters in Girei to the north and from 3meters at Jimeta to 51 meters at Sebore to the South has been observed with yields ranging from 2 to 15 liters/sec (with drawdown ranging from 6meters to 43.04 meters) in several hours of pumping from shallow boreholes tapping this aquifer. Borehole lithological logs reveal that the depth to the base of this aquifer varies from 21 meters at Bagalchi and Gokra to 75 meters at Girei to the north and from 21 meters at Limawa and Rumde to 75 meters in Yola town to the south with an average depth of about 48.40 meters. The base is made up of clay-shales, sandy clays and clayey sands respectively.

The water table is usually high ranging from 1meter at discharge areas to 51 meters maximum at recharge areas. Recharge is also sustained mostly by seepage from the Benue River. The shallow water is thus susceptible to the vagaries of the relatively sensitive climatic regime in the study area. Downward leakage into the underlying Lower semi-confined to confined aquifer system is promoted by heterogeneity within the Bima Sandstone Formation. It also contribute to the groundwater development and are largely exploited in rural areas by means of shallow wells for domestic, agricultural and livestock needs.

Lower Semi-Confined to Confined Aquifer System

It is located within a complex sequence of clastic sediments of Albian age called the Bima Sandstone, the thickness and lithology of which show significant lateral variation. It is a multilayer aquifer system which extends over a surface area of 450km^2. The aquifer consist of medium to coarse grained sands and silt alternating with thick layers of clay-shales and sandy-clays and clayey-sands with hydraulic heads varying from 0.0006 to 0.0173 to the north and from 0.0064 to 0.0493 to the south. Its depth of occurrence reaches locally up to 80 to 200m while its penetrated thickness ranges between 9.09m and 129m with an average of about 36.09m. The depth to static water levels in wells tapping these aquifers are generally relatively low and ranges from 3.01m to 33.6m whereas yields are relatively higher than in alluvial aquifer and ranges from 0.436 m/hr to 54m/hr. They are confined above by clay-shales, sandy- clays and clayey -sands respectively.

About 68 wells are exploited from this aquifer in the region of interest. The principal recharge to these aquifers is perhaps by direct percolation of rainfall along the foothills of the Bagale hills as well as on the groundwater mounds to the north and south of the study area. Though minor contributions are expected from the infiltration of rainfall along the foothills of the Bagale hills during the rainy season. To the northeast the main flow direction is to the southwest while to the south it is northeastwards and the flow rate is about 27m/hr, part of the recharge that is possible is from infiltration of the Benue River water. Even though the mean annual precipitation in the area of study is relatively high especially when compared with other semi-arid regions, all is lost largely through evapotranspiration in the basin since surface runoff is low. Consequently, it is estimated that mean annual recharge to the aquifer from the River Benue is 38 percent of the amount of the annual precipitation. This aquifer thus receives recharge from River Benue by downward percolation.

Most of the groundwater abstraction is from this aquifer. Pumping is mostly through privately owned boreholes used to supply hotels, schools and industries which are largely concentrated within the city centre.

Transmissivity and Hydraulic Conductivity Values

The hydraulic conductivity, K, values for the upper alluvial (largely unconfined) aquifer were determined employing both the pumping test and the granulometric methods.

Hazen (1893) method indicates k, values ranging from $9.03\text{x } 10^{-3}$ m/s to $3.36 \text{ x } 10^{-1}$m/s with a mean value of about $7.53 \text{ x}10^{-2}$ m/s.

Harleman et al. (1963) method gave k values that range between $1.60 \text{ x}10^{-3}$m/s to $2.16 \text{ x}10^{-1}$m/s with a mean value of about $4.83 \text{ x}10^{-2}$m/s.

Masch and Denny (1966) gave k values that varies from $3.77 \text{ x}10^{-5}$m/s to $2.83 \text{ x}10^{-4}$ with a mean value of about $1.20 \text{ x}10^{-4}$m/s

Uma et al. (1989) indicate k values that range from $9.50 \text{ x}10^{-5}$m/s to $1.28 \text{ x}10^{-2}$m/s with a mean value of about $2.86 \text{ x}10^{-3}$m/s. Finally Uma and Leohnert (1994) gave k values ranging from $6.8 \text{ x}10^{-7}$m/s to $1.13 \text{ x}10^{-5}$m/s with a mean value of about $3.46 \text{ x}10^{-6}$. It was observed that only Masch and Denny (1966) and Uma et al. (1989) methods gave the results that matched the aquifer samples of the study area when compared with other statistical grain-size methods and were therefore adopted. The values obtained from the Cooper-Jacob (1946) vary from 0.021m/day to 152.17m/day whereas those obtained from the Theis (1935) Recovery method varies from0.052m/day to 19.28m/day.

The Logan (1964) method gave k values that vary from 0.081m/day to 23.28m/day as against values of 0.072m/day to 3.50m/day obtained from the step-drawdown method. The k values generally fall within the range (10^{-2} m/day to 10^{2}m/day) which indicate moderate to good aquifers. The values obtained from the statistical grain-size (granulometric) methods are essentially uniform and compare favorably with values got from the pumping test methods. This is expected in view of the fact that the upper alluvial aquifer occurs at a relatively shallow depth (less than 80m) and are composed of recent sediments. It has been observed by Uma and Leohnert (1994) that both methods generally give similar values in recent sediments that have not been subjected to significant diagenetic alteration.

A summary of the hydraulic conductivity, k, and values obtained from both the pumping test and the statistical grain size methods for the upper unconfined alluvial aquifer indicate values ranging from 0.051m/day to 56.98m/day with a mean value of 2.54m/day.

The hydraulic conductivity, K, values for the lower semi –confined to confined aquifer was determined employing the pumping test methods. These are the Cooper- Jacob (1946) method, the Theis (1935) Recovery technique, the Step-draw-down method and the Logan (1964) Approximation method. The values obtained from the Cooper- Jacob (1946) method vary from 0.59m/day to 27.98m/day whereas those obtained from the Theis (1935) Recovery Technique varies from 0.99m/day to 12.81m/day.

The Logan (1964) method gave values that varies from 1.11m/day to 15.47m/day as against values of 0.08m/day to 2.33m/day got from the Step-drawdown method.

A summary of hydraulic conductivity values obtained for the lower semi-confined to confined aquifer employing the pumping test and the statistical grain-size methods indicate values ranging from 0.22m/day to 10.48m/day with a mean value of 3.81m/day.

The k, values obtained for this aquifer falls within the range of 10^{-2}m/day to 10^{2}m/day which indicate an aquifer system of moderate to good performance (Todd, 1995). The k values obtained from pumping test method for the lower semi-confined to confined aquifer which is composed of older and diagenetically altered sediments is much lower than those of statistical grain-size methods. This is thus due to a long burial history resulting in increased compaction and cementation undergone by the sediments of this aquifer system (Uma and Leohnert, 1994).

Transmisssivity is defined as the ease with which a saturated aquifer transmits water through its entire thickness. It is represented mathematically as;

$$T = kb \ (m^2/s) \tag{10}$$

Where;

k = hydraulic conductivity (m/s)

b = saturated thickness of the confined aquifer or the height of

the water table above the top of the underlying aquitards that bounds the aquifer from the unconfined aquifer (Freeze and Cherry, 1979).

The saturated thickness of the upper unconfined alluvial aquifer was estimated from borehole lithological analyses and surface exposures of the aquifer whereas those of the lower semi- confined to confined aquifer was estimated from screen length and borehole logs. The transmissivity values for the upper unconfined alluvial aquifer ranges from 1.52m^2/day to 349.86m^2/day with a mean value of 37.99m2/day whereas those obtained from granulometric methods vary from 9.97×10^{-3} m^2/s to 1.53m^2/s.

The transmissivity values for the lower semi-confined to confined aquifer varies from 9.18m^2/day to 349.93m^2/day with a mean value of about 103.51m^2/day (Table 8b). The variation in transmissivity values in the two aquifer systems are because of variations in the thickness of the aquifer rather than in hydraulic conductivity.

The pumping test indicates that the phreatic aquifer has an average hydraulic conductivity of 2.54 m/d, a transmissivity value of 37.99 m^2/d and a specific yield of 27% (obtained from Pumping Test method). The recharge area on either side of the study area is about 337.5km^2 and the expected average net abstraction rate is 172800 m^3/day by the year 2020. This would yield an infiltration rate of 0.000512 m/d on average. The coefficient of storage values indicate that confining conditions exists in the two aquifer systems.

Figure 3 indicates that to the northeast a localized recharge area of the aquifers occurred to the west whereas to the southwest it is northeastwards towards the Benue River. They discharge naturally at points or areas where the aquifer with its underlying relatively impermeable alluvial units such as clay-shales and mudstone intercepts the ground surface in river or stream valleys.

Multiple independent methods of recharge estimation such as chloride mass balance, water balance, darcy flux and hydrograph separation techniques indicate groundwater recharge ranging from 5.78 mm/yr to 331.02 mm/yr. These gave an average groundwater reserve of about for the two aquifer systems.

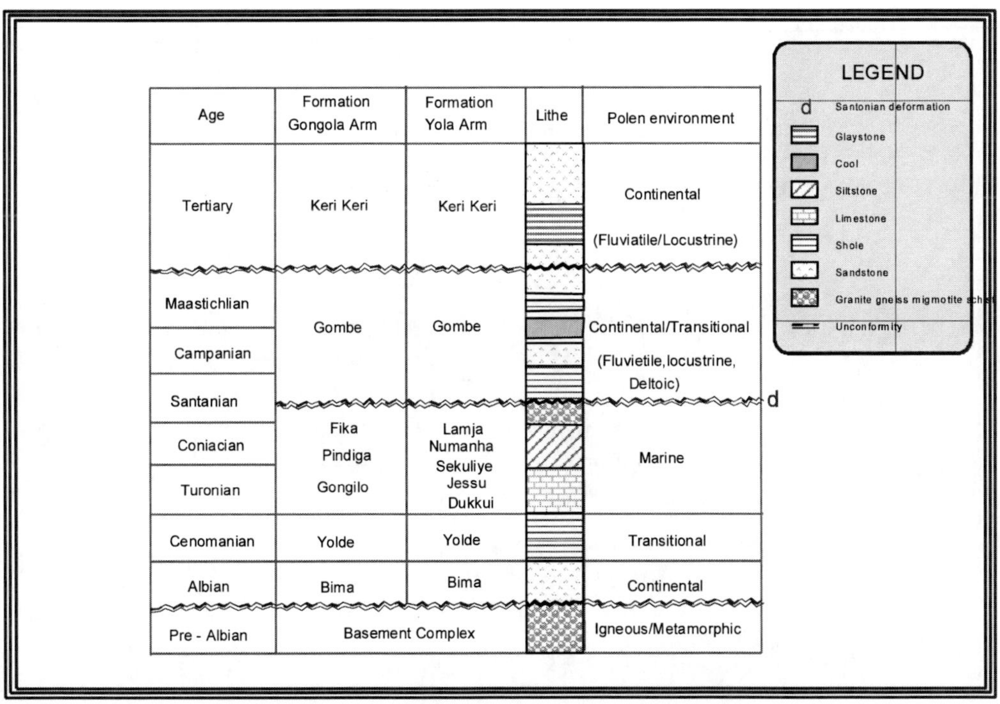

Figure 2a. Stratigrophic suceession in the Upper Benue trough (modified from Petters, 1982).

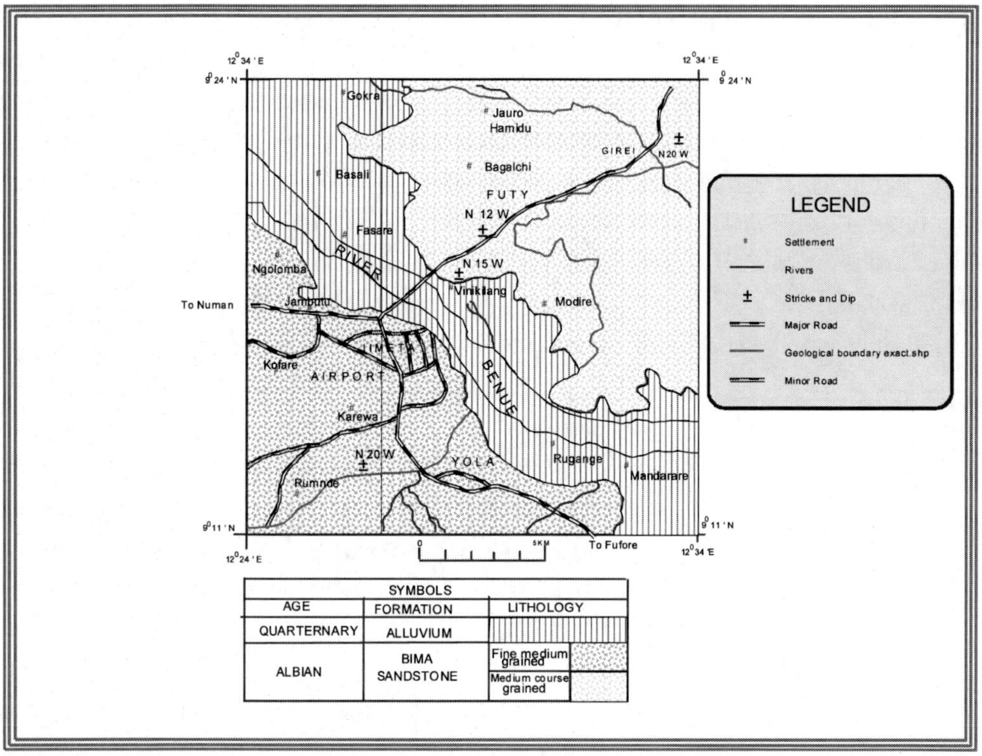

Figure 2b. Geological map of the study area.

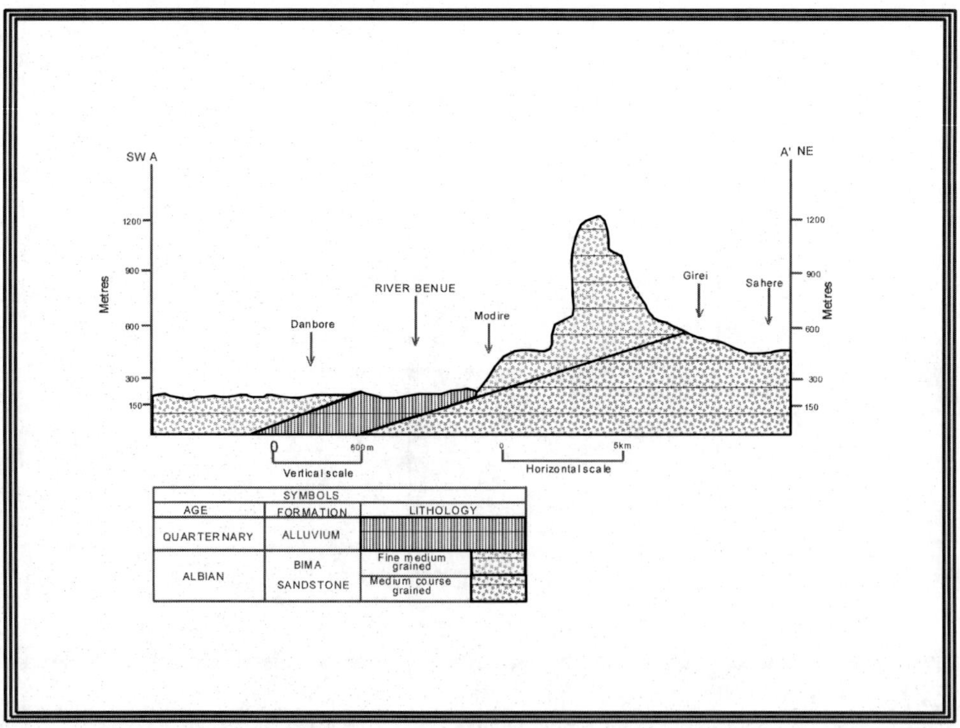

Figure 2c. Geological section A-'A'.

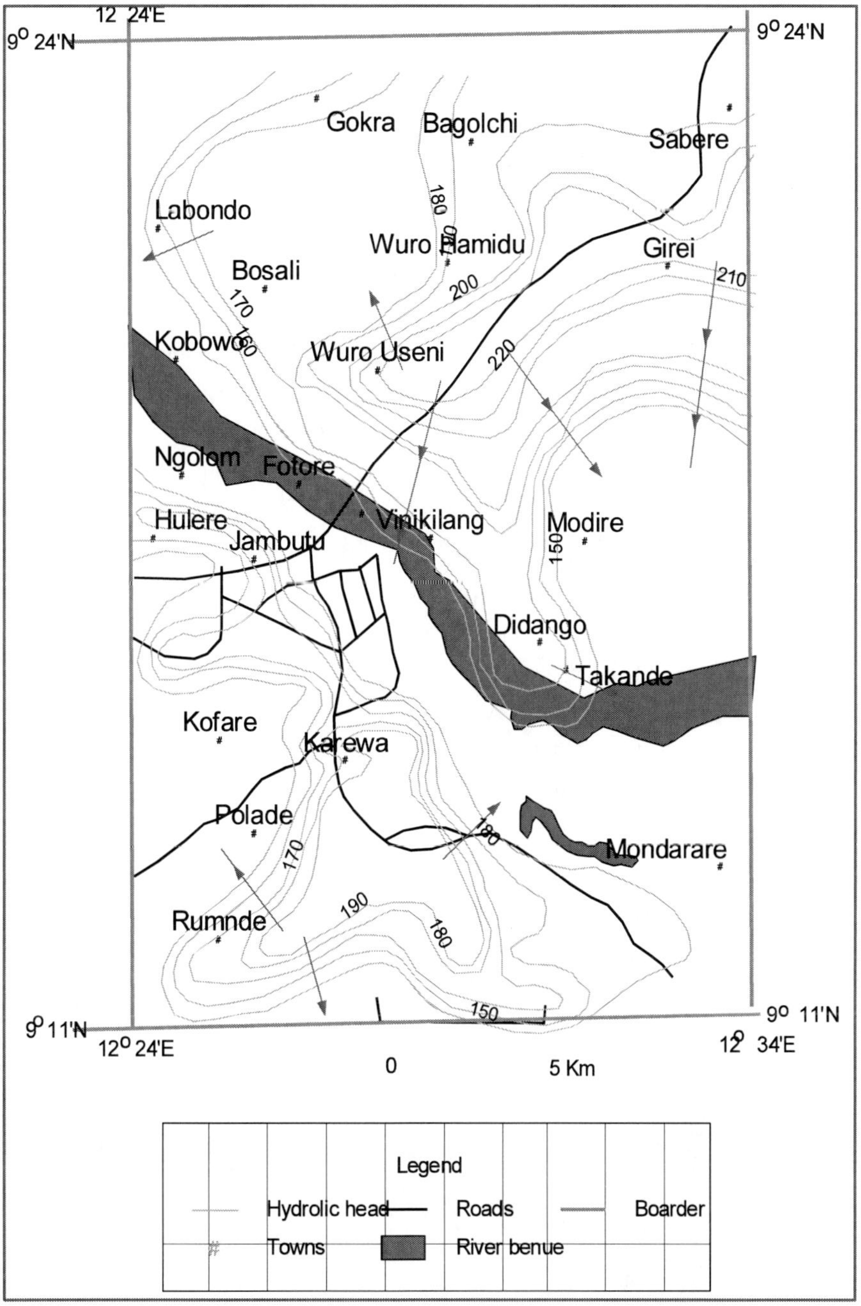

Figure 3. Hydraulic Head Distribution and Groundwater Flow Direction in the Upper Unconfined (Alluvial) Aquifer System.

RESULTS AND DISCUSSION

Details and a summary of the physicochemical parameters for the different water sources for therainwater, groundwater and surface samples are given in Table 1.

Surface Water/Precipitation Chemistry

The surface water temperature in the study area ranged from 26 to 34°C. The electrical Conductivity (EC) varies between 0.07 and 0.45 ds/m. The mean total dissolved solids (TDS) are 155 mg/l (Surface water) and 18 mg/l (precipitation).

The EC and TDS are low in comparison to the WHO (2006) standard of 1.4 ds/m for EC and 1000 mg/l for TDS indicating that the waters are fresh.

The mean pH for the surface water bodies and precipitation are 7.57 and 6.40 with ranges of 7.20 And 8.00 (Surface water) and 4.30 and 7.70 (precipitation). Considering the entire sample set, about 13 out of 16 samples had pH>7. In addition, all the surface water and precipitation samples have pH values that fall within the limits of the World Health Organization (WHO 2006), of 6.5-8.5. Most of the samples are thus neutral in nature which is attributed to the fact that the surface waters are neutralized as they flow through calcareous rocks in the process.

The dissolved oxygen (DO) varied from 1.10 to 7.40 for the surface water and from 4.81 to 7.20 for the precipitation. The surface water samples show moderately high (Standard deviation SD= 24.54) variation in terms of hardness with values ranging from 21 to 112 mg/l as CaCO3 for the surface water and low (SD=1.74) for the precipitation with values ranging from 1.95 to 5.80 mg/l as $CaCO_3$. However, all of the samples can be described as moderately hard to hard with hardness values of more than 61 mg/l (Freeze and Cherry 1979).

Table 1. Summary of physicochemical parameters in water samples for the study area

Parameter	Rainfall (n = 5)	Surface water (n = 11)	Shallow groundwater (n = 81)	Deep groundwater (11)
	Mean Min Max SD	Mean Min Max SD	Mean Min Max SD	Mean Min Max SD
Temp (°C)	27.62 26.70 29 1.03	29.56 26 34 3.02	30 30 31 0. 06	30.2030 32 0.60
pH	6.40 4.30 7.70 1.34	7.577.20 8.00 0.23	7.09 6.10 7.08 0.39	7.22 6.60 7.80 0.41
EC (ds/m)	0.03 0.02 0.05 0.11	0.22 0.07 0.45 0.11	0.090.05 0.21 0.03	0.09 0.04 0.18 0.04
DO	6.07 4.81 7.20 0.87	3.22 1.10 7.40 1.75	0.90 0.01 2.80 0.70	0.96 0.05 2.10 0.83
EH (volts)	0.83 0.76 0.96 0.13	0.76 0.73 0.83 0.07	0.72 0.46 0.82 0.10	0.75 0.60 0.85 0.10
TDS	18.19 13.40 30.60 7.04	155 41 298 72.60	60 33 137 18.83	63 27 121 26
TH	4.59 2.26 6.90 1.74	61 21 112 24.54	54 0.18 190 36.41	59 10 105 28
Ca	3.61 1.95 5.80 1.59	26 8 37 9.25	12.19 0.48 25 6.19	11.30 0.50 38 11.68
Mg	1.78 1.10 2.50 0.50	35 12 84 19.09	21 0.50 51 9	31 8 55 13.46
K	2.56 0.00 4.21 1.56	5 1.20 9 2.40	4 0 37 6.89	4.32 0 12 4.16
Na	1.23 0.00 1.98 0.80	73 2.23 211 89.04	34 0 157 33.74	34 0 91 34.30
HCO_3	17.93 16.20 19.20 1.17	137 49 273 63.27	93.4 19.90 251 41.26	103 50 207 44.41
NO_3	1.44 0.00 2 .23 0.83	35 9 59 17.97	4.36 0 23 4.63	6 0.66 31 8.63
SO_4	2.83 1.60 4.50 1.21	18 2 29 7.52	11 0 61 12.28	14.80 0 65 19.66
CI	0.65 0.50 0.80 0.11	146 28 455 138.28	68 0 239 56	76 0 172 61.82
Fe	0.21 0.00 0.39 0.17	0.30 0 0.73 0.26	0.35 0 1.80 0.41	0.09 0 0.32 0.10
Mn	0.00 0.00 0.01 0.00	0.00 0 0 .02 0.00	0.00 0 0.00 0.00	0.00 0 0.00 0.00
E coli (Per 100ml)	00 0 0	13.80 8 21 3.54	5.60 0 7 2.12	2 0 4 1.21

The mean concentration of cations in the surface water samples are Na (73 mg/l), Ca (26 mg/l), K (5mg/l), Mg (35 mg/l) and Na (1.23 mg/l), Ca (3.61mg/l), K (2.56 mg/l), Mg (1.78 mg/l) for the precipitation respectively. The mean values of chloride concentration are 146 mg/l and 0.65 mg/l for the surface water and precipitation respectively. The maximum and minimum values of bicarbonate in the surface waters are 273 and 41 mg/l and 19.20 and 16.20 mg/l for the precipitation. The amount of sulphate varies from 2 to 29 mg/l for the surface water and 1.60 to 4.50 mg/l for the precipitation. The concentrations of the sulphate are below the WHO (2006) limits of 250 mg/l.

In the surface water samples, the nitrate concentrations varied from 9 to 59 mg/l and 0 to 2.23 mg/l for the precipitation. The concentrations are lower than that are normally expected for unpolluted water, 0.124-9.92 mg/l (Custodio and Llamas 1983) and 11.3 mg/l (WHO2006) maximum recommended value for drinking purposes.

Shallow Groundwater Chemistry

Shallow groundwater temperatures ranged from 30 to 31°C. The values of EC vary from 0.05 to0.21ds/m whereas the total dissolved solids vary from 33 to 137 mg/l. The EC and TDS are relatively low in comparison to the WHO (2006) standard of 1.4 ds/m for EC and 1000 mg/l for TDS indicating that the waters are also fresh.

The pH values vary from 6.10 to 7.08 indicating slightly acidic to neutral water which could be Attributed to rainfall and the absence of buffering materials within the aquifers. The hardness values (as $CaCO_3$) ranged from 0.18 to 190 mg/l. Data from Table 1 shows that out of all the 81 samples considered for the study area only 25 samples can be described as hard with hardness values of > 60 mg/l. The data on chemical composition (Table 1) show that the mean concentrations of the cations in groundwater are Na (34 mg/l), Ca (12.19 mg/l), K (4 mg/l) and Mg (21 mg/l). The values of chloride concentrations vary from 0 to 239 mg/l whereas the value of sulphate concentration varied from 0 to 61mg/l. The relatively low concentration of both chloride and sulphate could be attributed to natural maximum concentrations derived from anthropogenic reactions. Nitrate values ranged from 0 to 23 mg/l with mean of 4.36 mg/l (as N) for the groundwater samples. Thus all the samples have lower concentrations reflecting littlepollution (e.g., Custodio and Llamas 1983) and less than 11.3 mg/l (WHO 2006) maximum recommended value for drinking purposes.

Deep Groundwater Chemistry

Deep groundwater temperatures ranged from 30 to 32°C. The mean values of EC vary from 0.05 to 0.21 ds/m whereas the mean total dissolved solids vary from 27 to 121 mg/l.

The EC and TDS are relatively low in comparison to the WHO (2006) standard of 1.4 ds/m for EC and 1000 mg/l for TDS indicating that the waters are also fresh.

The pH values vary from 6.60 to 7.80 indicating slightly acidic to neutral water which could be Attributed to rainfall and the absence of buffering materials within the aquifers. The hardness values (as CaCO3) ranged from 10 to 105 mg/l. Data from Table 4 shows that out of all the 11 samples considered for the study area only 6 samples can be described as hard with hardness values of > 60 mg/l.

The data on chemical composition (Table 1) shows that the mean concentrations of the cations ingroundwater are Na (34 mg/l), Ca (11.30 mg/l), K (4.32 mg/l) and Mg (31 mg/l). The values of chloride concentrations vary from 0 to 172 mg/l whereas the value of sulphate concentration varied from 0 to 65mg/l. The relatively low concentration of both chloride and sulphate could be attributed to natural maximum concentrations derived from anthropogenic reactions.

Nitrate values ranged from 0.66 to 31 mg/l with mean of 6 mg/l (as N) for the groundwater samples. Thus all the samples have lower concentrations reflecting little pollution (e.g., Custodio and Llamas 1983) and less than 11.3 mg/l (WHO 2006) maximum recommended value for drinking purposes. The pH values range from 7.20 to 8.00 for the shallow groundwater, 6.60 to 7.80 for the deep groundwater and 6.40 to 7.70 for the surface waters indicating slightly acidic to neutral waters. The conductivity varies from 0.05 to 0.21ds/m for the shallow groundwater, 0.04 to 0.18 ds/m for the deep groundwater and 0.07 to 0.45 ds/m for the surface water bodies indicating relatively low mineralization in the area and largely suitable for both drinking and irrigation purposes. The estimated values of TDS ranged from 33 to 137 mg/l for the shallow groundwater, 27 to 121 mg/l for the deep groundwater and 41 to 298 mg/l for the surface water respectively indicating TDS values below the desirable limit of 500 mg/l.

Table 2. Correlation coefficients among different water quality parameters in surface water of the study area

	pH	EC	TDS	Na	K	Mg	Ca	Fe	NO_3	Cl	HCO_3	SO_4^{2-}
pH	1.000											
EC(ds/m)	0.416	1.000										
TDS	0.429	0.964**	1.000									
Na^+	-0.235	-0.695*	-0.767**	1.000								
K^+	0.110	0.242	0.310	-0.545	1.000							
Mg^{2+}	0.350	0.264	0.277	0.191	-0.394	1.000						
Ca^{2+}	0.590	0.766**	0.808**	-0.731*	0.127*	0.449	1.000					
Fe	0.275	0.730*	0.768**	0.893**	0.488	-0.084	0.744**	1.000				
NO_3^-	0.204	0.583	0.677*	-0.463	0.184	0.654	0.647*	0.421	1.000			
Cl^-	-0.168	-0.539	-0.602	0.944**	-0.630	0.471	-0.539	0.818**	-0.179	1.000		
HCO_3^-	-0.381	0.214	0.230	-0.252	-0.011	-0.120	0.361	0.200	-0.254	-0.381	1.000	
SO_4^{2-}	0.08	0.04	-0.17	-0.36	0.63	0.30	0.37	-0.10	0.62	-0.26	-0.04	1.000

* Check the level of significance i.e.,

** 1% level of or correlation is significant at the 5% level

Table 3. Correlation coefficients among different water quality parameters of shallow groundwater

	pH	EC	TDS	Na^+	K^+	Mg^{2+}	Ca^{2+}	Fe	NO_3^-	Cl^-	HCO_3^-	SO_4^{2-}
pH	1.000											
EC (ds/m)	-0.045	1.000										
TDS	0.046	1.000	1.000									
Na^+	-0.007	0.126	0.122	1.000								
K^+	0.113	0.207	0.207	0.224*	1.000							
Mg^{2+}	0.108	0.092	0.096	0.089	-0.028	1.000						
Ca^{2+}	0.217	0.259*	0.257*	0.266*	-0.091	0.244*	1.000					
Fe	-0.451**	0.117	0.118	0.014	-0.027	-0.331**	-0.049	1.000				
NO_3^{-1}	0.001	0.132	0.134	0.106	0.097	0.055	0.019	0.057	1.000			
Cl^{-1}	-0.005	0.195	0.194	0.929**	0.223*	0.303**	0.259*	-0.042	0.068	1.000		
HCO_3^{-1}	0.285*	-0.199	0.197	0.209	0.003	0.508**	0.544**	-0.280*	0.109	0.170	1.000	
SO_4^{2-}	-0.28	-0.20	-0.20	0.03	0.11	0.14	0.19	0.23	-0.04	-0.04	-0.03	1.000

NB: All parameters are in mg/L except where otherwise stated ** correlation is significant at the 1% level
* Correlation is significant at the 5% level.

Table 4. Correlation coefficients among different water quality parameters of deep groundwater

	pH	EC	TDS	Na^+	K^+	Mg^{2+}	Ca^{2+}	Fe	NO_3^-	Cl^-	HCO_3^-	SO_4^{2-}
pH	1.000											
EC (ds/m)	0.390	1.000										
TDS	0.401	1.000**	1.000									
Na^+	0.075	0.067	0.075	1.000								
K^+	0.170	0.140	0.122	-0.180	1.000							
Mg^{2+}	-0.626*	0.192	0.176	0.107	0.120	1.000						
Ca^{2+}	-0.452	0.153	0.132	0.067	0.335	0.544	1.000					
Fe	-0.872**	-0.070	-0.077	0.088	-0.223	0.548	0.476	1.000				
NO_3^{-1}	0.100	0.045	0.035	-0.372	0.017	0.214	-0.015	0.019	1.000			
Cl^{-1}	-0.256	0.193	0.197	0.960**	-0.102	0.323	0.177	0.196	-0.355	1.000		
HCO_3^{-1}	-0.347	0.175	0.155	-0.315	0.38	0.536	0.846**	0.448	0.232	-0.205	1.000	
SO_4^{2-}	-0.33	0.14	0.12	-0.19	0.32	0.71	0.09	0.29	0.42	-0.04	0.76	1.000

NB: All parameters are in mg/L except where otherwise stated ** correlation is significant at the 1% level **Correlation is significant at the 5% level

The systematic calculation of correlation coefficients between water quality variables and Regression analyses provide an indirect means for rapid monitoring of water quality. The correlation matrix for different groundwater quality variables for different water sources are shown in Tables 2, 3 and 4.

It was evident that the distribution of hardness, calcium, magnesium, sodium, chloride, bicarbonate, nitrate and sulphate were largely poorly correlated (r <5) with electrical conductivity in both the shallow and deep groundwater. The similarity in correlation matrixbetween the two groundwaters thus indicates that mineral dissolution is the dominant processinfluencing water quality. However, the surface waters of the study area largely exhibitedsignificant positive correlation (r >5). The surface water samples disclosed a good correlationbetween EC and NO_3^- (r= 0.54) which further indicated that variation in EC concentration iscontrolled by both Cl^- and NO_3^- concentration. Another fairly good correlation also observedbetween Cl^- and NO_3^-(r= 0.95) confirm a common primary source which is both domesticeffluent/sewage and/or fertilizer application. The level of significance is taken at 1% and 5%respectively (Obiefuna and Orazulike 2010 a).

Perfect positive correlation coefficient was observed between EC and TDS (r = 0.96) whereasHigh positive correlation was observed between Na^+ and Cl^- (r >0.80) virtually in all thedifferent water sources (Obiefuna and Orazulike 2010 a).

Further statistical study of the shallow groundwater employing the Pearson correlation indicates Highly positive correlation between the following irrigation indices SAR and SSP (r= 0.90); KR and SSP (r= 0.95); KR and SSP (r= 0.89); and EC and TDS (r= 1.00) and low negative correlation among MAR and SAR (r= 0.21); MAR and SSP(r= -0.28); KR and MAR(r= -0.25);PI and MAR (r= -0.44); RSBC and SSP (r= -0.21); TDS and PI (r= -0.04); RSBC and TDS (r= -0.12) and EC and PI (r= -0.04). The deep groundwater also indicates a closely similar picture asfollows: SSP and SAR (r= 0.96); KR and SAR (r= 0.99); SSP and KR (r= 0.97) and EC and TDS (r= 1.00) whereas a strongly negative correlation exists between the following SAR and RSBC(r= -0.70); RSBC and SSP (r= -0.76); RSBC and KR (r= -0.70).

The surface water samples exhibit strong positive correlation among KR and MAR (r= 0.42); SAR and MAR (r = 0.56) SSP and SAR (r= 0.97); KR and SAR (r= 0.97); KR and SSP (r= 0.90) and TDS and EC (r= 0.94) whereas a strong negative correlation was found to exist between; SSP and TDS (r= -0.82); KR and TDS (r= -0.91); EC and SAR (r= -0.81); SSP and EC (r= -0.75); EC and KR (r= -0.87) and PI and EC (r= 0.76) (Tables 5, 6 and 7).

Thus based on the above data set it was concluded that the correlation studies of the water Quality parameters and irrigation indices have a great significance in the study of water resources. The relatively high positive correlation between some chemical parameters and irrigation indices signifies a common origin or progressive enrichment of both parameters.

Table 5. Correlation coefficient among different irrigation indices in surface water concentration in the study area

	SAR	SSP (%)	MAR (%)	KR (meq/l)	PI (%)	TDS (mg/l)	RSBC (meq/l)	EC (ds/m)
SAR	1.000							
SSP (%)	0.967**	1.000						
MAR (%)	0.594	0.659*	1.000					
KR (meq/l)	0.970	0.903**	0.421	1.000				
PI (%)	0.927	0.833**	0.555	0.926**	1.000			
TDS (mg/l)	-0.866**	-0.819**	-0.362	-0.908	-0.800**	1.000		
RSBC (meq/l)	-0.124	-0.186	-0.481	-0.102	-0.210	-0.018	1.000	
EC (ds/m)	-0.809*	-0.750**	-0.365	-0.871**	-0.764	0.939**	0.034	1.000

NB: *Correlation is significant at the 5% level ** correlation is significant at the 1% level

Table 6. Correlation coefficients among different irrigation indices in shallow groundwater concentration in the study area

	SAR	SSP (%)	MAR (%)	KR (meq/l)	PI (%)	TDS (mg/l)	RSBC (meq/l)	EC (ds/m)
SAR	1.000							
SSP (%)	0.983*	1.000						
MAR (%)	-0.555	-0.696	1.000					
KR (meq/l)	0.996**	0.993**	-0.623	1.000				
PI (%)	0.885	0.956*	-0.871	0.917	1.000			
TDS (mg/l)	-0.955*	-0.993**	0.766	-0.972*	-0.982*	1.000		
RSBC (meq/l)	-0.346	-0.169	-0.587	-0.267	0.126	0.062	1.000	
EC (ds/m)	-0.955*	-0.993**	0.76	-0.972*	-0.982*	1.000**	0.061	1.000

NB: *correlation is significant at the 5% level ** correlation is significant at the 1% level

Table 7. Correlation coefficients among different irrigation indices of deep groundwater concentration

	SAR	SSP (%)	MAR (%)	KR (meq/l)	PI (%)	TDS (mg/l)	RSBC (meq/l)	EC (ds/m)
SAR	1.000							
SSP (%)	0.960**	1.000						
MAR (%)	-0.340	-0.346	1.000					
KR (meq/l)	0.990**	0.972**	-0373	1.000				
PI (%)	0.016	0.209	0.276	0.049	1.000			
TDS (mg/l)	0.059	0.068	-0.070	0.033	-0.141	1.000		
RSBC (meq/l)	-0.696*	-0.759**	0.151	-0.695	-0.409	0.087	1.000	
EC (ds/m)	0.050	0.05	-0.081	0.024s	-0.154	1.000**	0.096	1.000

NB: * correlation is significant at the 5% level ** correlation is significant at the 1% level

Furthermore the relatively high negative correlation between some chemical parameters and Irrigation indices indicate evidences of groundwater mixing or pollution from anthropogenic activities (Obiefuna and Orazulike 2010 a,b).

The calculated irrigation indices such as SAR, MAR, SSP, RSBC, PI, EC, TDS thus indicate that the water samples are largely suitable for irrigation purposes.

The adjusted R^2 parameter was employed as a more desirable goodness- of- fit measure because as earlier mentioned it considers the degrees of freedom in estimating the desired parameters.

The R^2 uses variations instead of variances (variance equals variation divided by the degrees of freedom) (Joarder et al. 2008). The R^2 values of different water quality parameters with electrical conductivity for the three water sources are given in Tables 8, 9 and 10.

It follows the electrical conductivity is the most appropriate variable predicting or explaining the TDS values in the samples of both shallow and deep groundwater. It is also evident that the Electrical conductivity is the most appropriate variable predicting or explaining about 92.90%, 58.70%, 53.20%, 34%, 29.10%, and 7% values of the dependent variables such as TDS, Ca^{2+}, Fe, NO_3^-, Cl and Mg respectively in the surface water sample of the data.

However apart from TDS electrical conductivity poorly predicted the above parameters in both The shallow and the deep groundwater samples. It further explained the fact that the variance of The residual is small compared to the variance of the dependent variable. Thus while correlation Techniques do not involve an implicit assumption of casuality the regression techniques do.

Pindyck and Rubinfeld (1997) pointed out that while the dependent variable is a variable to be Explained the independent variable is a moving force. It thus follows that the choice of a dependent and independent variable in a regression model is therefore crucial.

Table 8. Regression equations for different water quality variables in surface water of the study

Regression Equation	R^2 value	t_{value}	p_{value}	f_{value}
Ca = 11.480 + 66.236EC	0.587	3.577	0.006	12.796
Mg = 23.996 + 47.119EC	0.070	0.823	0.432	0.677
Cl = 301.052 + 696.474EC	0.291	-1.920	0.087	3.689
NO_3 = 12.823 + 97.844EC	0.340	2.152	0.060	4.632
TDS = 9.145 + 653.777EC	0.929	10.853	0.000	117.783
Fe_{total} = -0.093 + 1.770EC	0.532	3.200	0.011	10.242
HCO_3 = 108.687 –126.577EC	0.046	0.658	0..527	0.432
TH = 35.888 + 113.453EC	0.245	1.708	0.122	2.916

NB: Indicates significance at the 5% level

Table 9. Regression equations for different water quality variables in deep groundwater

Regression Equation	R^2 value	t Value	P Value	F Value
Ca = 6.913 + 46.707EC	0.023	0.465	0.653	0.216
Mg = 24.513 + 67.701EC	0.037	0.588	0.571	0.346
Cl = 46.397 + 311.031EC	0.037	0.589	0.571	0.346
No_3 = 4.820 + 10.141 EC	0.002	0.135	0.896	0.018
TDS = -0.224 + 670.994 EC	0.999	99.202	0.000	9840.967
Fe^{Total} = 0.109 – 0.180 EC	0.005	-0.212	0.837	0.045
HCO_3 = 84.104 + 203. 349 EC	0.031	0.534	0.606	0.285
TH = 30.084 + 311.406 EC	0.181	1.409	0.192	1.985

NB: indicates significance at the 5% level

Table 10. Regression equations for different water quality variables in shallow groundwater

Regression Equation	R^2 value	t Value	P Value	F Value
Ca = 17.257-56.924EC	0.067	-2.365	0.021	5.592
Mg = 18.187 + 29.342EC	0.008	0.814	0.418	0.662
Cl =33.655 + 387.992EC	0.038	1.756	0.083	3.083
No_3 = 2.422 + 21.736 EC	0.017	1.176	0.243	1.383
TDS = 0.021 + 669.459EC	1.000	529.129	0.000	279977.0
Fe^{Total} = 0.192 + 1.722 EC	0.014	1.039	0.302	1.080
HCO_3 = 119.265 – 291.215EC	0.039	-1.789	0.078	3.200
TH = 51.352 + 26.397 EC	0.000	0.180	0.858	0.032

NB:indicates significance at the 5% level.

Two variable least squares approach was used to develop a relationship between electrical Conductivity as an independent variable and different water quality variables such as Ca^{2+}, Mg^{2+}, Cl^-, NO_3^-, TDS, HCO_3^-, TH, Fe as a dependent variable. The cross-section of the results obtained is presented in Tables 8, 9 and 10. The first column of the tables indicates the results of the ordinary least squares (OLS) regressions with only the EC as the controlling variable and a constant. The second column indicates the values of the R^2.

Regression results for Ca^{2+}, Mg^{2+}, Cl^-, NO_3^-, TDS, Fe, HCO_3^- and TH show that apart from both deep and shallow groundwater samples, they are all significant at 1%, 5% and 10% levels. All the regression equations are obtained for 96-2= 94 degree of freedom. The significance of the relationship is also supported by F test (Tables 8, 9 and 10). Furthermore Whiteneys test for heteroskedasticity in the residuals of the basic specification rejects them null of no heteroskedasticity and thus all standard errors of coefficients are calculated using whiteney test.

The robustness of the results obtained was checked which indicated a fairly satisfactory outcome between the observed and the computed values.

Assessment of Groundwater Quality for Agricultural Purposes

The study was done to assess the irrigation water quality of three water systems namely surface water, shallow groundwater and deep groundwater of Yola Area of Northeastern Nigeria. The surface water analyzed includes River Benue, Lake Geriyo and Lake Njuwa. The shallow groundwaters are tapped mainly by Hand-dug wells whereas deep groundwater is tapped largely by boreholes (Obiefuna and Orazulike 2010 a,b).

The range of values of analyzed parameters is given in Tables 11,12 and 13. On the basis of theparameters the suitability criteria for irrigation purpose were determined (Table 14).The important parameters affecting the suitability of water for irrigation purposes are

ElectricalConductivity (EC), Soluble Sodium Percentage (SSP), Total Dissolved Solids (TDS), Sodium Adsorption Ratio (SAR), Magnesium Adsorption Ratio (MAR), Permeability Index (PI), KellysRatio (KR) and the Residual Sodium Bicarbonate (RSBC) (Figures 4, 5 and 6).

Table 11. Summary of physico-chemical parameters of the surface water in yola area (After Obiefuna and Orazulike 2010 a, b)

Parameter	Range	Mean	Standard Deviation	WHO (2006)
PH (unit)	7.20 – 8.00	7.57	0.23	6-5 – 8.5
Temperature (0/c)	26 – 34	29.56	3.02	-
Conductivity (ds/m)	0.07 – 0.45	0.22	0.11	1400
Sodium (mg/l)	2.32 –210.80	74.79	88.90	200
Potassium (mg/l)	1.20 – 8.70	5.21	2.40	55
Calcium (mg/l)	8.30 – 37	26.64	9.25	75
Magnesium(mg/l)	12.40 – 84.90	34.95	19.01	50
Iron (mg/l)	0 – 0.73	0.30	0.26	0.3
Chloride (mg/l)	27.90 -455.20	145.87	138.28	250
Sulphlate (mg/l)	2– 29.11	17.79	7.52	400
Bicarbonate (mg/l)	48.50– 273	136.89	63.27	variable
Nitrate –N (mg/l)	8.50-58.50	34.53	18.02	50
Fluoride (mg/l)	0 – 0.21	0.057	0.068	2
Phosphate (mg/l)	0.45 – 24	10.80	6.63	-
Silica (mg/l)	0 – 18	7.53	6.90	-
TDS (mg/l)	41-298	154.82	72.60	1000
Total Alkalinity (mg/l)	84-305	184.73	65.86	-
Total Hardness (mg/l)	20.70 – 112.80	61.17	24.54	500
Odour	Odourless	Odourless	Odourless	Unobjectionable odour
Turbidity (NTU)	0.23 – 28	10.83	9.59	5

Table 12. Summary of physico-chemical parameters of the shallow groundwater in yola area (After Obiefuna and Orazulike 2010 a,b)

Parameter	Range	Mean	Standard Deviation	WHO (2006)
pH (unit)	6-1 – 7.8	7.08	0.39	6-5 – 8.5
Temperature (°C)	30 – 30.5	30	0.05590	-
Conductivity (ds/m)	0.05 – 0.205	0.08839	0.02837	1400
Sodium (mg/l)	0 – 156	34.3506	33.7417	200
Potassium (mg/l)	0 – 37	3.95775	6.89631	55
Calcium (mg/l)	0.48 – 24.84	12.11247	6.19152	75
Magnesium(mg/l)	0.50 – 50.9	20.39275	9.15513	50
Iron (mg/l)	0.001 – 1.80	0.31673	0.38812	0.3
Chloride (mg/l)	0 -238.5	67.48326	56.47236	250
Sulphlate (mg/l)	0 – 60.70	10.61037	12.24415	400
Bicarbonate (mg/l)	19.9 – 251	93.3505	41.25757	Variable
Nitrate –N (mg/l)	0 – 22.50	4.40544	4.64074	50
Fluoride (mg/l)	0 – 1.20	0.07952	0.27085	2
Phosphate (mg/l)	0 – 18.50	1.61678	3.53411	-
Silica (mg/l)	0 – 2.80	0.27039	0.57741	-
TDS (mg/l)	26.8 – 137.4	59.35345	18.80811	1000
Total Alkalinity (mg/l)	61- 274	102.90706	42.52973	-
Total Hardness (mg/l)	0.18 – 190	54.39	35.80507	500
Odour	Odourless	Odourless	Odourless	Unobjectionable odour
Turbidity (NTU)	0.005 – 10.2	0.96905	1.50	5

Table 13. Summary of physico-chemical parameters of the deep groundwater in Yola area (After Obiefuna and Orazulike 2010 a, b)

Parameter	Range	Mean	Standard Deviation	WHO (2006)
PH (unit)	6-8 – 7.8	7.21818	0.40943	6-5 – 8.5
Temperature (0/c)	30 – 32	30.18182	0.60302	-
Conductivity (ds/m)	0.04 – 0.180	0.0938	0.03827	1400
Sodium (mg/l)	0.001 –90.5	33.65555	34.31843	200
Potassium (mg/l)	0 – 12.10	4.32182	4.1608	55
Calcium (mg/l)	0.46 – 38.30	11.29545	11.67562	75
Magnesium(mg/l)	8.10 – 54.80	30.86455	13.4684	50
Iron (mg/l)	0 – 0.32	0.09173	0.09813	0.3
Chloride (mg/l)	0.004 -171.5	75.57782	61.82	250
Sulphlate (mg/l)	0 – 64.50	14.80818	19.66051	400
Bicarbonate (mg/l)	50– 207	103.18182	44.40923	variable
Nitrate –N (mg/l)	0.66-30.50	5.77182	8.63412	50
Fluoride (mg/l)	0 – 0.1	0.037	0.03604	2
Phosphate (mg/l)	0 – 5.80	1.53	2.59901	-
Silica (mg/l)	0 – 40	4.43545	11.89613	-
TDS (mg/l)	26.80-120.60	62.66818	25.63874	1000
Total Alkalinity (mg/l)	61- 119	92.63536	15.0218	-
Total Hardness (mg/l)	10 – 105	59.30	28.02906	500
Odour	Odourless	Odourless	Odourless	Unobjectionable odour
Turbidity (NTU)	0.005 – 1.50	0.329	0.53689	5

Electrical Conductivity (EC)

The most influential water quality guideline on crop productivity is the water salinity hazard as measured by electrical conductivity (EC). The primary effect of high EC water on crop productivity is the inability of the plant to compete with ions in the soil solution for water (physiological drought). The higher the EC, the less water is available to plants; even though the soil may appear wet. Thus because plants can only transpire water ("pure water"), usuable plant water in the soil solution decreases dramatically as EC increases (Joshi et al, 2009). On the basis of electrical conductivity, the irrigational classification is given in Table 14.

Table 14. Quality of irrigation water in relation to electrical conductivity (EC) after Richard 1954

S/No	Electrical conductivity (μmhos/cm)	Type of water	Suitability for irrigation	Remarks
1	Below 250	Low saline water	Entirely safe	All samples in the study area fall within this zone
2	250-750	Moderately saline (2)	Safe under practically all conditions	NIL
3	750-2250	Medium to high salinity (3)	Safe only with permeable soil and Moderate Leaching	NIL
4	Above 2250			
(i)	2250-4000	High salinity	Unfair for irrigation	NIL
(ii)	4000-6000	Very high salinity	Unfair for irrigation	NIL
(iii)	Above 6000	Excessive salinity class	Unfair for irrigation	NIL

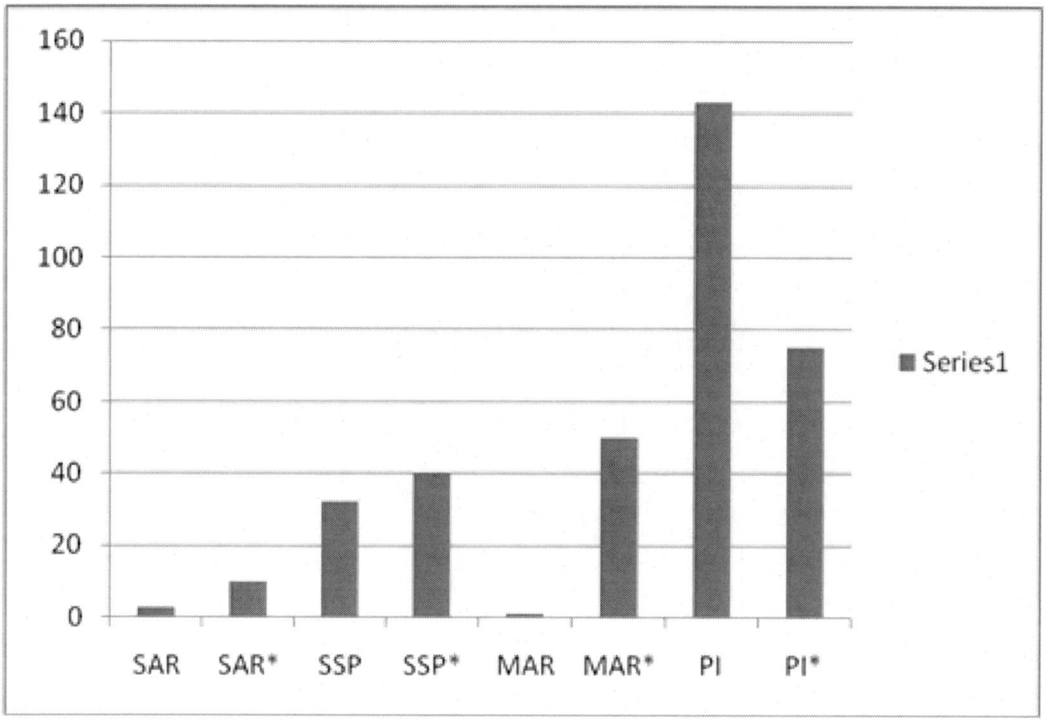

Figure 4. Summary Statistics of different irrigation indices of surface water.

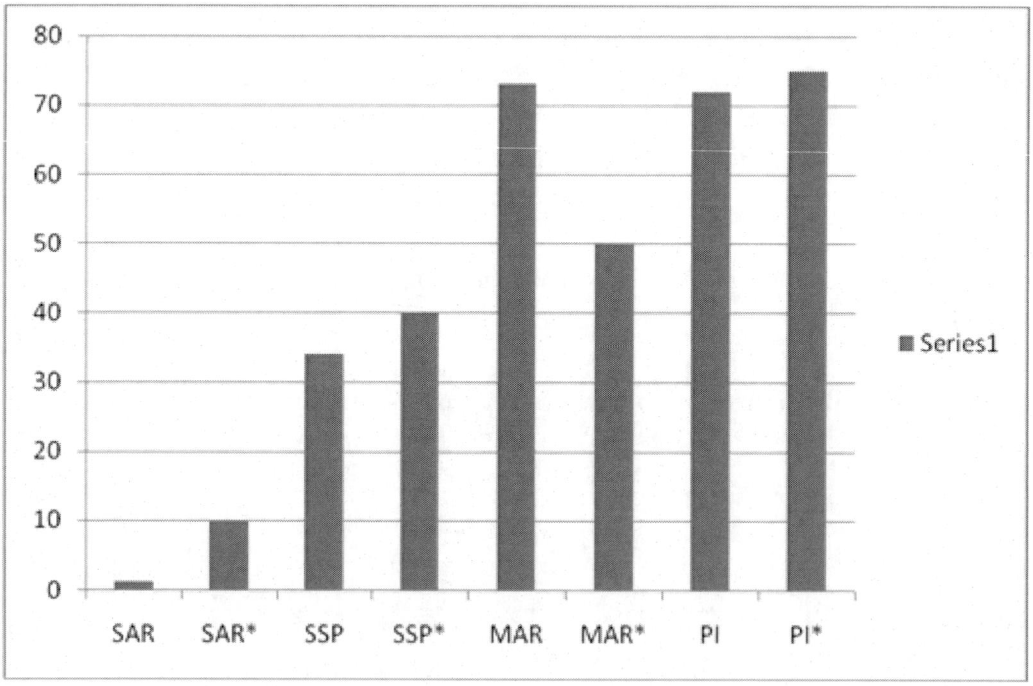

Figure 5. Summary Statistics of different irrigation indices of shallow groundwater.

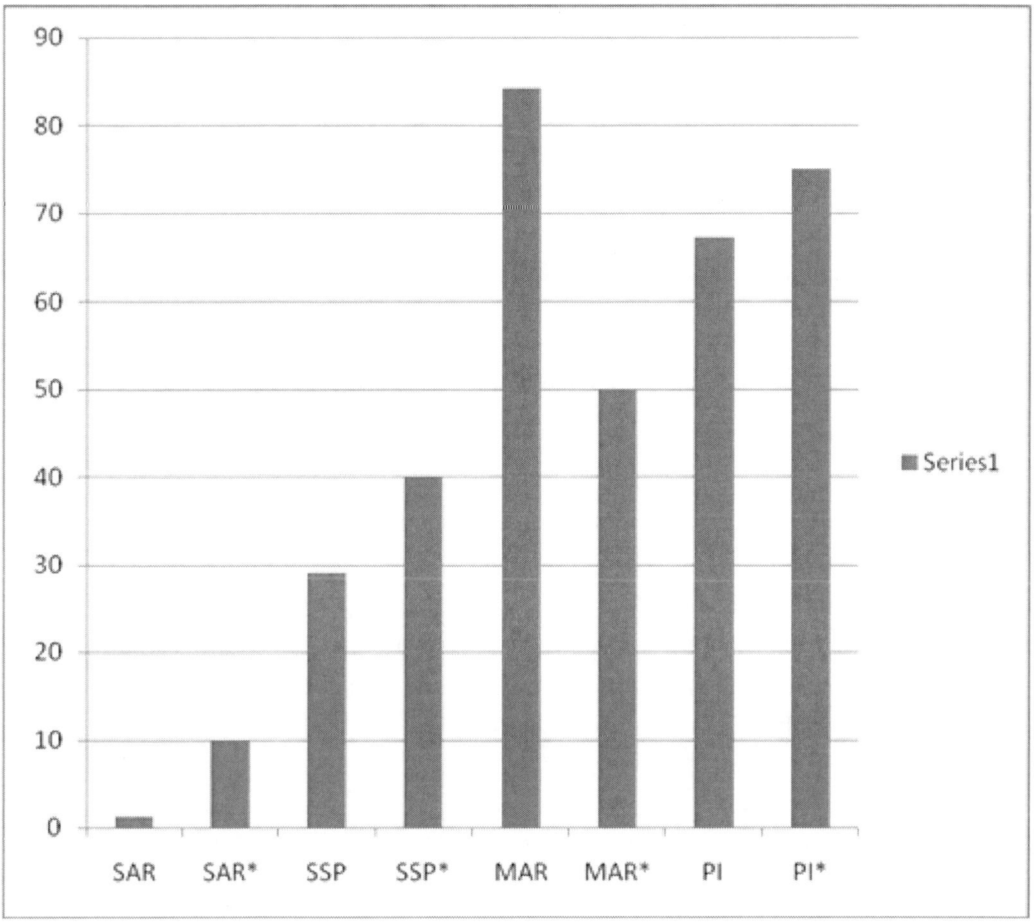

Figure 6. Summary Statistics of different irrigation indices of deep groundwater.

For the surface water the EC values are 0.194 ds/m and 0.25 ds/m (Lakes Geriyo and Njuwa) whereas its values in River Benue varies from 0.061 ds/m at Sw7/Sw8 to 0.266 ds/m at Sw11 with a mean value of 0.204 ds/m.

The electrical conductivity of shallow groundwater range from 0.04ds/m to .195ds/m whereasthose of deep groundwater range from 0.05ds/m to 0.18ds/m. They are thus generally less than 4ds/m and fall within the excellent categories for irrigation water and also fall below the WHO (2006) recommended limit of 1400 micro-mhos/cm.

Soluble Sodium Percentage (SSP)

Sodium percent is another important factor to study sodium hazard. It is calculated as the percentage of sodium and potassium against all cationic concentration. It is also used for adjudging the quality of water for agricultural purposes. The use of high percentage sodiumwater for irrigation purpose stunts the plant growth. Sodium reacts with soil to reduce its permeability. Sodium percent in water is a parameter computed to evaluate the suitability for irrigation (Joshi et al. 2009). The soluble sodium percentage values of lakes Njuwa and Geriyo are 52.39% and 58.32% whereas those of River Benue varies from 4.40% at Sw6 to 83.64% at Sw8 with a mean value of 32.50%. It range from 0.01% to 79.08% for shallow groundwater, and 1.31% to 57.12% for deep groundwater respectively indicating low alkali hazards and fair (ClassIII) to excellent (ClassI) irrigation quality (Wilcox, 1950).

Total Dissolved Solids (TDS)

Salts of calcium, magnesium, sodium, potassium present in the irrigation water may prove to beinjurious to plants. When present in excessive quantities, they reduce the osmotic activities of the plants and may prevent adequate aeration. The total dissolved solids in the waters can beestimated from their specific conductance values by multiplying with a factor of 0.64 which is taken as the conversion factor for most natural waters (US Salinity Laboratory 1954 Figure 7). For the surface waters the TDS value are 167.5 mg/l and 120.98 mg/l (Lakes Njuwa and Geriyo); Whereas its value in River Benue varies from 41 at Sw8 to 218 mg/l at Sw6 with a mean value of 142.95 mg/l (Obiefuna and Orazulike 2010 a,b).

In shallow groundwater it ranges from 26.8 mg/l to 137.4 mg/l and 33.50 mg/l to 120.6 mg/l for deep groundwater. They are generally less than 200 mg/l and can be classified as excellent irrigation water according to Robinove et al. (1958).

Sodium Adsorption Ratio (SAR)

The sodium adsorption ratio gives a clear idea about the adsorption of sodium by soil. It is theproportion of sodium to calcium and magnesium which affect the availability of the water to the crop. The sodium adsorption ratio of both surface water and groundwater examined in the present investigation are generally less than 9 and fall under the category of C2S1 indicating low alkali hazards and excellent irrigation water (Figure 7).

Magnesium Adsorption Ratio (MAR)

Magnesium content of water is considered as one of the most important qualitative criteria indetermining the quality of water for irrigation. Generally, calcium and magnesium maintain a state of equilibrium in most waters. More magnesium in water will adversely affect crop yields as the soils become more saline (Joshi et al. 2009).

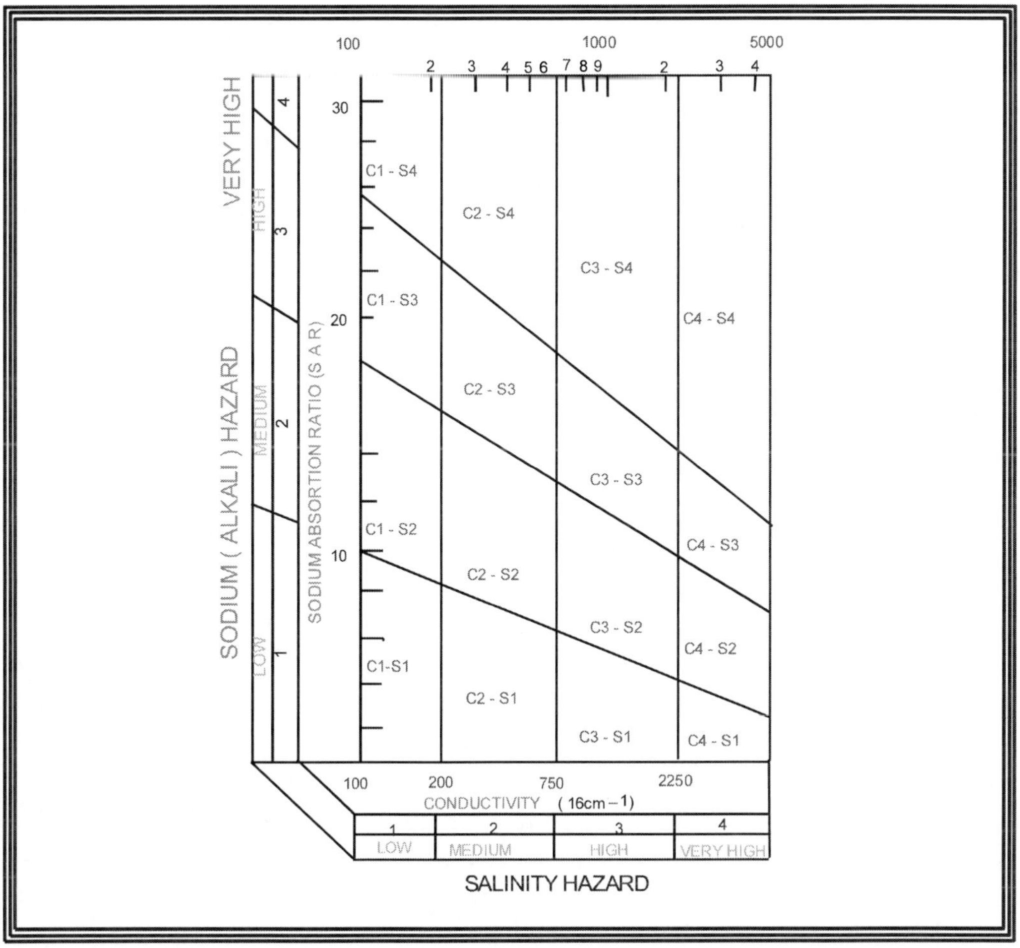

Figure 7. Classification of the analyzed water samples with respect to sodium adsorption ration and salinity hazard (After us salinity lab, 1954).

The magnesium adsorption ratio calculated for water samples from lakes Geriyo and Njuwa were71.18% and 83.32% while the values for water samples from River Benue varies from 57.09%at Sw10 to 71.11% at Sw7 with a mean value of 67.31% (Obiefuna and Orazulike 2010 a,b).

The values of shallow groundwater varies from 13.34% to 97.83% and 70.22% to 96.67% for deep groundwater indicating that they are above the acceptable limit of 50% (Ayers andWestcot, 1985). The waters are therefore considered unsuitable. This is because high magnesium adsorption ratio causes a harmful effect to soil when it exceeds 50%.

Permeability Index (PI)

The soil permeability is affected by the long term use of irrigated water and the influencing constituents are the total dissolved solids, sodium bicarbonate and the soil type. For the surface water samples obtained from lakes Njuwa and Geriyo the permeability index values are 57.75% and 77.31% whereas those of River Benue varies from 3.19% at Sw4 to 98.60% at Sw8 with a mean value of 25.64% (Obiefuna and Orazulike 2010 a,b).

For the shallow groundwater permeability index values range from 39.5% to 118% and 28.70%To 131.36% for deep groundwater. The above results therefore suggests that most of the water samples fall within ClassI and ClassII and can be categorized as good irrigation water (Doneen, 1964).

Residual Sodium Bicarbonate (RSBC)

The concentration of bicarbonate and carbonate also influences the suitability of water for irrigation purpose. One of the empirical approaches is based on the assumption that all Ca^{2+} and Mg^{2+} precipitate as carbonate. Thus based on this assumption Eaton (1950) proposed the concept of residual sodium bicarbonate (RSBC) for the assessment of high carbonate waters. The water with high RSBC has high pH and land irrigated with such water becomes infertile owing to deposition of sodium carbonate; as observed from black colour of the soil.

The residual sodium bicarbonate values of water samples from lakes Njuwa and Geriyo are -0.46 And 5.81 meq/litre whereas those of River Benue varies from -0.40 meq/l at Sw1 to 4.15 meq/l at Sw10 with a mean value of 1.44 meq/l.

The values for shallow groundwater vary from 0.06 meq/litre to 3.18 meq/litre and 0.77 meq/litre to 1.68 meq/litre for deep groundwater respectively (Obiefuna and Orazulike 2010 a,b). The residual sodium bicarbonate values are less than 2.5 meq/litre and are therefore considered safe for irrigation purposes.

Kellys Ratio (KR)

The Kellys Ratio (KR) values of lakes Njuwa and Geriyo are less than 1.40 while those of River Benue at Jimeta and Yola Treatment Plants are 4.98% and 5.63% respectively.

The values for shallow groundwater ranged between 0 and 3.46% but less than 1.25% for deepgroundwater (Obiefuna and Orazulike 2010 a,b).

These indicate that while the KR for water samples from the two lakes (1.09% for Njuwa and 1.39% for Geriyo) are close to the permissible limit; the KR values of water samples obtained from River Benue varies from 0.02% at Sw4 to 5.63% at Sw7 with a mean value of 1.27%.Most of the KR values for most surface water and groundwater samples however fall within thepermissible limit of 1.0 and are considered largely suitable for irrigation purposes.

Chemical Speciation of Selected Major Ions in Groundwater Samples

Analytical data are provided in Table 1. The pH values for tested samples varied from 6.1 to 7.8for shallow groundwater, 6.8 to 7.8 for deep groundwater and 7.20 to 8.00 in surface watersamples. Sodium concentrations range was between 0 and 157 mg/l in shallow groundwater0.001 91.00 mg/l in deep groundwater and 2.23 to 211 mg/l for the surface waters respectively.In shallow groundwater potassium concentration ranges from 0 to 37 mg/l, 0 to 12.10 mg/l indeep groundwater and 1.20 to 9.00 mg/l for then surface waters respectively. The concentrationof calcium in shallow groundwater samples are less than 25 mg/l, less than 39 mg/l in deepgroundwater samples and between 1.20 to 9 mg/l in the surface water samples respectively.Magnesium concentration in shallow groundwater is less than 51mg/l, less than 55 mg/l in deepgroundwater and between 12 and 84 mg/l in the surface water samples respectively. Ironconcentrations range was between 0.001 and 1.80 mg/l in shallow groundwater, 0 and 0.32 mg/lin deep groundwater and 0 to 0.73 in the surface water samples respectively.

Tabulated analytical values of major cations, anions, and pH were used to create the input datafile (Obiefuna and Orazulike 2010 a,b). Generated output by the model included the calculatedionic strength in the analyzed samples and activities of various metal ionic species whether freeor complexed ions. A variety of reactions may occur which influences the speciation andmobility of metal contaminants including acid/base, precipitation/dissolution, sorption and ionexchange reactions can retard the movement of metals in groundwater. The rate and extent ofthese reactions will depend on factors such as pH, Eh, complexation with other dissolvedconstituents, sorption and ion exchange capacity of the geological materials and organic mattercontent.

Groundwater flow characteristics also influence the transport of metal contaminants.Tabulated analytical values of major cations, anions, and pH were used to create the input data file. Generated output by the model included the calculated ionic strength in the analyzed samples and activities of various metal ionic species whether free or complexions (Tables 15, 16 and 17).

Table 15. Summary of PHREEQC output file containing activities and concentrations of some ionic species in surface water (Obiefuna and Orazulike 2010 d)

Species	Concentration mg/l	Activity mg/l	Species	Concentration mg/l	Activity mg/l
Ca^{2+}	5.955E-004	3.909E-004	MgF+	1.806E-007	1.619E-007
$CaHPO_4$	2.648E-0.005	2.655E-005	Na+	3.274E-003	2.916E-003
$CaHCO_3^+$	1.747E-0.005	1.573E-005	$NaHCO_3$(aq)	4.844E-006	4.856E-006
$CaSO_4$	8.112E-0.006	8.132E-006	$NaSO_4^-$	1.677E-006	1.504E-006
$CaCO_3$(aq)	4.067E-0.006	4.077E-006	$NaHPO_4$	7.221E-007	6.476E-007
$CaH_2PO_4^+$	5.854E-0.007	5.250E-007	$NaCO_3^-$(aq)	4.273E-007	3.832E-007
$CaPO_4$	2.832E-006	2.540E-006	NaOH	7.138E-010	7.156E-010
CaOH+	2.687E-0.009	2.409E-009	NaF	4.438E-009	4.449E-009
CaF+	1.118E-0.008	1.002E-008	$Fe(HS)_2$	1.562E-010	1.567E-010
$CaHSO_4$+	1.517E-0.12	1.361E-012	Fe + 2	1.912E-007	1.265E-007
K+	1.332E-0.004	1.192E-004	$FeHPO_4$	5.723E-008	5.738E-008
KSO_4^-	1.010E-0.007	9.058E-008	$FeHCO_{3+}$	4.179E-008	3.748E-008
$KHPO_4^-$	2.952E-0.008	2.647E-008	$FeCO_3$	1.704E-008	1.708E-008
KOH(aq)	1.531E-011	1.535E-011	$FeH_2PO_4^+$	3.405E-009	3.054E-009
Mg^{2+}	1.289E-003	8.522E-004	$Fe(HS)_{3-}$	2.242E-014	2.010E-014
$MgHPO_4$	7.805-005	7.825E-005	$FeSO_4$	2.436E-009	2.443E-009
$MgHCO_3^+$	3.367E-005	3.020E-005	FeCl+	7.177E-010	6.436E-010
$MgSO_4$	2.237E-005	2.242E-005	FeOH+	2.318E-009	2.079E-009
$MgH_2PO_4^+$	1.625E-006	1.458-006	FeF+	3.741E-012	3.355E-012
$MgPO_4^-$(aq)	8.328E-006	7.469E-006	$FeHSO_4$+	4.912E-016	4.405E-016
$MgCO_3$	4.895E-006	4.907E-006	$Fe(OH)_3$	4.088E-006	4.098E-006
MgOH+	1.922E-007	1.724E-007	$Fe(OH)_2$+ Fe(OH)4	7.852E-007 1.854E-007	7.042E-007 1.663E-007

Table 16. Summary of PHREEQC output file containing activities and concentrations of some ionic species in Shallow Groundwater Samples (After Obiefuna and Orazulike 2010 d)

Species	Concentration mg/l	Activity mg/l	Species	Concentration mg/l	Activity
Ca^2+	2.923E-004	2.145E-004	Na+	1.492E-003	1.380E-003
$CaHCO_3$+	6.089E-006	5.636E-006	$NaHCO_3$ (aq)	1.490E-006	1.492E-006
$CaSO_4$	3.161E-006	3.165E-006	$NaHPO_4$	4.535E-008	4.189E-008
$CaHPO_4$	2.005E-006	2.008E-006	$NaSO_4$-	5.455E-007	5.039E-007
$CaCO_3$ (aq)	4.892E-007	4.898E-07	$NaCO_3$- (aq)	4.346E-008	4.014E-008
CaH2PO4+	1.295E-007	1.197E-07	NaF	1.472E-009	1.474E-009
$CaPO_4$-	6.941E-008	6.411E-008	NaOH	1.120E-010	1.121E-010
CaF+	4.212E-009	3.890E-009	Fe+2	2.292E-006	1.690E-006
CaOH+	4.741E-010	4.379E-010	$FeHCO_3$+	3.518E-007	3.250E-007
CaHSO4+	1.742E-012	1.609E-012	$FeHPO_4$	1.046E-007	1.048E-007
K+	1.010E-004	9.320E-005	$FeCO_3$	4.937E-008	4.943E-008
KSO_4-	5.456E-008	5.040E-008	Fe(HS)2	3.371E-008	3.375E-008
$KHPO_4$-	3.064E-009	2.830E-009	FeSO4	2.319E-008	2.322E-008
KOH (aq)	3.970E-012	3.975E-012	FeH2PO4+	1.818E-008	1.608E-008
Mg^{2+}	8.201E-004	6.041E-004	FeOH+	1.028E-008	9.492E-009

Species	Concentration mg/l	Activity mg/l	Species	Concentration mg/l	Activity
$MgHCO_3$+	1.509E-005	1.394E-005	FeCl+	4.483E-009	4.141E-009
MgSO4	1.134E-005	1.135E-005	FeF+	3.397E-011	3.137E-011
MgHPO4	7.636E-006	7.645E-006	$Fe(HS)_3$	1.884E-011	1.704E-011
$MgCO_3$	7.581E-007	7.591E-007	$FeHSO_4$+	1.372E-014	1.268E-014
MgH_2PO_4+	4.646E-007	4.292E-007	$Fe(OH)_3$+	2.158E-006	2.160E-006
MgPO4- (aq)	2.637E-007	2.436E-007	$Fe(OH)_2$ +	1.191E-006	1.100E-006
MgF+	8.771E-008	8.102E-008	$Fe(OH)_4$ -	3.197E-008	2.953E-008
MgOH+	4.554E-008	4.207E-008	FeOH +2	3.079E-010	2.241E-010

Table 17. Summary of PHREEQC output file containing activities and concentrations of some ionic species in deep groundwater samples (After Obiefuna and Orazulike 2010 d)

Species	Concentration mg/l	Activity mg/l	Species	Concentration mg/l	Activity
Ca^2+	2.700E-004	1.934E-004	$NaHCO_3$ (aq)	1.528E-006	1.530E-006
$CaHCO_3$+	5.823E-006	5.358E-006	$NaSO_4$-	6.953E-007	6.381E-007
CaSO4	3.704E-006	3.709E-006	$NaCO_3$- (aq)	6.124E-008	5.620E-008
CaHPO4	1.720E-006	1.722E-006	NaHPO4-	4.220E-008	3.873E-008
CaCO3 (aq)	6.304E-007	6.313E-007	NaF	1.050E-009	1.051E-009
CaH_2PO4+	8.284E-008	7.602E-008	NaOH	1.472E-010	1.474E-010
$CaPO_4$-	8.110E-008	7.443E-008	Fe+2	2.939E-007	2.116E-007
CaF+	2.807E-009	2.576E-009	Fe(HS)2	2.609E-007	2.613E-007
CaOH+	5.802E-010	5.325E-010	FeHCO3+	4.666E-008	4.282E-008
$CaHSO_4$+	1.527E-012	1.401E-012	FeHPO4	1.243E-008	1.244E-008
K+	1.104E-004	1.012E-004	FeCO3	8.800E-009	8.813E-009
KSO_4-	7.766E-008	7.127E-008	FeSO4	3.781E009	3.786E-009
$KHPO_4$-	3.177E-009	2.916E-009	FeOH+	1.770E-009	1.624E-009
KOH (aq)	5.816E-012	5.825E-012	$FeH2PO_4$+	1.610E-009	1.478E-009
Mg^{2+}	1.212E-003	8.719E-004	$Fe(HS)_3$-	1.154E-009	1.059E-009
$MgHCO_3$+	2.308E-005	2.118E-005	FeCl+	6.222E-010	1.5710E-010
MgSO4	2.133E-005	2.136E-005	FeF+	3.132E-012	2.874E-012
MgHPO4	1.049E-005	1.050E-005	FeHSO4+	1.671E-015	1.533E-015
MgCO3	1.562E-006	1.564E-006	Fe(OH)3	6.859E-007	6.870E-007
$MgPO_4$- (aq)	4.934E-007	4.528E-007	Fe(OH)2+	2.805E-007	2.574E-007
$MgH2PO_4$+	4.757E-007	4.366E-007	Fe(OH)4-	1.390E-008	1.276E-008
MgF+	9.350E-008	8.581E-008	FeOH+2	5.444E-010	3.862E-011
MgOH+	9.065E-008	8.319E-008	FeH_2PO4+2	1.412E-015	1.002E-015
Na+	1.464E-003	1.345E-003	Fe+3	5.336E-16	2.671E-016

Chemical Speciation of Sodium

In natural water system dissolved sodium occurs in +1 oxidation state. The results indicate thatNa^+ is the most soluble chemical species of sodium in shallow groundwater and representabout 99% of the sodium. Complexed species indicate that 0.09% occurs as bicarbonate,0.03% as sulphates and less than 0.002% as other anions. In deep groundwater Na+ represent99.84% of the sodium whereas complexed species revealed that 0.10% occur as bicarbonates,0.05% as sulphate and 0.004% as carbonate respectively. In surface water Na+ represent99.82% of the sodium while complexed species disclosed that 0.12% occur as bicarbonate,0.04% as sulphate, 0.02% as $NaHPO_4$ and 0.01% as $NaCO_3$.Sodium can be removed fromnatural water by ion exchange and sorption to mineral surfaces. Removal by these mechanismsincreases as pH increases. Sorption is also influenced by the cation exchange capacity (CEC)of clays, carbonate minerals and organic matter present in soils and sediments. Ion exchangeand sorption of Na+ controls the mobility of sodium. In surface water and groundwater systems,a significant fraction of sodium is undissolved and occurs as precipitates ($NaHCO_3$, Na_2SO_4),sorbed ions or surface coatings on minerals, or as suspended organic matter

In natural water system dissolved sodium occurs in +1 oxidation state. The results indicate thatNa^+ is the most soluble chemical species of sodium in shallow groundwater and representabout99% of the sodium. Complexed species indicate that 0.09% occurs as bicarbonate, 0.03%as sulphates and less than 0.002% as other anions (Figure 8).

In deep groundwater Na^+ represent 99.84% of the sodium whereas complexed species revealedthat 0.10% occur as bicarbonates, 0.05% as sulphate and 0.004% as carbonate respectively(Figure 9). In surface water Na^+ represent 99.82% of the sodium while complexed speciesdisclosed that 0.12% occur as bicarbonate, 0.04% as sulphate, 0.02% as $NaHPO_4$ and 0.01% as$NaCO_3$ (Figure 10).

Chemical Speciation of Potassium

Potassium like sodium has only one significant oxidation state K+ and it tends to be soluble inmost types of natural water as sodium. Chemical speciation of Potassium in shallowgroundwater with different anions is shown in Figure 5. It indicate that 99.94% of potassiumexists as free (K+), about 0.05% as sulphate, 0.003% as $KHPO_4$ and about 0.0005% as KOH.

In deep groundwater the chemical speciation of potassium with different anions indicate that99.93% of potassium exists as free ions (K+), 0.07% as sulphate, 0.003% as $KHPO_4$ and 0%as KOH respectively. In surface water samples an estimated 99.93% of potassium occurs infree ionic form (K^+), 0.06% occurs as KSO_4, 0.02% as $KHPO_4$ and 0% as KOH respectively.

Potassium like sodium has only one significant oxidation state K^+ and it tends to be soluble inmost types of natural water as sodium. Chemical speciation of Potassium in shallowgroundwater with different anions is shown in Figure 11. It indicate that 99.94% of potassiumexists as free (K^+), about 0.05% as sulphate, 0.003% as $KHPO_4$ and about 0.0005% as KOH.

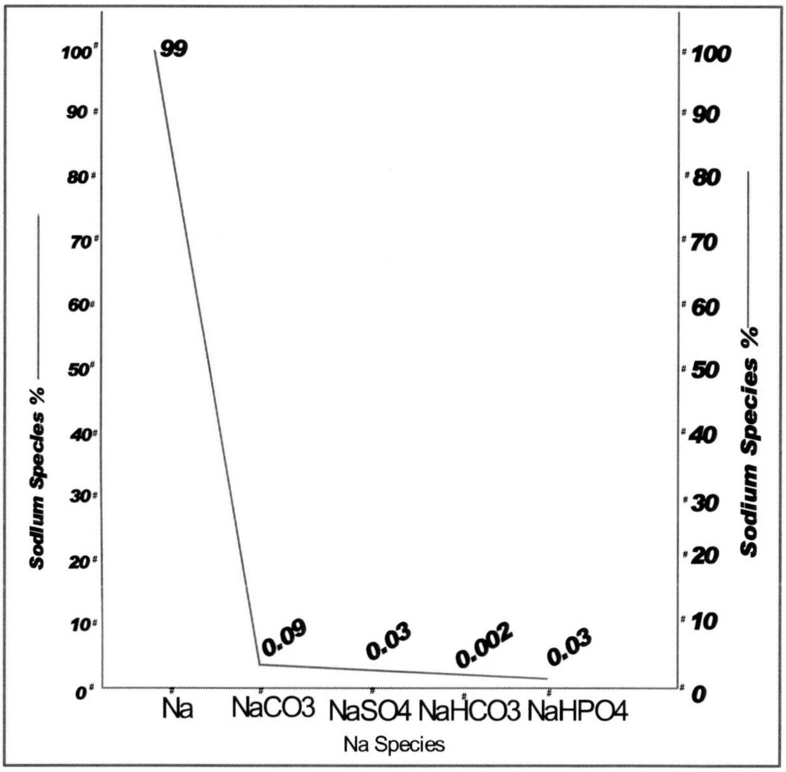

Figure 8. Percent Distribution of Sodium Species in Shallow Groundwater.

In deep groundwater the chemical speciation of potassium with different anions indicate that 99.93% of potassium exists as free ions (K^+), 0.07% as sulphate, 0.003% as $KHPO_4$ and 0% as KOH respectively (Figure 12).

In surface water samples an estimated 99.93% of potassium occurs in free ionic form (K^+), 0.06% occurs as KSO_4, 0.02% as $KHPO_4$ and 0% as KOH respectively (Figure 13).

Speciation of Magnesium

In natural water system dissolved magnesium occurs in +2 oxidation state. Chemical speciation of magnesium in shallow groundwater samples indicate that 95.83% of magnesium is in the free ionic form (Mg^{2+}), 1.76% occurs as bicarbonate, 1.33% as sulphate, 0.89% as $MgHPO_4$ and 0.09% as carbonate. The deep groundwater disclosed that 96.27% of magnesium exists as free ions (Mg^{2+}), 1.83% as bicarbonate, 1.69% as sulphate, 0.12% as carbonate and 0.04% as phosphates. The surface water samples revealed similar trends as 92.45% of magnesium occur in free ionic state, 3.94% as phosphates, 1.83% as bicarbonates, 1.21% as sulphate and 0.19% as carbonates. Under most conditions Mg^{2+} and magnesium-hydroxide complexes are the most stable forms of lead. Low solubility compounds are formed by complexation with inorganic Cl^-, CO_3, SO_4, PO_4. In surface water and groundwater systems, a significant fraction of magnesium is undissolved and occurs as precipitates ($MgPO_4$, $MgCO_3$, $MgSO_4$), sorbed ions or surface coatings on minerals, or suspended organic matter.

In natural water system dissolved magnesium occurs in +2 oxidation state. Chemical speciation of magnesium in shallow groundwater samples indicate that 95.83% of magnesium is in the free ionic form (Mg^{2+}), 1.76% occurs as bicarbonate, 1.33% as sulphate, 0.89% as $MgHPO_4$ and 0.09% as carbonate (Figure 14). The deep groundwater disclosed that 96.27% of magnesium exists as free ions (Mg^{2+}), 1.83% as bicarbonate, 1.69% as sulphate, 0.12% as carbonate and 0.04% as phosphates (Figure 15). The surface water samples revealed similar trends as 92.45% of magnesium occurs in free ionic state, 3.94% as phosphates, 1.83% as bicarbonates, 1.21% as sulphate and 0.19% as carbonates (Figure 16).

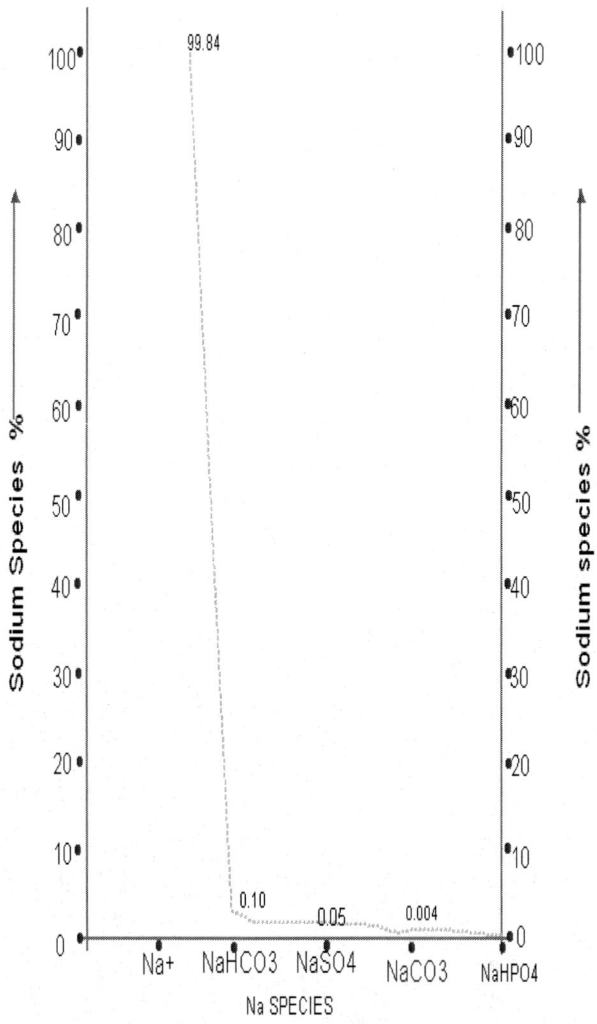

Figure 9. Percent Distribution of Sodium Species in Deep Groundwater.

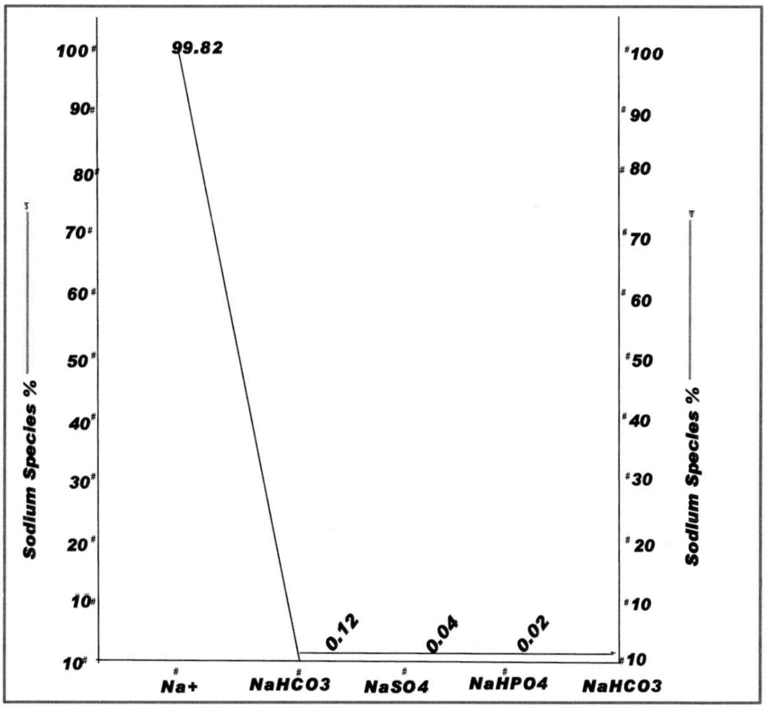

Figure 10. Percentage Distribution of Sodium Species in Surface Water.

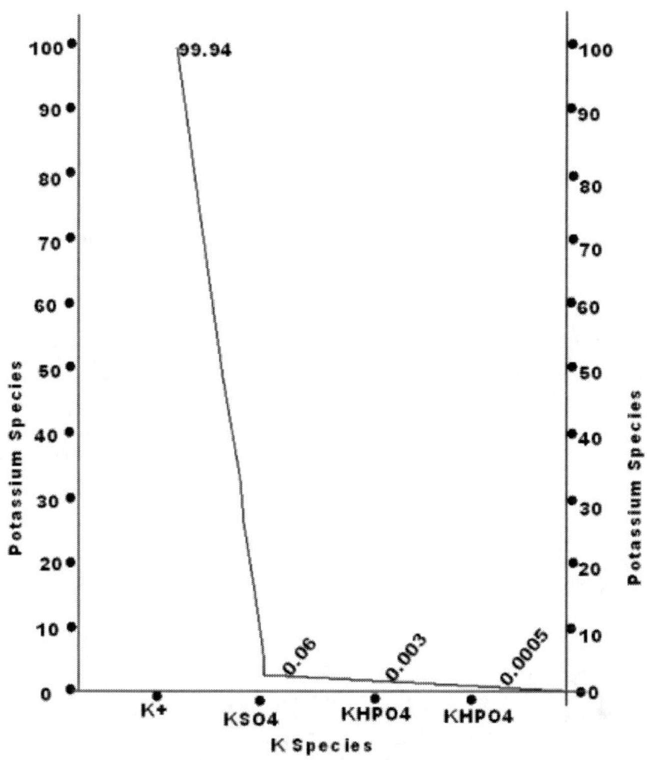

Figure 11. Percentage Distribution of Potassium species in shallow Groundwater.

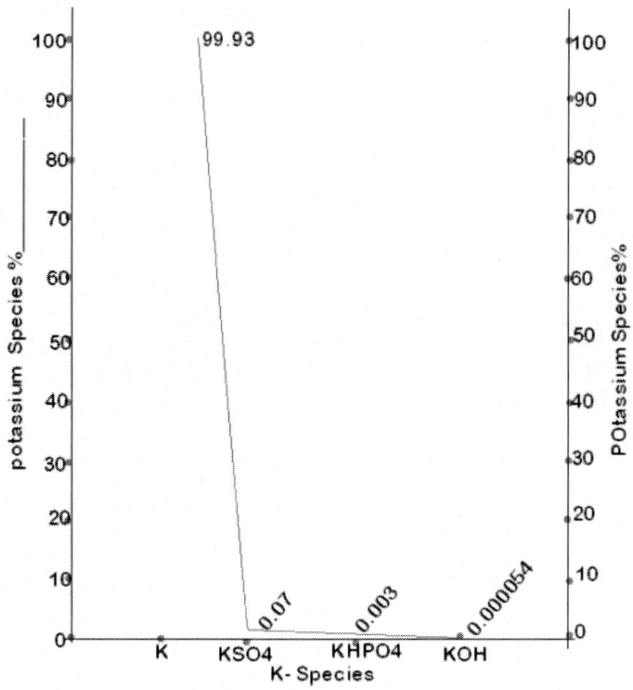

Figure 12. Percent Distribution of Potassium Species in Deep Groundwater

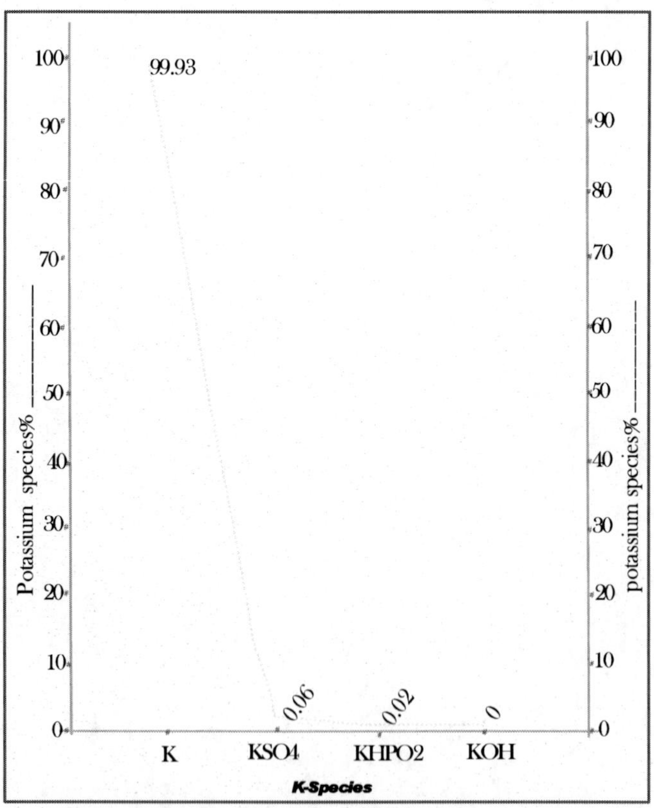

Figure 13. Percentage Distribution of Potassium Species in Surface Water.

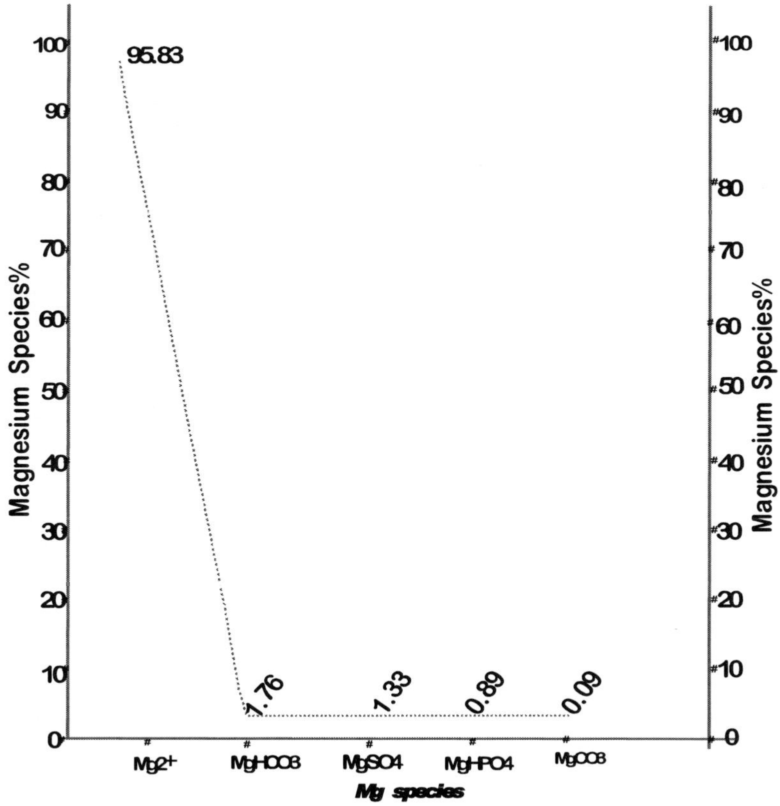

Figure 14. Percentage distribution of Magnesium Species in shallow Groundwater.

Chemical Speciation of Calcium

Calcium like magnesium has only one oxidation state Ca^{2+} and is largely soluble in most types of natural waters. Chemical speciation of calcium with different anions in shallow groundwater show that 96.07% exists as free ions (Ca^{2+}), 2% as bicarbonates, 1.04% as sulphates and 0.66% as phosphates. In deep groundwater samples 95.79% of calcium occurs as free ion (Ca^{2+}), 2.07% as bicarbonates, 1.31% as sulphates and less than 0.90% as phosphates and carbonates respectively. The surface water samples revealed values of 93.48% of calcium occurring in free ionic form (Ca^{2+}), and less than 7% occurring as phosphates, bicarbonates and sulphates respectively. Calcium can also be removed from natural water by ion exchange and sorption to mineral surfaces. Removal by these mechanisms increases as pH increases. Sorption is also influenced by the cation exchange capacity (CEC) of clays, carbonate minerals and organic matter present in soils and sediments. Ion exchange and sorption of Ca^{2+} controls the mobility of calcium. In surface water and groundwater systems, a significant fraction of sodium is un-dissolved and occurs as precipitates ($CaCO_3$, $CaSO_4$), sorbed ions or surface coatings on minerals, or as suspended organic matter.

Calcium like magnesium has only one oxidation state Ca^{2+} and is largely soluble in most types of natural waters. Chemical speciation of calcium with different anions in shallow groundwater as shown in Figure 17 show that 96.07% exists as free ions (Ca^{2+}), 2% as bicarbonates, 1.04% as sulphates and 0.66% as phosphates. In deep groundwater samples 95.79% of calcium occurs as free ion (Ca^{2+}), 2.07% as bicarbonates, 1.31% as sulphates and less than 0.90% as phosphates and carbonates respectively (Figure 18).

The surface water samples revealed values of 93.48% of calcium occurring in free ionic form (Ca^{2+}), and less than 7% occurring as phosphates, bicarbonates and sulphates respectively (Figure 19).

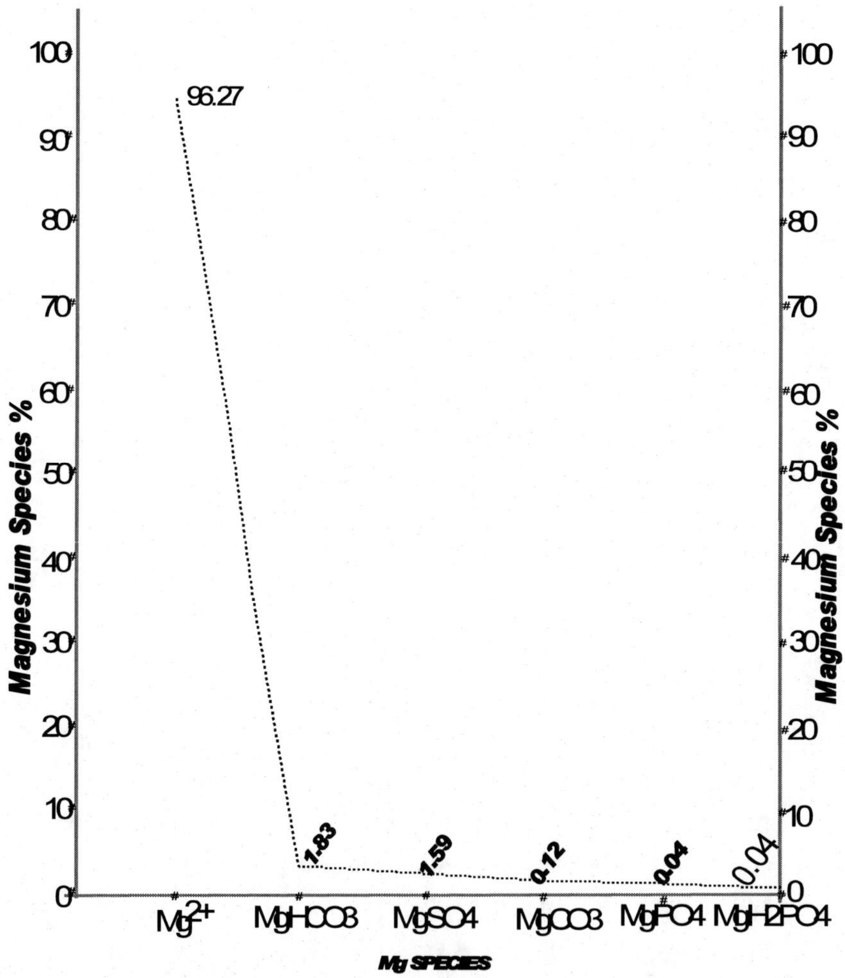

Figure 15. Percentage Distribution of Magnesium Species in Deep Groundwater.

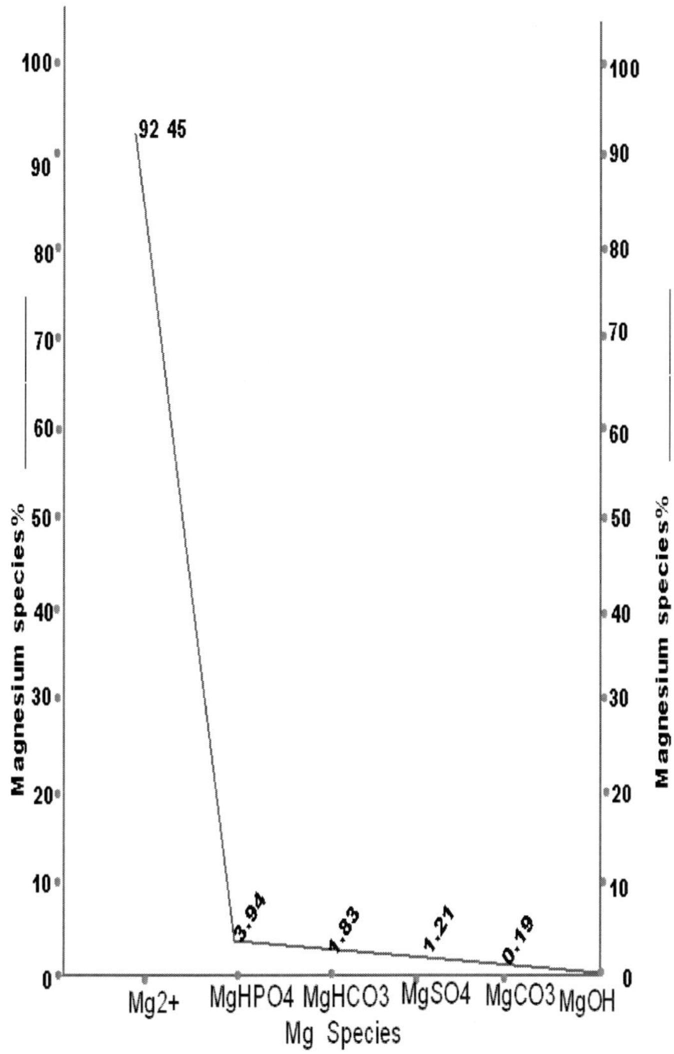

Figure 16. Percentage distribution of magnesium species in surface water.

Chemical Speciation of Iron

Iron may occur insolution as Fe^{+2} or Fe^{+3} oxidation state. The redoxconditions in oxygenated Waters favour the moreoxidized form (Fe^{3+}). Ferrous ions (Fe^{2+}) formcomplexes with a number of ligands, a strong$FeHCO_3$ and $FeCO_3$ appears likely to be the majorform in naturalwater containing dissolved CO_2.Chemical speciation of iron in shallow groundwatersamplesindicated that 79.37% of iron is in the freeionic form (Fe^{+2}), 12.18% as bicarbonate and lessthan 8% as phosphates, carbonates, sulphides andsulphates respectively. In deep groundwater46.53%exists as free Fe^{2+} ions, 41.31% as sulphides, 7.39%as bicarbonates and less than 4% asphosphates, carbonates and sulphates. In surface water samples an estimated 22.58% of ironoccurs in free ionic form (Fe^{2+}), a predominant

67% occur as sulphides while the remaining 10.42% occur as phosphates, bicarbonates and carbonates respectively (Obiefuna and Orazulike2010 d). Iron is an extremely mobile ion and its removal from groundwater is in most cases anextremely difficult tasks. Under mild oxidizing conditions iron is in a +3 valence state and forms an insoluble precipitates, Fe_2O_3 while under reducing conditions is converted to +2 valence statesand is present as soluble Fe^{2+}cation. Increased iron concentrations in water samples are due tore-dissolution of previouslyformed unstable precipitates or due to leaching of iron from theadjacent rocks that bounds the aquifers of the study area.

Iron may occur in solution as Fe^{+2} or Fe^{+3} oxidation state. The redox conditions in oxygenatedwaters favour the more oxidized form (Fe^{3+}). Ferrous ions (Fe^{2+}) form complexes with a numberof ligands, a strong $FeHCO_3$ and $FeCO_3$ appears likely to be the major form in natural watercontaining dissolved CO_2. Chemical speciation of iron in shallow groundwater samplesindicated that 79.37% of iron is in the free ionic form (Fe^{+2}), 12.18% as bicarbonate and lessthan 8% as phosphates, carbonates, sulphides and sulphates respectively (Figure 20). In deepgroundwater 46.53% exists as free Fe^{2+} ions, 41.31% as sulphides, 7.39% as bicarbonates andless than 4% as phosphates, carbonates and sulphates (Figure 21).

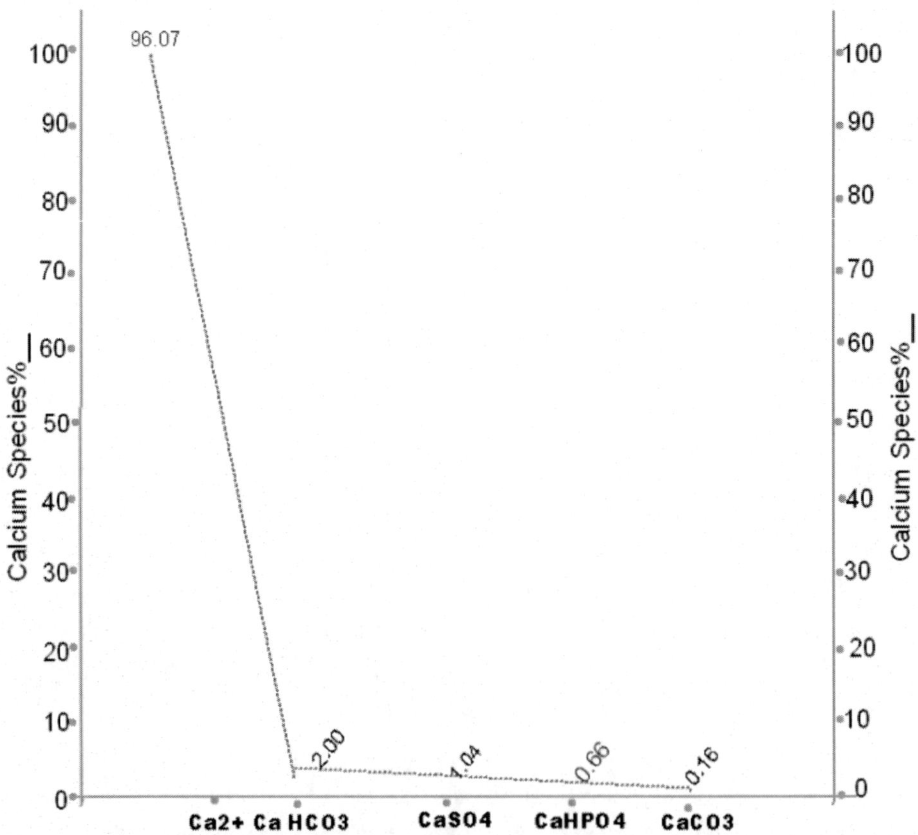

Figure 17. Percentage Distribution of Calcium species in shallow Groundwater.

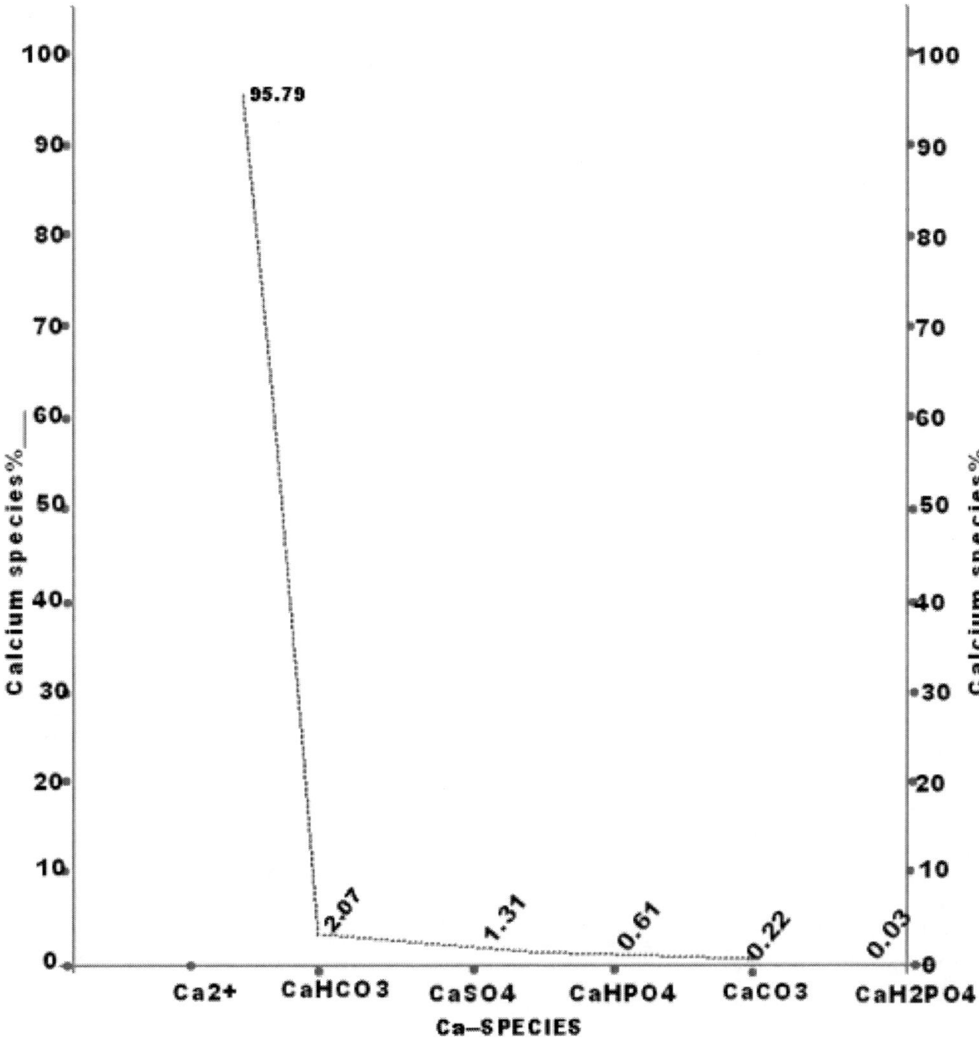

Figure 18. Percent distribution of Calcium Species in Deep Groundwater.

In surface water samples an estimated 22.58% of iron occurs in free ionic form (Fe^{2+}), a predominant 67% occur as sulphides while the remaining 10.42% occur as phosphates, bicarbonates and carbonates respectively (Figure 22).

Main iron precipitates are hematite (αFeO_3) goethite (α-FeOOH) and to a lesser amounts ferrihydrite ($Fe_5HO_8.4H_2O$) or iron hydroxide [$Fe(OH)_3$]. Geochemical speciation calculations suggest that at pH values between 5 and 6 copper precipitates as cupric and cuprous ferrite. Iron precipitates can be quite massive and often cause serious clogging problems of drinking water wells.

The study findings based on model calculations indicated that Na^+, Ca^{2+}, Mg^{2+} and K^+ are highly soluble and consequently will be bioavailable at the pH ranges of the tested wells. Iron is also expected to be bioavailable as 46.53% to 79.37% of the total iron is presented as Fe^{2+}. The lower values of Fe^{2+} are related to the organic complexation of the ions. Identifying ionic distributions is of particular importance as pollutants affect the groundwater environment by the chemical behavior of the ionic species and transformation of species than by total concentrations. The adverse effects of highly soluble free Fe^{2+} are important to groundwater chemistry because their inherent toxicity is related to their bioavailability. The result suggests that Na^{+}, Ca^{2+}, Mg^{2+} and K^{+} at such level would be available for uptake by plants, animals and humans. It is also highly possible that such conditions may constitute a negative health impact. The model findings also suggest that Fe^{2+} levels represent a health risks in some water samples.

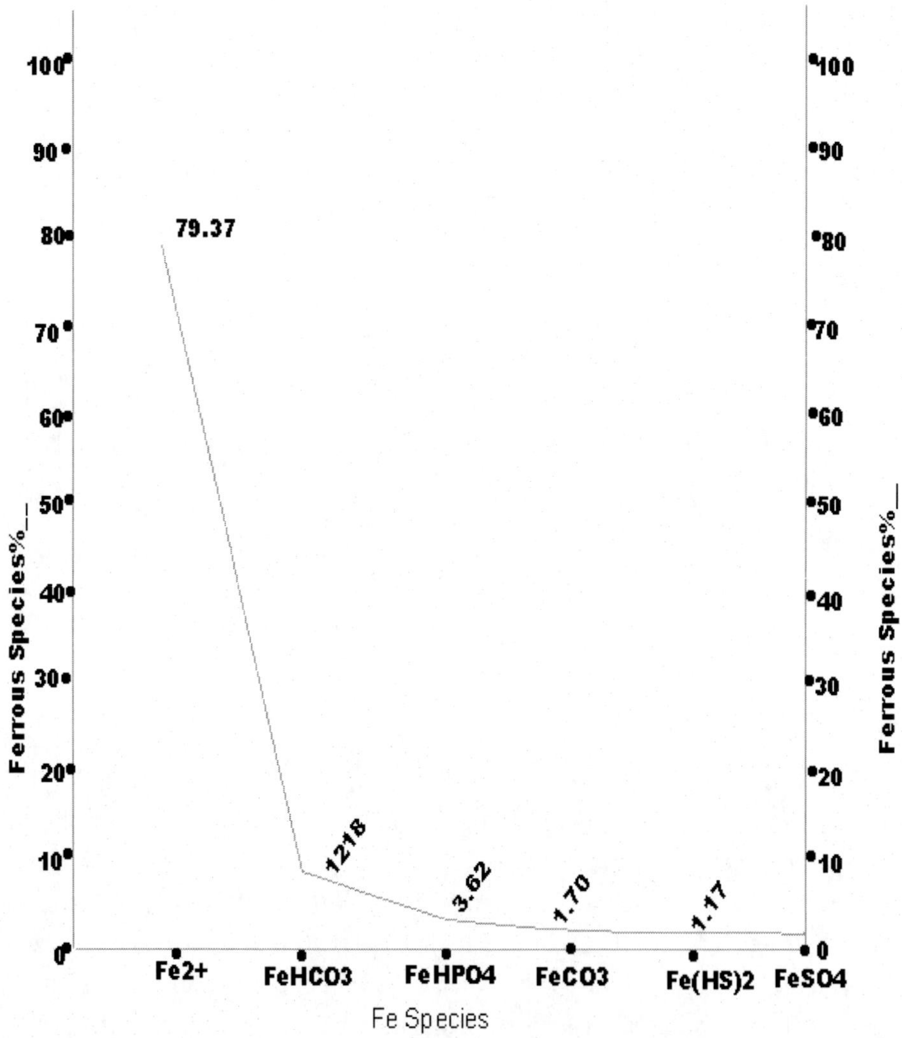

Figure 19. Percentage distribution of Iron (Ferrous) in shallow Groundwater.

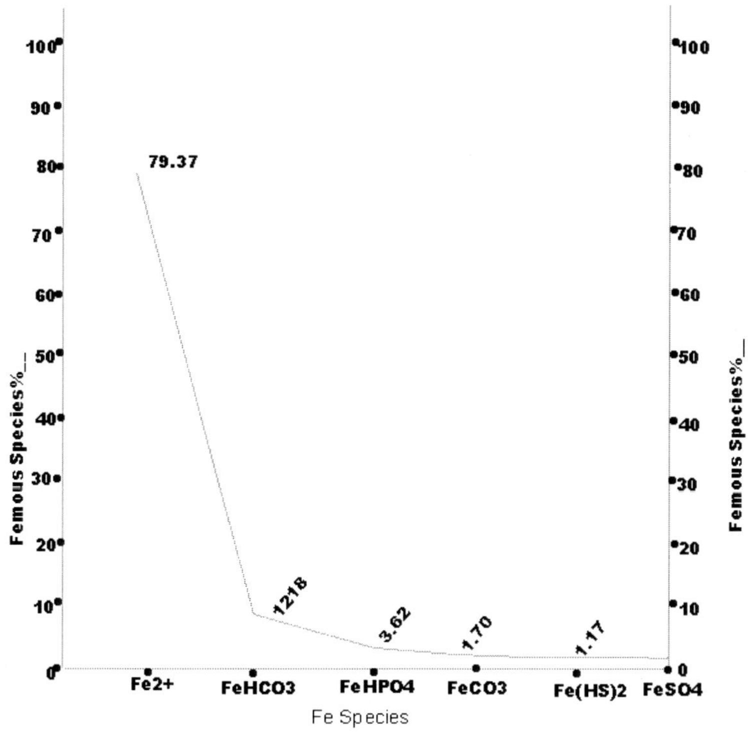

Figure 20. Percentage distribution of Iron (Ferrous) in shallow Groundwater.

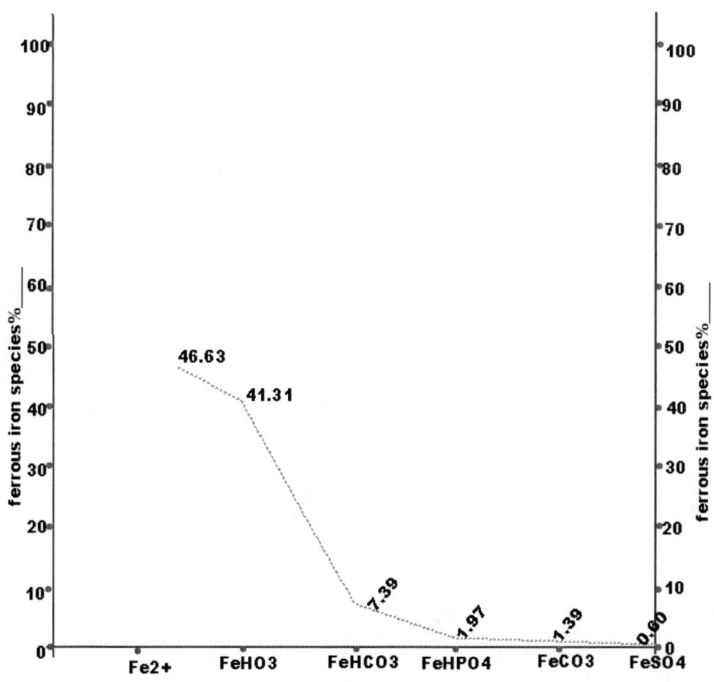

Figure 21. Percentage Distribution of Ferrous Iron in Deep Groundwater.

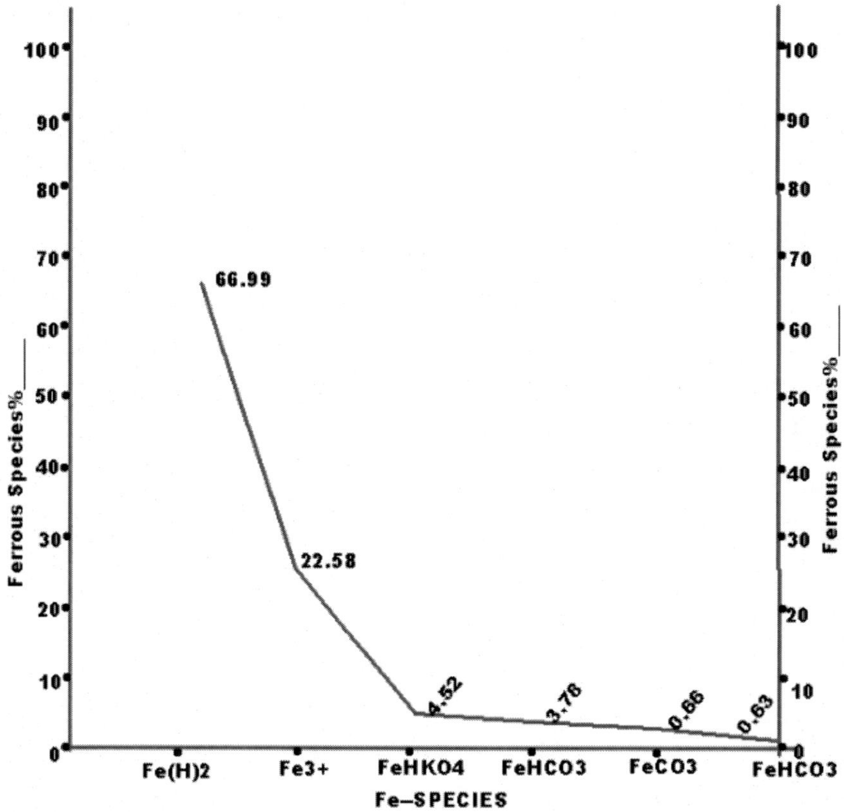

Figure 22. Percentage distribution of Ferrous Iron Species in Surface Water.

Conclusion

A total of one hundred and seven water samples from hand-dug wells and boreholes tapping these aquifers, surface water and rain water were chemically analyzed employing standard methods. Groundwater in the study area is of enormous importance, as it constitutes more than 90% of potable water. The assessment of some contamination indicators, such as pH, dissolved organic matter, dissolved oxygen, phosphate, chloride, sulphate and nitrate indicates that the water is largely suitable for domestic purposes.

However nitrate and iron contamination have been detected in some wells. Possible sources of pollution are likely from fertilizer application. It was found that the correlation coefficient of common parameters such as hardness, Ca^{2+}, Mg^{2+}, Na^{+}, Cl^{-}, HCO_3^{-}, NO_3^{-} and SO_4^{2-} indicate poor correlation with electrical conductivity (EC) in groundwater. The similarity in correlation matrix between both shallow and groundwater was attributed to mineral dissolution as the dominant process influencing the water quality.

Good correlation between electrical conductivity (EC) and nitrate (NO_3^{-}) (r=0.80) and EC and Cl- (r=0.95) suggest that variation in EC is controlled by both Cl^{-} and NO_3^{-} concentration. This confirms common primary sources which are both domestic effluent/sewage and/or fertilizer application. Furthermore the values of total dissolved solids (<250 mg/l), electrical conductivity (<3 ds/m), soluble sodium percentage (2.22% to 85%), permeability index (28.7% to 131.36%), residual sodium bicarbonate (< 1.0) and sodium adsorption ratio (<9) obtained for most of thewater samples were found to be within the safe limits for irrigation. In addition, most of the other irrigation indices of the sampled water also fall within the permissible level indicating low sodic waters. The groundwater will neither cause salinity hazards nor have an adverse effect on the soil properties and thus largely suitable for irrigation purposes.

The linear regression equations for predicting the concentration of different parameters based on electrical conductivity was successfully applied in the study area with fairly reasonable certainty. In some of the water sources the problems of bicarbonate, chloride, nitrate and iron were noticeable. It is however less appropriate variable for predicting values of the dependent variable in the groundwater samples. It further explained the fact that the variance of the residual is small compared to the variance of the dependent variable.

ACKNOWLEDGMENT

The author will like to thank the Upper Benue Basin Development Authority Yola, Adamawa State Water Board Yola for their assistance during field work. The chemical analysis of the water samples was done in the chemical laboratory of the Adamawa State Water Board Yola Nigeria employing standard methods (Atomic Absorption Spectrophotometry for cations and conventional titration for anions). The tireless efforts of the water chemist in the Laboratory Mohammed Ahmed are highly acknowledged. Finally I will like to thank Ibrahim Ahmed for drafting the figures.

REFERENCES

Abubakar, M.B., Obaye, N.G., Luterbacher, H.P., Dike, E.F.C. & Ashraf, A. R. (2006). A Report on the occurrence of Albian-Cenomanian elater-bearing pollen in Nasara-1 well, Upper Benue Trough Nigeria: Biostratigraphic and Paleoclimatological Implications. *Jour of African Earth Sciences*, Vol. *45*, Issue 3, 347-354.

Alfarrah, N.& Walraevens, K. (2018). Groundwater overexploitation and seawater intrusion incoastal areas of arid and semiarid regions. *Water*, *10*(2), 143.

American Public Health Association, American Water Works Association, Water Pollution Control Federation. (1998). *Standard methods for the examination of water and wastewater*. 20th edn, American Public Health Association, Washington, DC.

Ayers, R.S.& Westcot, D.W. (1985). *Water quality for agriculture FAO Irrigation and Drain.*, Paper No 29(1), 1-109.

Bano, N. (2017). Assessment of physicochemical characteristics of groundwater quality used for drinking water supply of firazabad City, *India using Index Method Jour Appl Sci and Environ Mgt, 21* (4), 728-731.

Bhunta, G.S., Keshavarzi, A., Shit, P.R., Omran, E.W.& Bagherzadeh, A. (2017). Evaluation ofgroundwater quality and its suitability for drinking and irrigation using GIS andgeostatistical techniques in a semiarid regions of Neyshabur, *Iran Appl. Water Sci.*, (8), 168.

Carter, J.O., Barber, W., Tart, E.A.& Jones, G.P. (1963).The Geology of parts of Adamawa, Bauchi and Bornu Provinces in Northeastern Nigeria. *Geol Surv. Of Nig. Bull, 30*, 35-53.

Carter, T.A.& Sneed, R.E. (2001). *Drinking water quality for poultry*, publication No. PSAnd T No 42 North Carolina Cooperative Extension Services, Raleigh, NC, USA.

Cooper, H. H. & Jacob, C. E. (1946). A generalized graphical method for evaluatingFormation constants and summarizing well- field history. *Trans Amer. Geophysical Union, 27*, 526-534.

Custodio, E.& Llamas, E. (1983). *Hydrologia Subterranea (Hydrogeology)* 2nd Edn Omega Barcelona, Spain.

Doneen, L.D. (1964).*Notes on water quality in agriculture*. Published as a water science and Engineering paper 4001 Dept of water science and engineering, Univ. of California.

Eaton, F.M. (1950). Significance of carbonate in irrigation waters. *Soil Sci., 67* (3), 128-133.

Felmy, A.R., Garvin, D.C.& Jenne, E.A. (1984). *MINTEQ A computer program for Calculating aqueous geochemical equilibrium.* NTIS PB 84-157148(EPA-600/3-84032) Nat. Tech. Inf. Ser. Springfield.

Freeze, R.A.& Cherry, J. A. (1979).*Groundwater* Prentice-Hall, Englewood Cliffs, NJ, 604pp.

Garcia, M.C., Hidalgo Model, V., Apella, M.C.& Blessa, M.A. (1998a). *Chemical quality of deep groundwater in Yerba Buena and San-Miguel de Tucuman Cities (Abstract),*vol 2, School of Environment Sciences and Technology, Buenos-Aires 498p.

Garcia, M.G., Hidalgo Mdol, V., Apella, M.C.& Blesa, M. (1998b).*Chemical Characteristics of deep groundwater in the cities of San Miguel de Tucuman and Yerba Buena Province of Tucuman in Proc 10th Latin-American Geological Congress and 6th Natural Congress of Economic Geology*, Buenos-Aires, Pub.,*3*, pp. 342 – 347.

Galindo M.C., Vece, M.B., Perondi, M.E., Moserrate Araoz, M., Garcia, M.G., Hidalgo-Medel, V., Apella, M.C.& Blesa, M.A. (2001). Chemical behaviour of the Sali-River Province of Tucuman, Argentina. *Environ Geol., 40* (7), 847 – 852.

Gupta, S.K.& Gupta, I.C. (1987). *Management of saline soils and water*. Oxford and IBH Publication Coy New Delhi India, pp. 399.

Hamlat, A.& Guidoum, A. (2018). Assessment of groundwater quality in a semiarid region of Northwestern Algeria using water quality index (WQI) *Appl Water Sci.*, (8), 220. http://doi.org/10.1007/s13201-018-0363-y.

Harleman, D.R.E., Mehlhorn, P.E.& Rumer, R.R. (1963). Dispersion-Permeability corelation in porous media *Jour Hydraulic Div. Amer Soc. Civ Eng. NO HYZ*, Vol. *89*, 67-85.

Hazen, A. (1893). *Some physical properties of sands and gravels.* Mass State Board of Health, 24th Annual Report.

Jenne, E.A. (1979). Chemical modeling-goals, problems, approaches and priorities. *American Chemical. Society.*, 3-12.

Jezersky, Z. (2007). Hydrogeochemistry of a deep gas-storage cavern, Czech Republic. *Hydrogeology Journal*, Volume *15*(3), 599–614.

Joarder, M.A., Raihan, F., Alam, J.B. & Hasanuzzaman, S. (2008). Regression analysis of Groundwater quality data of Sunamjang District, *Bangladesh International Jour. Environ Research*, *2*(3), 291-296.

Joshi, D.M., Kumar, A.& Agrawal, N. (2009). Assessment of the irrigation water quality of River Ganga in Haridwar District India. *Jour Chem*, vol *2* (2), 285-292.

Kelly, W.P. (1963). Use of saline irrigation water. *Soil Sci.*, *95*(4), 355-391.

Logan, J. (1964). Estimating transmissivity from routine production tests of water wells. *Groundwater*, vol *2*, no 1, 35-37.

Masch, F.D.& Denny, K. J. (1966). Grain-size distribution and its effect on the permeability of unconsolidated sands *Water resources Research*, Vol *2*, 665-677.

Mclean, W., Jankowski, J.& Lavitt, N. (2000). Groundwater quality and sustainability in analluvial aquifer, Australia. In: Silio, O (eds) *Groundwater, Past achievements and future challenges.* AA Balkema, Rotterdam, pp. 567 – 573.

Millero, J. (2001). Speciation of metals in natural waters *Geochemical Transactions, 8.*

Obiefuna, G.I.& Orazulike, D.M. (2010a). Physicochemical characteristics of groundwater Quality from Yola area Northeastern Nigeria *Journal of Applied Sci. and Env. Mgt.*, *14* (1) www.bioline.org.br/ja.

Obiefuna, G.I.& Orazulike, D.M. (2010b).Assessment of the groundwater quality of Yola Area for irrigation purposes Water Resources *Journal of the Nigerian Association Of Hydrogeologists*, vol *20*, Nos 1-2.

Obiefuna, G.I.& Orazulike, D.M. (2010c). Geology and hydrogeology of groundwater Systems of Yola Area, Northeast, *Nigeria Journal of Mining and Geology*, vol *46*, Nos 1-2.

Obiefuna, G. I.& Orazulike, D. M. (2010d). Chemical Speciation of Some Metal Ions in Groundwater of Yola Area, Using Geochemical Model *Journal of Applied Sciences and Environmental Management*, Vol. *14*(2), 65-70.

Pannatier, E. G., Broers, H. P., Venema, P.& Van Beusekom, G. (2000). *A new Process – Based Hydrogeochemical classification of Groundwater: application to the*

Netherlands National Monitoring System. INO report, NITG 00 – 143-B, 38, NITG,Utrecht.

Parkhurst, D.L.& Appelo, C. A.J. (1999). User's guide to PHREEQC (version 2) - A Computer program for speciation, batch-reaction, one-dimensional transport, and inverse geochemical calculations. *Water Resources Investigation*, Report 99-4259, Denver, USA.

Peters, S.W. (1982).*Central West African Cretaceous Tertiary benthic foraminifera and stratigraphy Palaeontographica Abt, 179*, 104.

Pindyk, R.S. & Rubinfeld, D.L. (1997). *Econometric models and economic forcasts*, 4th ed. Irwin McGraw-Hills Coy USA.

Raghunath, I.M. (1987). *Groundwater.* Second edition Wiley Eastern Ltd New Delhi India.

Raihan, F.& Alam, J.B. (2008). Assessment of ground water quality in Sunamganj Bangladesh *Irianian Jour. Of Env. Health Sci. and Eng.*, Vol6 (3), 155-166.

Richards, L.A. (1954). Diagnosis and Improvement of saline and alkali soils. *Agric Handbook, 60*, USDA and IBH Pub. Coy Ltd New Delhi India, pp. 98-99.

Robinove, C.J., longfort, R.H.& Brook, J.W. (1958). Saline water resource of NorthDakota US Geol. *Surv. Water Supply Paper, 1428*, p72.

Saatsaz, M., Sulaiman, W.N.A., Eslamian, S & Mohammadi, K (2013).Hydrogeochemisry and groundwater quality assessment of Astaneh-Kouchesfahan Plain, *Northern Iran Int. Jour. Water* Vol 6 (2) 44-65.

Seluakumar, S., Chandrasekar, N.& Kumar, G. (2017).*Hydrogeochemical characteristics and groundwater contamination in a rapid urban Water Res. Ind., 17*, 26-33.

SPSS Inc. (2005). *Website for SPSS Inc.*, http://www.spss.com. Assess December 2009.

Stuyfzand, P.J. (1986). A new Hydrochemical classification of water Types: Principles andapplication to the coastal dunes aquifer system of the Netherlands. *Proceedings of the 9th Salt water Intrusion Meeting*, Delft pp. 641 – 655.

Tawfic, T.A., Blaylock, M.J.& Vance, G.F. (1997). Selenite adsorption and solution Complexes in reclaimed mine soils: A comparative study using geochemical models *International conference on: Water management, salinity and pollution control towards sustainable irrigation in the Mediterranean region*, Bari Italy.

Theis, C. V. (1935). The relation between the lowering of the piezometric surface and the rate And duration of discharge of well using ground-water storage. *Trans AM Geophys Union, 16*, 519-524.

Todd, D. K. (1995).*Groundwater hydrology*. J. Wiley and Sons Singapore.

Uma, K.O., Egboka, B.C.E.& Onuoha, K.M. (1989). New Statistical Grain-size Method of Evaluating the Hydraulic Conductivity of Sandy aquifers of Ajali Sandstones Formation *Jour of Hydrology, 108*, 343-366.

Uma, K.O.& Leohnert, E.P. (1994). Hydraulic conductivity of shallow sandy aquifers: Effects of Sedimentologic and diagenetic differences. Environmental Geology

Springer Verlag USEPA (2003) *National primary drinking water regulations USEPA* Washington, DC.

U S Salinity Laboratory Staff. (1954). *Diagnosis and Improvement of Saline and Alkali Soils U S Dept of Agric. Handbook, 60.*

Walraevens, K. (1990). Hydrogeology and Hydro-geochemistry of the Ledo-Paniselian semi-confined aquifer in East and West Flanders. *Acad Analecta*, *52* (3), 11 – 66.

Walraevens, K., Mahauden, M. & Coetsiers, M. (2003). Natural background Concentrationof trace element in aquifers of the Flemish region, as a reference for thegovernment sanitation policy. FZK/TNO Conference on contaminated soil (Consoil), *Ghent*, pp. 215-224.

WHO. (2006). *Guidelines for drinking-water quality: recommendations*, 1st addendum to 3rd edn, vol *1*, World Health Organization, Geneva.

Wilcox, L.V. (1950.) *Classification and use of Irrigation waters*. United States Dept. of Agric circular, No 696, Washington D.C. pp16.

In: Advances in Hydrogeochemistry Research
Editors: S. Saeid Eslamian and Faezeh Eslamian
ISBN: 978-1-53617-829-6

Chapter 10

IMPACT ON WATER RESOURCES BY HYDRAULIC FRACKING ASSOCIATED WITH COALBED METHANE PRODUCTION

David Steffy*

Jacksonville State University, Alabama, US

ABSTRACT

The regional aquifers in the vicinity of coalbed methane (CBM) production can be impacted by the improper injection of hydraulic fracturing (fracking) fluid. The activity leds to an unsustainable extraction and contamination of fresh water in a watershed. This investigation of the Cahaba River Watershed in Alabama is presented as a case study. The watershed currently experiences an average of 200,000 barrels (35,000 m^3) of fracking fluid per production well that is injected, and an estimated 4,000,000 barrels (358,000 m^3) of wastewater disposed of per well in the same watershed. The composition of the fluid released through the permit with the Oil and Gas Board of Alabama (OGB) is given in broad generic terms rather than specific chemical composition and concentration. The retrieval of only 60% of the injected fluid (flowback water) leaves 40% of the fluid remaining in the subsurface. Wastewater generated with the fracking process is disposed by either a permitted Type II injection well or a surface water release permit. These disposal sites are utilized by a company's accumulated wastewater from multiple wells. The unknowns are the composition of the wastewater and the assurance that the injection point is isolated from local water resources utilized by humans. Both federal and many state guidelines are lacking or are vague that govern these activities. Finally, the ongoing CBM production sites impose increasing stress on the remaining freshwater resources in the watersheds by the withdraw of millions of gallons of groundwater and surface waters.

Keywords: coalbed methane, wastewater injection, hydraulic fracturing

* Corresponding Author's Email: dsteffy@jsu.edu.

INTRODUCTION

Thirty-seven nations were evaluated for coalbed methane production potential by the U.S. Environmental Protection Agency [2015]. Eleven countries presently produce 1.635 Trillion cubic meters (TCM) per year from virgin (unmined) coalbeds, and 24 countries have an estimated 154.3 TCM of proven coalbed methane potential. Fracking of the coalbed is used to improve CBM production is reportedly being used in 3 countries, and the use of this technology is being adapted in a handful of other countries. The potential for expanding use of this technology will definitely grow, and the known environmental problems will grow with its use.

The environmental problems caused by fracking are many including the contamination of air, water, soil, and humans. This study will focus on the impact of water resources of small regional watersheds impacted by fracking coalbeds to improve the extraction of methane trapped in the coal.

The Energy Information Agency (EIA, 2017] states that Alabama has 6,044 wells producing natural gas in 2015. This accounts for 1.1% of total U.S. production in 2015. Alabama extracted 5.8% of the 1.269 billion cubic feet (BCF) or 35.9 (BCM) of methane produced nationally from CBM in 2015 (Figure 1). Alabama's largest producer of CBM is the Black Warrior Basin, which has an estimated proven reserve of 1,893 trillion cubic feet (TCF) or 53.6 TCM [EIA, 2007].

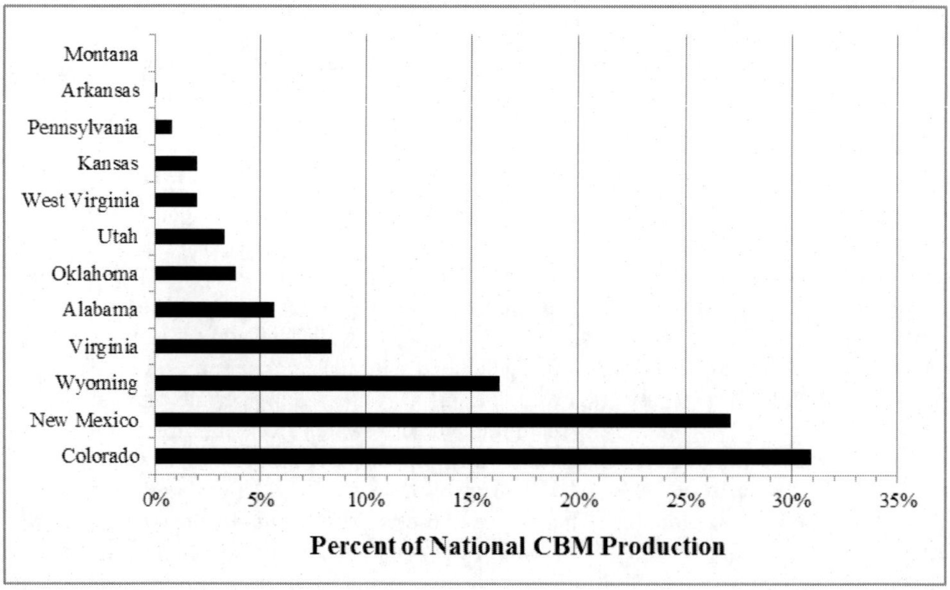

Figure 1. Percent of CBM national production for the top twelve states in 2015 [BALLOTPEDIA, 2017].

Water and wastewater management of fracking of the source rock layers in the oil and gas industry is an ongoing problem that generates risk to humans and the watershed environment. Not fully understanding and quantifying these risks generated by the extraction and disposal are part of the concerns. Regulations established by federal and state agencies that are intended to alleviate impact of these water resources are unevenly applied and led to future unsustainable water requirements. These problems are presented by a case study of fracking coalbeds present in the Cahaba River Watershed, Alabama.

Fracking involves injecting a chemical solution and a proppant under high pressure through perforations of the well casing into the rock layers. The proppant that maintains the openings generated providing an increased permeability for the rock layer that is judged to contain carbon-based resources. The proppant can be either: natural sand; man-made sand; resin-coated sand; or sintered bauxite [Smith and Montgomery, 2015]. Roostaei [2017] describes the injection of proppant into a fracture a three-phase process: (1) fluid is injected through well perforations into the desired rock layer under high pressure, which will extend or generate new fractures; (2) proppant materials and designed chemicals are injected to maximize the filling of the fractures to maintain the openings; and (3) removal of proppant and fluids remaining in the drill stem, and the injected hydraulic fluids. The efficiency of this process, in part, can be measured by the amount of backflow produced by stage 3.

Alabama's CBM wells are primarily vertical and directional wells designed to improve the extraction of methane gas associated with layers of coal. After the producing wells are drilled, and the well casings were then perforated and fracked to access known coal seams to improve the CBM production [Smith and Montgomery, 2015; Thakur, 2017]. The depths below surface of the perforations reported on the OGB of Alabama permit are assumed to be the vertical extent of the coal seams encountered by the CBM wells.

The focus of this study is the CBM production from the Pottsville Formation deposited in the Black Warrior Basin, which is also present in the Ridge and Valley Province. These coalbeds are found have an average permeability of 10 to 25 mD (millidarcy) [Thakur, 2017]. In contrast, oil and gas wells in the basin commonly have a horizontal well component, which extends through the potential unconventional and conventional hydrocarbon reservoir rocks with lower permeabilities (< 1 mD) [King, 2012].

The delivered fracking fluids are designed to optimize this process for a particular rock formation. This optimization results in a wide variety of chemical mixtures, which combines with the carbon-rich source rocks. After the injection is complete and fractures are formed, much of the injected water is recovered from the rock and is handled as flowback wastewater [Arnauld, 2015]. Flowback wastewater is a combination of chemicals added to promote the fracking process and natural compounds such as salts, metals, and radioactive materials. Subsequently, a smaller amount of "produced water" is also recovered as wastewater with hydrocarbon extraction. Improper handlings of these wastewaters have caused environmental problems.

Table 1. Fracking fluids used, their purposes, and resulting activity which were identified by FracFocus [2017] are presented

Additive	Purpose	Downhole Result
Acid	Help dissolve minerals and initiate cracks in the rock	Reacts with minerals present in the formation to create salts, water, and carbon dioxide (neutralized)
Acid/Corrosion Inhibitor	Protect casing from corrosion	Bonds to metal surfaces (pipe) downhole. Any remaining product not bonded is broken down by micro-organism and consumed or returned in produced water.
Biocide	Eliminates bacteria in the water that can cause corrosive by products	Reacts with micro-organism that may be present in the treatment fluid and formation. These micro-organism break down the product with a small amount of the product returning in produced water.
Base Carrier Fluid (water)	Create Fracture Geometry and Suspended Proppant	Some stays in formation while remainder returns with natural formation water as "produced water" (actual amounts returned vary from well to well)
Beaker	Allows a delayed break down of gels when required.	Reacts with the "crosslinker" and "gel" once in the formation making it easier for the fluid to flow to the borehole. Reaction produces ammonia and sulphate salts which are returned in produced water.
Clay and Shale Stabilization/control	Temporary or Permanent Clay Stabilizer to lock down clays in the shale structure	Reacts with clays in the formation through a sodium - potassium ion exchange. Reaction results in sodium chloride (table salt) which is returned in produced water. Also replaces binder salts like Calcium, Chloride helping to keep the formation in tact as the Calcium Chloride dissolves.
Crosslinker	Maintains viscosity as temperature increases	Combines with the "breaker" in the formation to create salts that are returned in produced water
Friction Reducer	Reduces Friction effects over base water in pipe	Remains in the formation where temperature and exposure to the "breaker" allows it to be broken down and consumed by naturally occurring micro-organisms. A small amount returns with produced water.
Gel	Thickens the water in order to suspend the proppant	Combines with the "breaker" in the information thus making it much easier for the fluid to flow to the borehole and return in produced water.
Iron Control	Iron chelating agent that helps prevent precipitation of metal oxides	Reacts with minerals in the formation to create simple salts, carbon dioxide and water all of which are returned in produced water
Non-Emulsifier	Used to break or separate oil/water mixtures (emulsion)	Generally returned with produced water, but in some formations my enter the gas stream and return in the produced natural gas.
pH Adjusting Agent/Buffer	Maintain the effectiveness of other additive such as crosslinkers	Reacts with acidic agents in the treatment fluid to maintain a neutral (non-acidic, non-alkaline) pH. Reaction results in mineral salts, water and carbon dioxide which is returned in produced water.
Propping Agent	Keeps Fractures Open allowing for hydrocarbon production	Stays in formation, embedded in fractures (used to "prop" fracture open)
Scale Inhibitor	Prevent Scale in Pipe and Formation	Product attached to the formation downhole. The majority of product returns with produced water while remaining reacts with microorganisms that break down and consume the product.
Surfactant	Reduce Surface tension of the treatment fluid in the formation and helps improve fluid recovery from the well after the frac is completed	Some surfactants are made to react with the formation, some are designed to be returned with produced water, or, in some formations they may enter the gas stream and return in the produced natural gas.

A controversial method of disposal for this wastewater is by the use of injection wells. The intent is to inject the wastewater into an isolated rock layer that will presumably not contaminate future water supplies. CBM production companies tend use one saltwater disposal (SWD) well for disposal of their wastewater generated from multiple producing wells in the coal field, or a surface discharge under a National Pollution Discharge Elimination System (NPDES) permit issued by the OGB of Alabama. Regulations are established by some states that place restrictions on the final disposal method allowed.

The coal industry utilizes an EPA Type II injection well in which retrieved fluids and wastewater are injected into the subsurface rock layer for disposal. The EPA requires that Type II injection wells be drilled to the deepest extent to assure that the most toxic waste is injected far below drinking water aquifers. The words "far below" are not quantified in the regulations. This type of well is supposed to be walled with multiple layers of steel casing and cement grout which will be regularly monitored for cracks.

In a ProPublica study of Type II injection wells, Lustgarden [2012] reported after a "review of well records, case histories and government summaries of more than 220,000 well inspections found that structural failures inside injection wells are routine." He states that 17,000 violations of regulatory requirements for Type II injection wells occurred between 2007 to 2010, which is a 7.7% rate of violation. Also, one integrity violation was noted for every 6 deep injection wells. A 17% rate of possible leakage of the injection well for the designed system overseen by the EPA. The study did not discriminate between those wells are supporting CBM activity and other oil and gas waste disposal activity.

This study examines the fracking activity associated with CBM production in Alabama and its potential risk of damaging the water resources of the Cahaba River Watershed. Between 2000 and 2010, approximately 8,401 wells have been fracked in the Black Warrior Basin known to produce CBM and oil and natural gas [Gallegos and Varela, 2014]. This investigation focused on CBM production from the Gurnee Coal Field which is a portion of the Black Warrior Basin located within the Cahaba River Watershed.

Fracking promotes the access to and extraction of natural gas from the Gurnee Coal Field. Production of hydrocarbons from "tight" sedimentary rock deposits or unconventional shale deposits is improved by drilling and then fracking a well which penetrated the rock layers that serve as reservoirs of the hydrocarbons. According to King [2012], "tight" is a petroleum engineering term applied to sedimentary rock that has restricted permeability measuring 0.0001 to 1 mD. The permeability of a deposit is its capacity to transmit fluid, which is dependent on pore size and shape, and their interconnections. Shale rock is an unconventional natural gas source with permeability less than 0.0001 mD. By fracking the sedimentary rock deposit, man-made fractures are generated to provide additional pathways for the movement of the hydrocarbons towards the future production well. In contrast, coal is a sedimentary rock containing varying amounts of biogenic carbon with adsorbed methane gas, which needs to be extracted before mining of the coal is commenced. If sufficient quantities of gas are present, then the gas is sold as a natural resource.

Fracking of coalbeds in the Black Warrior Basin commenced in the 1990's as a way to increase the methane production from the Pottsville Formation [Milici, 2002]. The formation is of lower and middle Pennsylvania age (320 to 300 mya (million years ago)) sedimentary rock including 3 to 14 coal seams of various thicknesses. The basin extends across two physiographic provinces: the Appalachian Plateau; and the Ridge and Valley [Pashin and Groshung, 1998]. The Gurnee Coal Field is located in the Ridge and Valley Province and has a distinct tectonic history and a resulting structural character, which influences the CBM production efficiency.

The fracking fluids include a variety of chemicals each having a specific purpose, which could include: friction reducer, surfactant, biocides, thickeners, scale inhibitor, hydrochloric acid, and a corrosion inhibitor [King, 2012]. Table 1 is a compilation of all possible generic forms of fracking chemicals used.

EPA [2015] in a preliminary draft assessment determined that fracking does not affect the water resources of the U.S. In this report, EPA's analysis from a voluntary database of the petroleum industry's fracking activity results in water pollution in a small number of wells drilled. This small number of problem-well sites is infrequent, and EPA determined that fracking should not be considered as a regional source of water pollution, or indicative of the petroleum industry's standard.

Fracking fluids are a mixture of water, chemicals, and proppant (sand). The sand maintains the fractured openings that are generated by the fracking process [King, 2012]. The exact composition of the fluid is dependent on the company and their understanding of the geologic formation to be fractured. The solution includes a variety of chemicals each having a specific purpose, which could include: friction reducer, surfactant, biocides, thickeners, scale inhibitor, hydrochloric acid, and a corrosion inhibitor [King, 2012].

Geological and Hydrological Settings

Five geological provinces listed in Table 2 are located in Alabama (Figure 2). The provinces are distinguished by the occurrence of either conventional, and unconventional reservoir rock, which include coalbeds having a natural gas resource potential. These provinces are separate from the traditional or historically developed oil and gas reservoirs along the Gulf Coast (Table 2). These provinces are drilled and explored to different levels depending upon their level of success.

The following rock sequences are important gas and oil producers found in the Black Warrior Basin province.

Table 2. Regions of Alabama are classified by geological provinces and management areas of the Alabama Oil & Gas Board

Physiographic Region	Geological Formation*	Oil & Gas Board Management Designation
Black Warrior Basin	Devonian Chattanooga Shale	Black Warrior
	Mississippian-Devonian Floyd-Chattanooga Shale Pennsylvanian Pottsville Coal	Coal Bed Methane
	Shale Basin Defined by ETA	
Valley & Ridge	Pennsylvanian Pottsville Coal Shale Basin Defined by ETA	Appalachian
	Cambrian Conasauga Formation	

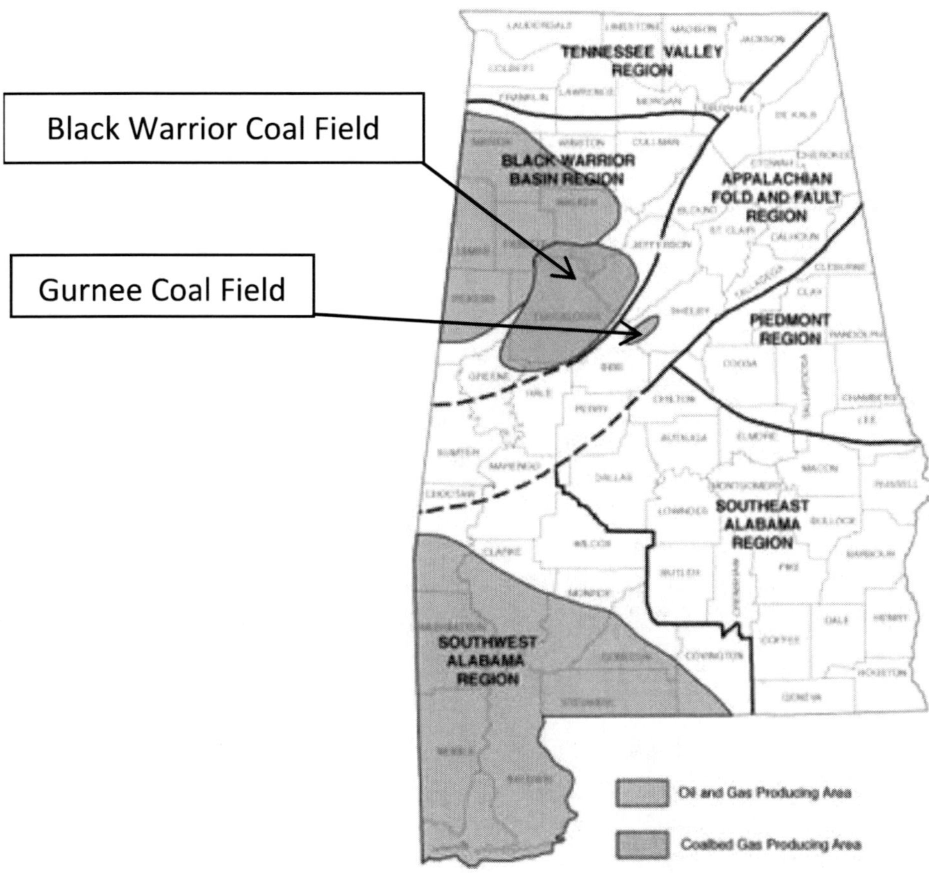

Figure 2. Oil and Gas Regions of Alabama, the location of Black Warrior Coal Field, and the Gurnee Coal Field [Alabama Oil and Gas Regions, n.d.].

Chattanooga Shale

Also known as the Ohio Shale, the Chattanooga Shale is generally targeted at a depth of 3,000 to 5,000 feet. It covers a large area from Kentucky to New York, and it has an estimated total resource of 12.2 Tcf [trillion cubic feet] of natural gas. It was initially believed that the Chattanooga was only in Tennessee and Kentucky but was later found to be an extension of the Appalachian Basin. In Alabama, it is referred to as the Black Warrior Basin. This shale has thicknesses in excess of 1,000 feet in Kentucky but thins to the south as much as 80 feet in some areas of Tennessee. Companies investing in the Chattanooga Shale are Atlas Energy, GeoMet, Range Resources, Energen, and Chesapeake.

Floyd Shale

The Black Warrior Basin Region north and west of Birmingham is a region which is underlain by the Lower Pennsylvanian Pottsville Formation and the Mississippian age Floyd-Neal shale. This shale is located in the Black Warrior Basin in Mississippi and northwest Alabama. This is still considered a natural gas prospective play.

Neal Shale

Located in the lower part of the Floyd Shale, the Neal Shale is a prospective natural gas play in Alabama and Mississippi. Given its lack of thickness, it may only be possible for horizontal wells drilled to provide commercial wells in this play.

Pottsville Formation

The Pottsville Formation is a Lower and Middle Pennsylvania age (318 to 298.8 mya) and is the source of CBM in the Gurnee Coal Field (Figure 2). The Pottsville is formed by up to 10 cyclothems of siltstones, shales, claystones, coals, and massive sandstones, which were deposited in the foreland Black Warrior Basin located west of the Southern Appalachian Mountains [Pashin, 2015, et al.; EPA, 2004]. There are more than five coal groups which produce CBM. This study will focus on the Mary Lee Coal Group which is a CBM producer in the Gurnee Coal Field located in Bibb and Shelby Counties, Alabama (Figure 3). The Mary Lee Coal Group consists of Newcastle, Mary Lee, Blue Creek, and Jagger seams (Karacan, 2012). Milici [2000] reports that Shelby County produced 3,874,186 MCF [million cubic feet] (110,000 MCM) (3.5%) and Bibb County produced 583 MCF (16.5 MCM) (0.0005%) of Alabama's 1.076 TCF (30.5 MCM) of CBM up to the year 2000. By the year 2000, however, there were only 25 producing wells of the 450 wells drilled in Shelby County, and there were no actively producing wells of the 3 drilled in Bibb County by the year 2000.

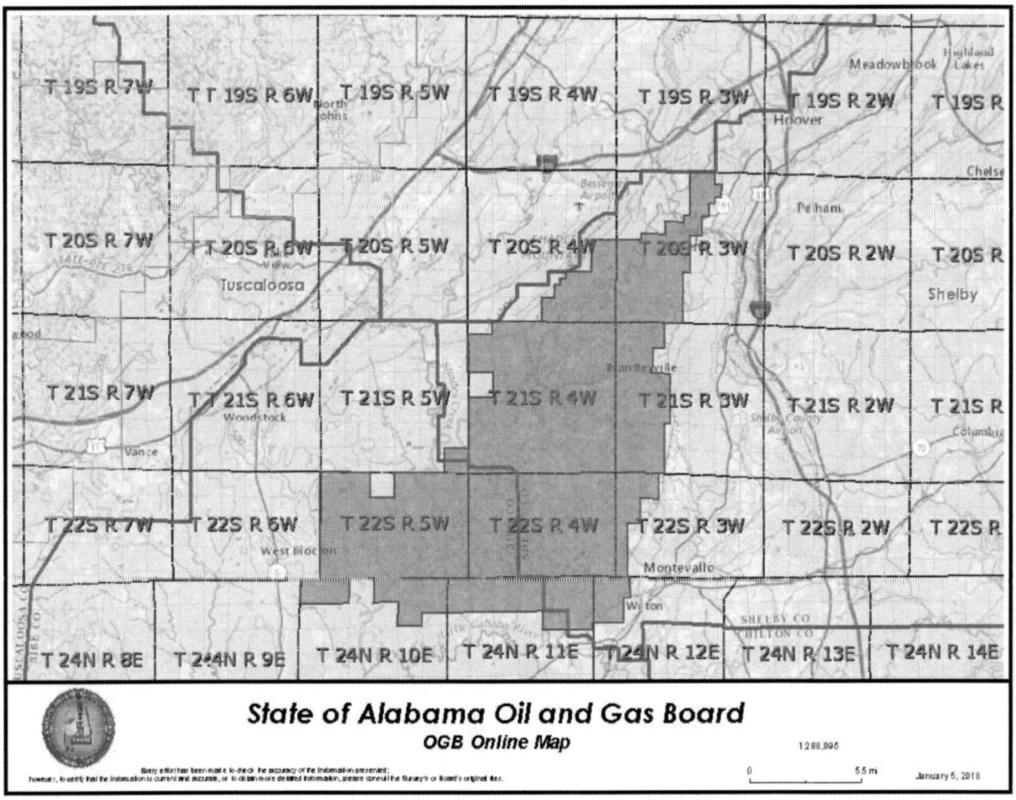

Figure 3. An Oil and Gas Board of Alabama location of the regulated Gurnee Coal Field.

The Gurnee Coal Field is the eastward extension of the Black Warrior Basin that occurs within the physiographic Appalachian Ridge and Valley Province of central Alabama. Figure 3 depicts the active OGB permits that cover the field. These 80-acre lease blocks issued for the construction of an exploration/production well targeting the underlying Pottsville Formation coal seams.

The layers of Pennsylvanian-age sedimentary rock interbedded with the coal seams were originally deposited in a foreland basin west of the Southern Appalachian Mountains [Pashin, et al., 2015]. These deposits were structurally deformed by compressional forces during the Appalachian orogeny between 325 to 260 mya into broad open folds and faults, which seems to have controlled the permeability and the gas and water production [Pashin, et al., 2015]. The coal seams may have acted as decollement surfaces at depths of 10,000 ft, which thrust faults deformed the layered coal seams [Pashin, et al. 2005, Figure 4]. The Gurnee Coal Field is located in the northeast-southwest trending Cahaba synclinorium.

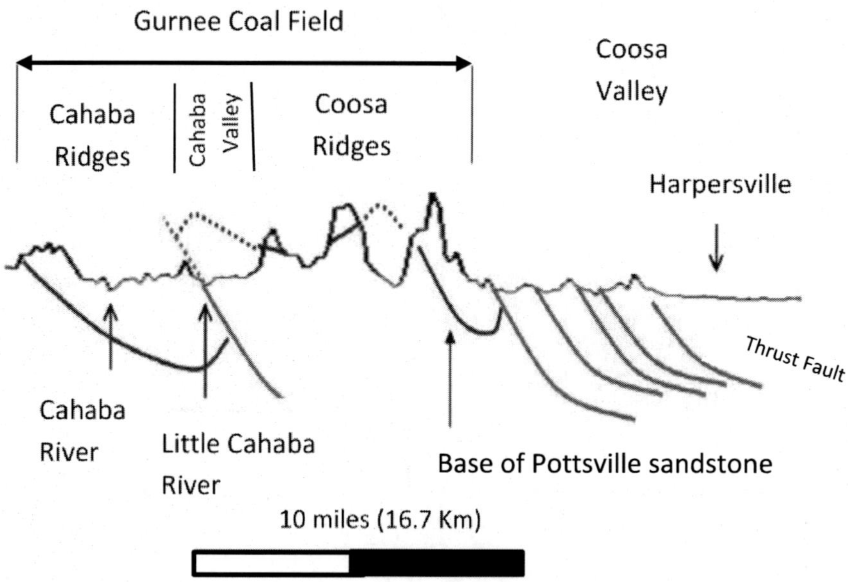

Figure 4. A northwest-southeast generalized cross-section illustrating the Gurnee Coal Field location in the Ridge and Valley Province, which is adopted from Neilson [2016]. The figure has an unknown amount of vertically exaggerated. Thrust faults (red) directed towards the northwest were formed by compressional tectonic force.

Water Resources of Bibb and Shelby Counties

The Gurnee Coal Field lies within the Cahaba River Watershed that drains southward for approximately 190 miles to its confluence with the Alabama River Subbasin. The Alabama River Subbasin is a part of the Mobile River Basin which empties into the Gulf of Mexico. The Cahaba River Watershed is a narrow (10 – 15 miles wide) and occupies 1,827 square miles of a southwest-trending valley bounded by mountain ridges formed by thrust faulting of Paleozoic rocks.

Kopaska-Merkel, et al. [2005] describe the Cahaba River Watershed as underlain by the Valley and Ridge aquifer system. The system consists of discontinuous rock layers that are faulted and folded limestone and sandstone that range in age from Cambrian to Pennsylvanian (545 to 300 mya). They describe the Pennsylvanian-age Pottsville Formation as a significant groundwater source in the watershed. Recharge areas for the aquifer system are the surface exposures of the Pottsville Formation; and the overlying upper Cretaceous Tuscaloosa Group of poorly sorted clay, sand, and gravel to its west; and by Miocene-Pliocene age rocks to the east [Tew, 2018]. These Cretaceous sedimentary rocks (145 to 66 mya) are disconformably overlain by Quaternary sand, gravel, and clay (2.6 mya to present).

Recharge of the aquifers is by local rainfall that ranges up 54 inches/yr (Chaney, 2013). The surface exposed Pottsville Formation rocks provide most of the 79.21 million gallons a day (Mgad) (300,000 m^3/d) for public water supply, 0.40 Mgad (1500 m^3/d) for industrial, 0.42 Mgad (1600 m^3/d) for livestock, and 3.91 Mgad (14,800 m^3/d) for irrigation [USGS, 2005].

Water resource management in Alabama is overseen by the Office of Water Resource (OWR) in the Department of Economic and Community Affairs (ADECA). Groundwater contamination and special cases such as wastewater injection are controlled by the Alabama Department of Environmental Management (ADEM). The approved injection wells are registered with the OGB of Alabama as part of the fracking wells design and operation plan.

Alabama adopted "The Reasonable Use Rule" when addressing groundwater issues [McDonald, 2010]. The rule limits a landowner's use of water to those uses that have a reasonable relationship to the use of the overlying land. The rule is essentially the rule of absolute ownership with exceptions for wasteful and off-site use.

Alabama as of 2015 has permitted 264 Type II injection wells, some of which can be used to inject fracking fluids, or for storage of waste fluids produced by the fracking process [EPA, 2015]. The OGB of Alabama guidelines for Type II injection wells meets the minimum requirements set by EPA Safe Drinking Water Act (SDWA) Sections 1422 and 1425 [Geraci, M. et al., 2017]. The Gurnee Coal Field has been developed for CBM since the 1980s with numerous small coal mines continue to operate in the basin as well. Several CBM developers operate within the Cahaba River Watershed with GeoMet, Inc. and CDX Gas being two of the largest. This development crosses over the watershed boundary between the Cahaba River Watershed and to the west the largerBlack Warrior River Watershed. For example, GeoMet, Inc. utilized a discharge wastewater pipeline originating in the Gurnee Coal Field for disposal in the Black Warrior River Watershed (ADEM NDPEL Permit #AL0076252), which was retired in April 2017. Currently, GeoMet, Inc. and CDX both operate CBM pipelines which join the SONAT Bessemer Calera Pipeline and the Enbridge Pipeline.

Gurnee Coal Field CBM Production Rock Properties

Gas contained in coal beds is methane with trace quantities of ethane, nitrogen, carbon dioxide, propane, and hydrogen sulfide. Naturally formed properties of coal determine the amount of gas that can be recovered. Reservoirs of CBM porosities related to cleats and matrix porosity are responsible for the storage of gas. The matrix porosity of coal beds can vary from 10% - 20%; the cleat (fracture) porosity of the reservoir is estimated to be in the range of 0.1% - 1%.

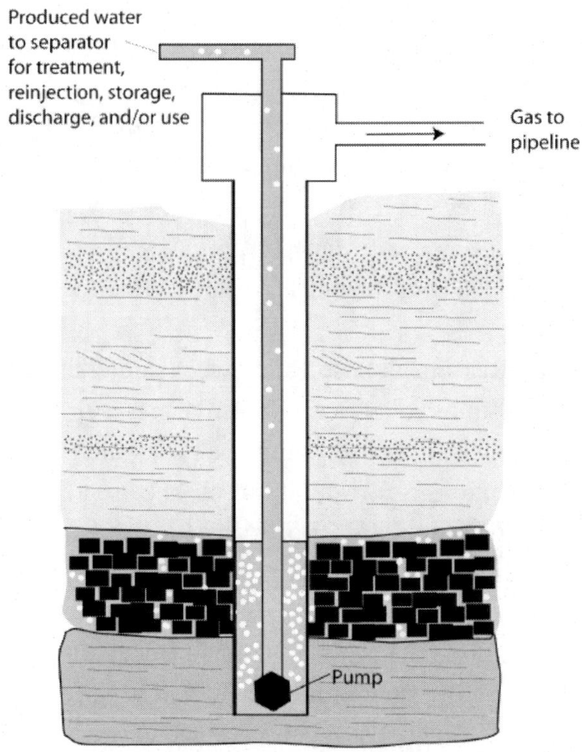

Figure 5. The separation of the adsorbed methane from the coalbed (black layer) occurs the when the pore water pressure is reduced as the water enters the well. The methane gas molecules (white bubbles) desorbs and enters the annular space and retrieved at the surface. The water is pumped to the surface and collected and is referred to as produced wastewater. Figure adopted from National Research Council [2010].

The adsorption capacity of coal depends on the type and its carbon fixation level. The range is 100 - 800 SCF/ton (standard cubic feet of gas measured at standard pressure and temperature per ton of coal) for most coal seams found in the US. Most of the gas in coal beds is in the adsorbed into the coal. When the reservoir is put into production, water in the fracture spaces is pumped off first. This leads to a reduction of pressure enhancing desorption of gas from the matrix.

The fracture permeability acts as the major channel for the gas to flow. The higher the permeability, the higher the gas production. For the Mary Lee – Blue Creek coal seams found in the Black Warrior Basin: the average depth is between 1200 – 2800 ft; the average gas content between 400 – 500 SCF/ton; and the permeability lies in the range of 10 - 50 mD [Thakur, 2017]. Coal formations in the eastern United States tend to have a lower water content resulting from, in part, a lower porosity than western coal formations. They also have more methane per ton of coal than coal from western states and can require fracturing to improve CBM production [EPA, 2010].

The permeability of fractured reservoirs changes with the stress applied by tectonic activity or by hydraulic fracturing. Coal displays a stress-sensitive permeability and this process plays an important role during stimulation and production operations. Fracture permeability in CBM reservoir tends to increase with depletion of gas; in contrast to conventional reservoirs. This unique behavior is because of shrinking of coal, when methane is released from its matrix, which results in opening up of fractures and increased permeability. It is also believed that due to shrinkage of coal matrix at lower reservoir pressures, there is a loss of horizontal stress in the reservoir which induces in-situ failure of coal.

METHODS

The data was collected from public online databases from both governmental and non-governmental organizations (NGOs). The data will be analyzed by developing spreadsheets that allow easy and comparable information that characterize the fracking activity used by company's production in the Gurnee Coal Field in the Cahaba River Watershed, Alabama. Comparison will be made between state and federal government databases about fracking fluid composition, and similar data from Australian coal fields.

To limit the number of variables in this study I focused on active CBM production wells from the Pottsville Formation in the Cahaba River Watershed. The completion of the well, stimulation (fracking), and production of gas and produced-water history of the wells are available on the OGB of Alabama. The amount of information provided for each well does vary, which makes comparisons less than complete. Similar limitations are provided by the EPA voluntary website concerning the chemical composition used for fracking.

Produced wastewater management by the oil and gas industry is considered the greatest risk to freshwater resources [Roswell and Reaven, 2012]. Therefore, this study focused on the volume of fracking fluid used, the flowback water, and produced water from these active. Particular attention was made on their composition and disposal methods of flowback and produced waters. This focus on the disposal method leads to our concern about environmental contamination of the water resources in the Cahaba River Watershed.

The management approach of hydraulic fracking was evaluated using the following information: location and depth of coal seams number of coal seams; gas generation; produced water; fracking fluid volume; chemical constituents of fracking fluids; and wastewater disposal using Type II SWD (saltwater disposal) injection wells. This evaluation considers aspects of gas generation near communities and their water resources which are at risk.

Data Collection

This study used several online databases containing fracking information about drilling activities in Alabama.

Alabama Geological Survey (https://www.gsa.state.al.us/) is charged with mapping the geological resources of the state such as energy and water. Exploration and production well permits are issued by the State Oil and Gas Board of Alabama (https://www.gsa.state.al.us/ogb/wells), which is an operational unit of the Alabama Geological Survey. Regulations promulgated by the Board approve well location, design of the well, and production units, and the production royalties to be required. Nationally, there are no federal standards for regulating fracking activity. Oil and Gas Board maintains a database of all permits issued, specific drilling designs, fracking undertaken, and the history of production.

FracFocus (http://fracfocus.org) is a NGO that provided an online database of fracking operation. The database is maintained by the Groundwater Protection Council and the Interstate Oil and Gas Compact Council. Its formation was funded by the U.S. Department of Energy, and oil and gas trade groups in 2011. The database is meant to compile all fracked well information including the composition of the fracking fluids being used [EPA, 2015]. Participation in this voluntary database varies between companies [Cook, 2015]. The exact composition of the fluid is dependent on the company and their understanding of the geologic formation to be fractured. By June 2017 there were 138 wells registered by FracFocus [2017] that are drilled in Alabama. The site was deactivated during 2018.

FracTracker Alliance (https://www.fractracker.org/) is a NGO supporting environmental studies of fracking and related operations in the U.S. [FracTracker, 2017].

The U.S. Environmental Protection Agency (EPA) (https://www.epa.gov/hfstudy) is the federal agency responsible for the collection, monitoring, and interpretation of water resource information needed to develop regulations that meet the requirements of congressional policy. This includes water and wastewater activities involving fracking of coalbeds for methane production. The regulations established are considered the minimum standards that all states have to meet in their regulations.

The U.S. Geological Survey maintains an extensive database on the occurrence; composition; and exploration and development of coal throughout the country. COALQUAL is the database that can accessed online at https://energy.usgs.gov/Coal/AssessmentsandData/CoalQuality.aspx.

TEDX (https://endocrinedisruption.org/) is a NGO research institute focused on the production and use of chemicals that interfere with healthy hormone function [TEDX, 2018]. They achieve their purpose by the systematic review of the U.S. National Toxicology Program to answer questions about the hazards chemicals pose. By partnering with other nonprofit organizations, they distribute information with the public, policy makers from the local to global level, government agencies, and health conscious businesses. Their primary focus is endocrine disruption, including chemicals utilized the oil and gas industry as part of their fracking process.

RESULTS

The voluntary listing of wells on the website FracFocus started in 2011. A review of the 186 wells in Alabama, indicated a significant drop in the 7 Alabama counties in this participation (Figure 6). This correlates with the total number of wells permitted by the Oil & Gas Board, which could be interpreted as FracFocus participation is dependent on the number of wells permitted. However, the percent of wells permitted that are registered on FracFocus decreased from 50% to 10% between 2011 to 2016 (Figure 7).

EPA [2004] described the relationship between the type of fractures developed by fracking as being dependent on the depth of fracturing. The shallower the depth of the coalbed being fracked, the more horizontal component of lithostatic pressure will be present. As a result, the more likely a vertical fracture will be present, and fracking will enlarge these fractures rather than producing horizontal fractures through the desired coal bed. These vertical fractures could endanger the overlying aquifer by inducing high-pressure injection of fracking fluids. In contrast fracking of deeply buried coalbeds will have a greater horizontal component as a result.

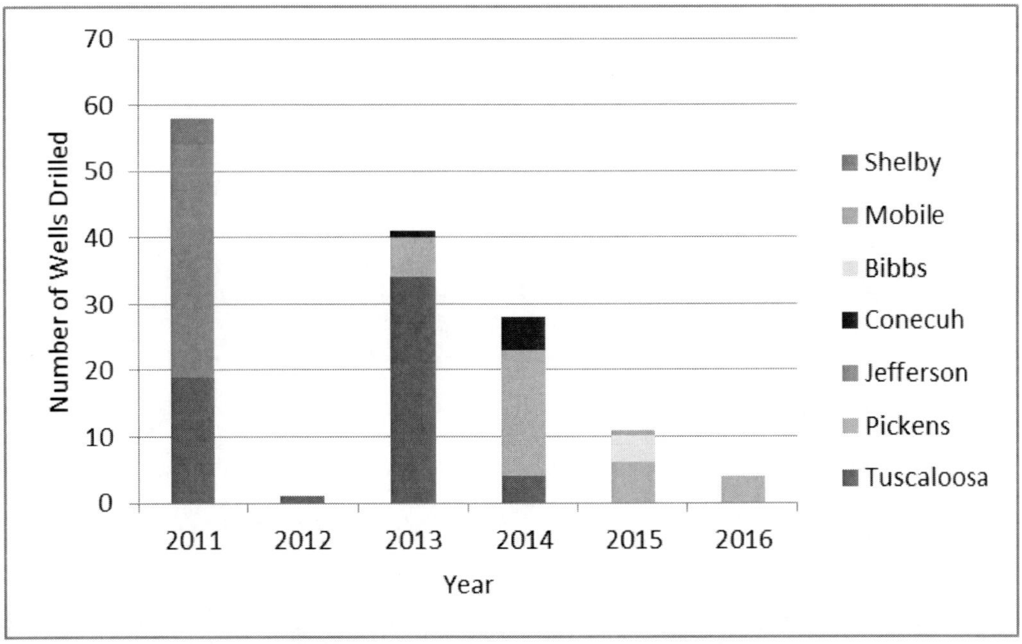

Figure 6. The number of wells drilled and fracked in 7 counties in Alabama listed by FracFocus.

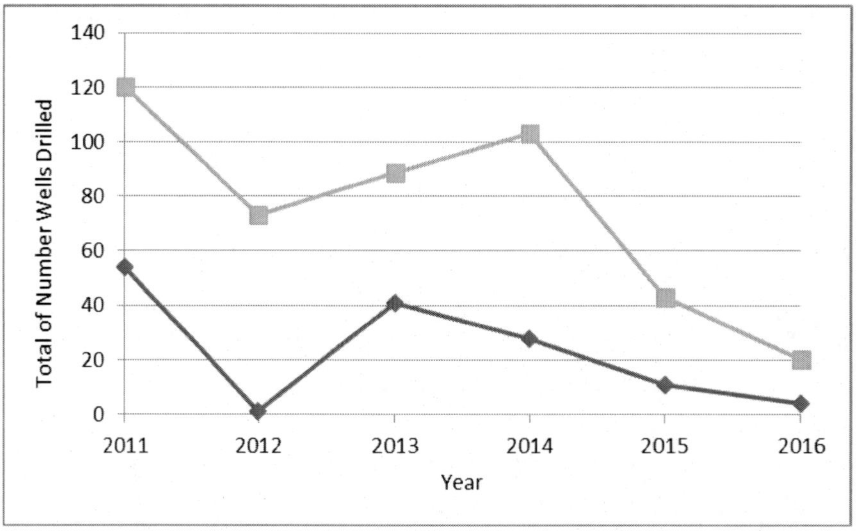

Figure 7. A comparison between the number of wells drilled and fracked in various counties in Alabama registered by FracFocus (diamond) and registered by the Alabama OGB (square).

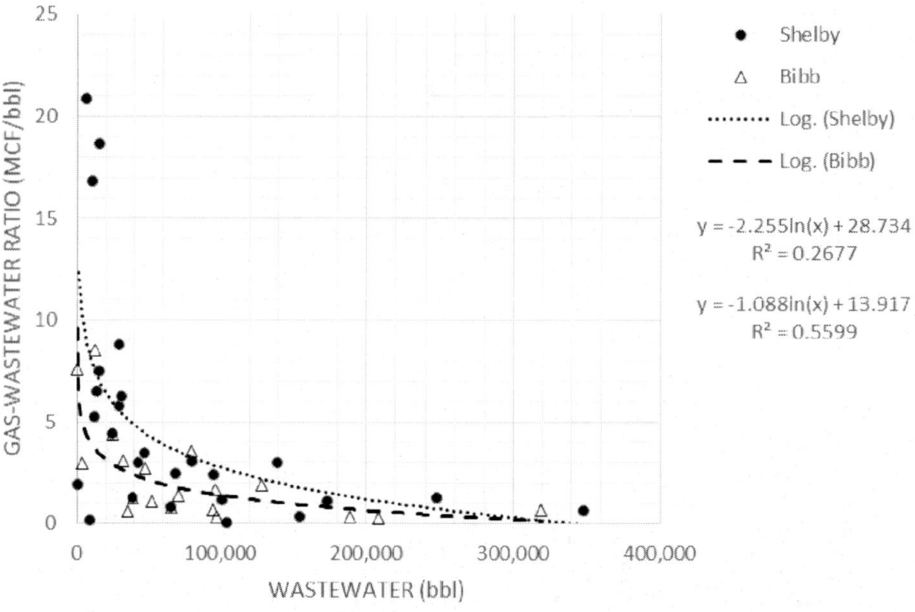

Figure 8. The CBM gas generation-produced wastewater ratio versus produced wastewater generated.

The Wastewater-Gas production ratio (bbl/MCF) from both Shelby and Bibb Counties demonstrate a logarithmic relationship to amount of wastewater produced (Figure 8). Although the relationship for neither is statistically strong. This study calculated that the Gurnee Field has an average ratio of 3.66 bbl/MCF (0.59 m^3/MCF). USGS (2000) states that the Black Warrior Basin, of which Gurnee Coal Field lies within, has an average ratio of 0.55 bbl/Mcf (0.08 m^3/MCF), and generates 58 bbl/day/well (9.22 m^3/day/well)

The average depth of currently producing CBM wells is 3,492 ± 460 ft below the surface. The cumulative thickness of the producing coal seams in a well is 194 ± 72 ft. The 52-active producing CBM wells in the Gurnee Coal Field have produced 3.8 million bbl (604,000 m^3) of wastewater based on data extracted from the OGB of Alabama online database.

Thakur [2017] reviewed the development of mathematical models of CBM production decline from multiple coal seams, which is the status seen in the Gurnee Coal Field. He states that exponential models are the most common for gas production wells and has a form of $q_t = q_i e^{-Dt}$, where q_t is the production rate at time t, q_i is the initial production rate, and -D is the rate of decline. This study calculated the -D value for each producing well in the Gurnee Coal Field for both Shelby and Bibb Counties.

Figure 9 displays the plot of CBM gas production as a natural log function of gas decline (D) for both Shelby and Bibb Counties. Production wells for Shelby County demonstrate a good linear relationship (R^2 = -0.8798), whereas Bibb County wells show no such correlation. The cause of the difference between the county's CBM production decline is not known. There are several possible reasons that could be involved such as: the production companies' approach to CBM development; the Gurnee Coal Field is tectonically complicated because of the thrust faulting history; and the lateral variation of the Pottsville Formation's coal seams. This investigation found no statistical differences between the counties with respect to: the Pottsville Formation's number of coal seams; total thickness of the coal seams being produced; the depth of coal seams; and the coal seams' permeabilities.

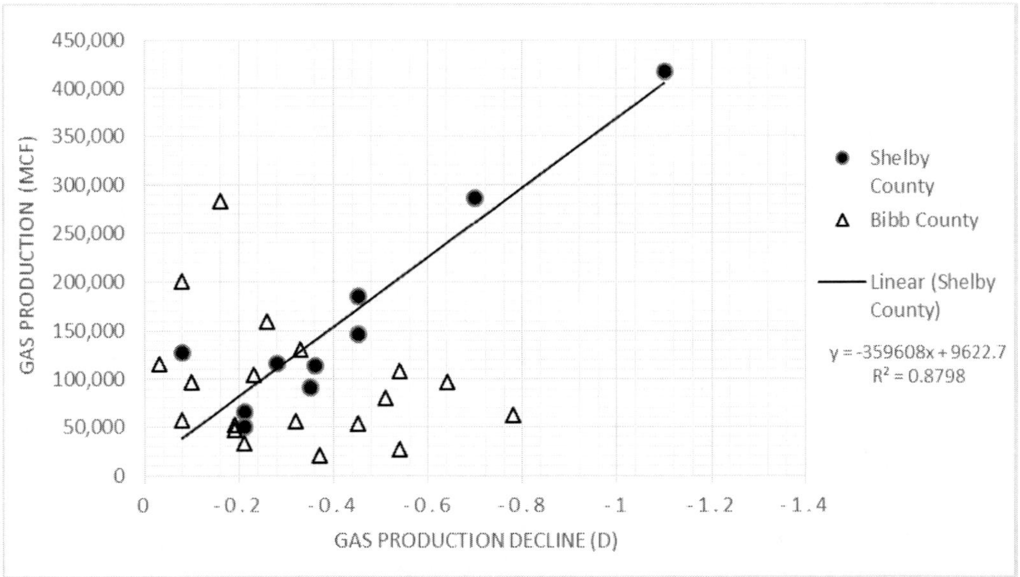

Figure 9. CBM gas production versus gas production decline (D) for both the Shelby and Bibb Counties. The linear correlation only applies to Shelby County wells.

Characteristics of Fracking Fluids Used in the Active Gurnee Coal Field

A problem common to all states faced with the regulation of fracking activity is how to manage the use and disposal of wastewater and retrieved fracking fluids. When mismanaged these fluids are known to contaminate the water resources.

The fracking of oil and natural wells is a process that involves injecting up to 3 to 5 million gallons (11,000 to 19,000 m^3) of high-pressure fluids into the subsurface deposit to generate openings for fluids and a proppant to enter. In contrast, information gathered from FracFocus retrieved from four CBM production wells drilled in Bibb County targeting Gurnee Coal Field (Table 6). The injection of 217,750 bbl (34,600 m^3) at high-pressure is the average volume of fracking fluid used, and an average volume of produced water retrieved during production is reported as 3,975,000 bbl (631,800 m^3).

The lack of participation by Alabama's oil and gas production companies in the federal voluntary disclosure database for fracking fluid use limits an accurate environmental assessment. In addition, there is also a limited public disclosure of fracking fluid composition released by the OGB of Alabama permits. Also, the release of wastewater generated by the fracking operation is also limited in both scope of quantity, quality, and method of disposal.

It is estimated that only 60% of the fluids injected are retrieved [WaterWorld, 2015]. The FracFocus database for current producing wells in the Gurnee Coal Field use fracking (Tables 4 and 5). Knowing the composition of fluids residing in the subsurface, and their location with respect to regional groundwater flow patterns are key characteristics required to assess risk.

Table 3. General components of fracking fluids listed on FracFocus used in 4 CBM wells, and the same wells registered with the OGB. The wells are drilled into three coal seams of the Pottsville Formation located in Bibb County

Component	FracFocus Percent Composition (%)		OGB Volume
	Mean	Range	Mean
Water	88.70	86.45 - 90.54	-
Proppant	11.34	9.51 – 13.63	59,700 Lbs (27,100 Kg)
Bulk Acid	0.03	0.01 – 0.05	260 Gal (0.98 m^3)
Diverting Agent	0.008	0.006 – 0.012	-
Corrosion Prevention	0.0007	0.0 – 0.0013	-
Biocide	0.016	0.012 – 0.022	-
Friction Resistance	0.012	0.001 – 0.036	-
Iron Control	0.00035	0.0002 – 0.0005	-
Gel Break	0.00076	0.0001 – 0.0022	11,600 Gal (43.9 m^3)
Other	0.042	0.017 – 0.094	-

Table 4. General composition of fracking fluids components listed in Table 4 from FracFocus data for wells listed in Table 6

Component	Constituents	Percent Composition Range (%)
Water	Water	86.41-90.51
Proppant	Crystalline Silica Quartz* Aluminum Oxide* Titanium Oxide* Iron Oxide*	2.72-7.00 0.07-0.08 0.006-0.007 0.006-0.007
Bulk Acid	Hydrochloric Acid	0.01-0.05
Diverting Agent	Zinc Oxide* Di (2-ethylhexyl) phthalate** Phthalic Anhydride**	0.004 0.001 0.0003
Corrosion Prevention	Ethylene Glycol** N, N-Dimethylformamide** 2-Butoxyethanol** Cinnamaldehyde** Ethoxylated Non-phenol** 1-Dencanol** Isopropyl Alcohol** 1-Octanol** Triethyl Phosphate** Propargyl Alcohol** Glycol Esthers** Tar Bases**	0.00056 0.00028 0.0001 0.00008 0.00012-0.00007 0.00007 0.00036-0.00004 0.00004 0.00004 0.00031-0.00036 0.00031-0.00036 0.00008-0.00009
Biocide	Polyethylene-Glycol** 2,2-dibromo-3-nitrilropionamide**	0.00853-0.01581 0.00341-0.00633
Friction Reducer	Hydrotreated Light Distillates**	0.0021-0.0336
Iron Control	Citric Acid Anhydrous	0.0003-0.0005
Gel Break	Aliphatic Acid** Ferrous Sulfate*	0.00007-0.00019 0.00007-0.00019
Other		0.017-0.095

* Inorganic, ** Organic

A review of the only 4 registered wells located in the Gurnee Coal Field on FracFocus are used to characterize the composition of fracking fluids. These 4 wells were drilled in 2014 by SAGA Petroleum LLC. A summary of the fracking fluid components is given in Table 4. The composition of the components in given in Table 5.

The following characteristics have been gathered and tabulated for comparison. The composition of the fracking fluids used in the permitted wells is obtained from online databases available from the websites: OGB of Alabama; and public available Fracfocus. The composition will be categorized into components of: water; proppant; bulk acid; diverting agent; corrosion prevention; biocide; friction resistance; iron control; gel break; and other components. The injection volume injection pressure for fracking; location beneath ground surface; and the number and total thickness of coal seams injected.

Table 5. Quantities in barrels (bbl) of fracking fluid listed in FracFocus used in 4 CBM wells drilled, and produced water listed by OGB from the Pottsville Formation located in Bibb County

Well	Fracking Fluid Injected (bbl)	Produced Wastewater (bbl)	Well Depth (ft)	Coal Seams	
				Number	Depth Range (ft)
16954-C	133,000	7,913,178	2401	3	1370 - 2366
16982-C	310,000	4,452,504	2360	3	1190 - 2243
16955-C	230,000	2,212,308	2518	3	1077 - 2029
16953-C	198,000	1,322,664	2429	3	1266 - 2246
Average	217,750	3,975,164	2427	3	1226 - 2221

Table 6. Disposal injection wells Type II (SWD) for produced water from fracked Pottsville Formation CBM wells located in Bibb and Shelby Counties

Well	County	Activity History	Produced Wastewater (bbl)	Well Depth Injected (ft)	Geologic Formation	Wastewater Reported Components
12943-SWD-03-01	Shelby	1/2004 – present	6,635,160	9810 – 11,512	Ord-Cam. Knox Group	Salt Water
7674-SWD-90-3	Shelby	6/1990 - 2/2007	177,364,656	4130 - 4596	Penn. Pottsville Fm.	Salt Water
4530-SWD-86-2	Bibb	1/1986 – 12/1986	627,107	11,185 - 11,197	Miss. Undiff.	Salt Water
6848-SWD-90-1	Bibb	7/1992 – PA*	0	7318-8572	?	Salt Water

* Plugged and abandoned.

Also, a comparison is provided between four registered wells on FracFocus with the same well permitted by the OGB of Alabama. Registration on FracFocus commenced during year 2011, whereas wells permitted by the OGB started by the 1940's. These OGB permits did list some fracking activity (depth of injection, pressure, and generalized fluid types) for some wells, but were not as extensive as those listed on FracFocus. Table 3 provides a striking difference between what was used, and what was released by the regulatory authority.

Type II SWD injection wells were utilized to dispose of produced water from the Gurnee Coal Field. Table 6 provides a history of this injection which started in 1992 and still continues. The OGB files indicate a substantial amount of injection of 177,365,000 bbl (28,198,000 m^3) between the years of 1990 and 2007 into the Pottsville Formation. This injection averaged 10,433,200 bbl/yr (1,658,700 m^3/yr) at a depth between 4130 – 4596 ft in the same formation, which is about 1700 – 2169 ft deeper than the coalbeds at an average depth of 2427 ft (Table 6). This injection well is located 1600 ft from the Cahaba

River Watershed in Shelby County. Freshwater is provided to communities in the Cahaba Watershed from this formation. Since 2004 a different SWD injection well in the watershed, but at a deeper and older Ordovician-Cambrian Knox Group which is an undifferentiated, presumably at a safer sedimentary rock layers, of sandy dolomite, dolomitic limestone, and limestone [USGS, 2019].

In all cases of Type II SWD injection, the wastewater is given a generic term Salt Water. The OGB of Alabama does not require injected wastewater to be monitored. No evidence is given on the projected movement of the wastewater.

Characteristics of Produced Water from CBM Wells

Fracking requires excessive amounts of water to generate the fractures in the coal beds. After the fracturing process, the well operator retrieves as much of the injected fracking water as possible, which is known as flowback water. An estimated 60% of the fracking fluid is retrieved. Continued pumping yields produced water along with the CBM being produced. Alabama does not require and list the composition of the flowback and produced waters. In accordance with EPA regulations issued under Sections 1422 and 1425 of the Safe Drinking Water Act (SDWA), the OGB requires water quality samples to be monitored for salinity to be collected of the produced water, which will be disposed. Based on the average OGB data, it an estimate of the quantity of produced water from the Gurnee Coal Field estimated from 51 production wells currently active since 1990 average 74,529 bbl (11,850 m^3). Collectively, these wells produce an average of 124,848 MCF/yr (35.353 MCM/yr) of methane gas. There is no documentation of flowback water listed for permitted wells and its disposal method.

Fracking fluid composition disclosure as part of the permitting process is determined by the individual state. According to Morrison [2015], when fracking is on state land, 25 of the states do require the components of the fracking fluid be released, 8 of the states are about to pass regulation requiring the components to be disclosed, and 17 states have no fracking fluid regulation because the industry activity is non-existent. Only Kentucky which is the sixth largest producer of fossil fuels (Figure 1) has pending disclosure regulations, and only New Mexico which is the ninth largest producer does not have active fracking occurring. Colorado is considered to have the most current fracking regulation [Colorado Rule 205A] that requires industry to register the fracking fluid composition, but there is a trade secret protection clause [Peiserich, 2012].

EPA [2015] compiled and analyzed 37,796 disclosures in FracFocus for 20 states between January 1, 2011, and February 28, 2013. An additional 734 disclosures did not meet criteria for inclusion in their analysis. The analyzed wells injected 91.8 billion gallons of fracking fluid with an average of 1.5 million gallons of fluid per fracked well. However, the Railroad Commission of Texas reported an increase of 46 million barrels of underground wastewater injection in 2005 to 3.6 billion barrels of fluid injection in 2011 [Cook, 2015]. A 7,726% increase in wastewater injection. The Commission also registered 17,685 wells as of December 4, 2014 in the Late Cretaceous Eagle Ford Shale Formation just south of San Antonio [Shade and Roest, 2015]. This is a 3,000% registration increase in one year, and an amazing 69,000% since 2008. This uncontrolled growth cannot be sustainable in terms of managing the environmental consequences that result from fracking.

Furthermore, analysis of 18,158 permitted wells recorded by 8 state agencies during a 9-month period in 2011 found 85% of the wells were fracked, but only 45% of these wells were registered on FracFocus [Haas, et al., 2012]. That indicates that 8,489 fracked wells have not been listed revealing the composition using this open forum. It should be noted that all listed wells on FracFocus do not fully list their composition claiming some level of proprietary in their fracking fluid makeup. If each of the 8,489 fracked well injected approximately 1.5 million gallons of fracked fluid [EPA, 2015], then upwards to 12 billion gallons of fluid (water and chemicals) were used during the fracking processes in the undisclosed wells. Some estimates place chemical components at 1% of the fluid mass injected [King, 2012]. As a result, approximately 120 million gallons of undisclosed chemicals were involved in the fracking during this 9-month period. An uncertain amount of fluid recovery is realized as flowback wastewater. However, approximately 60% of the fracking fluids injected are estimated to be recovered as flowback wastewater [WaterWorld, 2015]. This would indicate that 48 million gallons of chemicals remain undisclosed and could still reside in the rock formations in which they were injected during 2011. This unintended chemical contamination will continue to migrate through the subsurface through time, and could threaten water resources in the future. The fracking fluids need to be limited or at least disclosed by the permitting process of the state.

Given as a frame of reference, Table 7 lists components of produced water derived from coalbeds in the U.S. [Echchelh, et al., 2018], the Black Warrior Basin [EPA, 2015], and an average for Australian water produced from coalbeds [Commonwealth of Australia, 2014]. Radioactive elements, which are known to be released from other nonconventional oil and gas resources such as shale are not addressed. The age and type of coal from which produced water are being extracted were not identified in either the U.S. and Australian reports. Both reports provide a wide range of measurements for most parameters. Missing values were neither reported or measured.

We focused on one environmental issue which deals with the problem of wastewater disposal. We did not address other environmental concerns dealing with fracking: air pollution generated by methane flaring; worker health and safety issues; solid waste management issues; and local earthquake occurrences. These should also have the same amount of regulatory complexity.

Table 7. CBM produced water chemical composition in the U.S. [Echchelh, et al., 2018], Black Warrior Basin [EPA, 2015], and Australia [Commonwealth of Australia, 2014]

Chemical Parameter	Black Warrior	U.S.	Australia
EC* (mS/cm)	-	9-40	200-16,000
SAR*	-	4-1567	16-567
pH	-	5.4-10.4	7-9.1
TDS* (mg/L)	589-61,733	150-177,000	200-10,000
Cl^- (mg/L)	11-42,800	0.8-110,000	150-2500
HCO^{3-} (mg/L)	-	19-43,310	-
S^{2-} (mg/L)	-	-	0.1-1.0
SO_4^{2-} (mg/L)	0-302.00	BDL-1800	1-10
NO_3^- (mg/L)	0-127.50	0.01	-
PO_4^{3-} (mg/L)	-	BDL-9199	-
Ar (mg/L)	0.0-0.085	-	0.001-0.0065
Na (mg/L)	126-16,700	2.6-51,700	35-4500
K (mg/L)	0.46-74.00	0.1-20,100	1-300
Ca (mg/L)	0-1,640	0.42-13,900	0.5-49
Mg (mg/L)	0.18-414.00	0.01-15	0.7-16
Al (mg/L)	0-0.099	0.01-3	0.01-0.3
B (mg/L)	0-0.54	0.05-10	0.05-3.1
Be (mg/L)	0-0.008	-	0.001
Br (mg/L)	-	-	1-12
Cd (mg/L)	0.0-0.015	0.0001-1.4	0.0001-0.0002
Cr (mg/L)	0-0.351	0.001-3.7	0.005-0.3
Cu (mg/L)	0-0.098	0.002-4.6	0.001-0.2
Fe (mg/L)	0.045-93.1	0.005-4180	1-25
Hg (mg/L)	0.0	-	0.0001-0.001
Li (mg/L)	0-8.94	BDL-36	-
Mn (mg/L)	0.18-4.84	0.0018-6	0.004-0.3
Ni (mg/L)	0-0.358	0.0001-19.2	0.0001-0.003
Pb (mg/L)	0-0.25	0.001-0.2	0.001-0.2
Se (mg/L)	0-0.063	-	0.001-0.01
TOC (mg/L)	0.0-103.00	-	2000-3900
Zn (mg/L)	0-0.278	0.001-51	0.005-0.15
Oil & Grease (mg/L)	-	2.2	-
C6-C9 fraction (mg/L)	-	-	20
C10-C14 fraction (mg/L)	-	-	50
C15-C28 fraction (mg/L)	-	-	100
C29-C36 fraction (mg/L)	-	-	50-113
1,2-Dichloroethane-D4 (mg/L)	-	-	118-120
Toluene-D8 (mg/L)	-	-	94.6-98.22
4-Bromofluorobenzene (mg/L)	-	-	99.2-102.9

* SAR is sodium adsorption ratio, TDS is total dissolved solids, EC is the electrical conductivity.

Wastewater Disposal Management Approaches

The USGS [2000] provides a review of produced wastewater disposal practices utilized by coal fields that have undergone CBM production. The produced wastewater can be treated then is either reused or disposed. Reuse of wastewater is performed by irrigation, released to wetlands, or used as a water supply. Disposal of wastewater can be by injection into the subsurface, surface discharge, or stored in surface ponds. A review of OGB records indicate that subsurface injection has been the preferred method in Gurnee Coal Field (Table 6).

The high injection pressure required for wastewater disposal has been proposed to mobilize natural contaminants to migrate from the oil and gas producing formation [Arnauld, 2015]. This study found that CBM wastewater well injected 184.6 million bbl (29.4 million m^3) into the subsurface (Table 7). These include the remaining fracking chemicals present in the wastewater. The probability of the receiving rock layer of a failure to retain all wastewater injected and the release of natural contaminants are hard to quantify. An adoption of the probabilistic risk assessment approach by Bolster, et al. [2009] and Tartakovsky [2007] are deemed applicable.

Regulation by the EPA issues vague guidelines on the disposal of wastewater from produced water through Underground Injection Control (UIC regulations in 40 CFR Sections 1422-1425). As a result, the states of the U.S. are required to implement their own regulations that meet the minimum set by the EPA. These federal regulations are meant to guide the installation of an UIC: construction; operation; monitoring and testing; reporting; and closure requirements. Under section 1425 the operation requires: permitting; inspections; monitoring; record-keeping; and reporting.

Sadasivam [2014] cited a Governmental Accounting Office [GOA] 2014 report indicating that EPA failed to enforce regulations that guide the use of Type II Injection Wells used to inject waste generated by the oil and gas industry. This study cited that Alabama has as of 2010 permitted 240 Type II injection wells which are either plugged and abandoned or active. Since 1988 all oil and gas exploration production waste, which includes CBM production, is considered non-hazardous [EPA, 2002].

The quantity of produced water injected causing pollution derived from by CBM activity is determined, in part, by the hydraulic communication between the coalbeds and the surrounding aquifers. Hydraulic communication affects the recharge of the coalbeds and the surrounding sedimentary rocks [NASEM, 2017]. This report states that the CBM produced water may be treated, disposed, stored, or put to beneficial use. The report identifies several variables that could influence the economic decision of disposal: sustained production of water over time; treatment costs; proximity of location of produced water to the proposed beneficial use; costs and infrastructure for water transport and storage; compatibility between produced water with potential receiving landscapes or water bodies; availability of suitable storage and disposal sites; and the legal framework for application of produced water to beneficial uses.

CONCLUSION

There are multiple water resource concerns that need to be managed properly to minimize the impact of hydraulic fracking. The risk to an important regional Cahaba River Watershed, Alabama, was investigated. This case study of risks caused by fracking associated with CBM production in the Gurnee Coal Field has identified these potential problems:

- The excessive withdraw of freshwater needed to formulate and inject fracking fluid could lead to a permanent change in the hydraulic character of the watershed's groundwater-flow patterns and sustainability;
- The vertical extent of fractures formed in the coal beds by fracking could penetrate the surrounding sedimentary rock layers. This could allow contamination by the movement between coalbeds being fracked and nearby freshwater aquifers;
- The unknown amount and chemical composition of fracking fluids injected into the coalbeds;
- The impact on water quality resulting from the injection of fracking fluid which is not recovered;
- The amount of produced wastewater and its disposal associated with CBM production;
- The high-pressure injection of produced wastewater at known depths that still communicate with aquifers used as water supplies.

Water and wastewater management add to the cost of CBM production, which could make the production of the methane gas uneconomical. The extent of impact of groundwater withdraw during CBM production depends on the overall volume of water in the coalbed and hydrogeology of the watershed, and its capacity to meet the rate of water extraction and recharge to the surrounding aquifers.

Another problem is the poor participation in the voluntary disclosure forum FracFocus by the CBM production industry in Alabama. This indicates that a more rigorous approach for acquiring the components being used in a fracking fluid is required. Colorado Rule 205a [State of Colorado, 2015] is a comprehensive disclosure regulation and should become the new model going forward [Peiserich, 2012]. This state rule requires companies to file any use of a fracking fluid and its composition. If a company claims its use is a trade secret, then the composition is filed on Form 41, which is searched for a unique composition and intended use that makes it proprietary information, a decision is then made at that point by the state.

The history of wastewater disposal generated by CBM production and its disposal by SWD Type II wells meets the minimum requirements specified by the federal EPA and the State of Alabama OGB. The regulations, however do not require the monitoring of any component of the original fracking chemicals injected present in the wastewater to be before it is disposed. This monitoring activity is an important missing to ensure the damage to the regions water resources is minimal. Also, a rigorous investigation is needed to determine if the depth of the Type II Injection well for wastewater is at quantifiable safe depth for disposal. Hypothetically, there is no such depth that is safe. The safe solution is to treat the wastewater so that it can be used elsewhere in the watershed. Thus, keeping the water balance at a sustainable level, and treating water as a nonrenewable resource.

REFERENCES

American Oil & Gas Historical Society. "*Shooters – A "Fracking" History*," Accessed May 14, 2017. http://aoghs.org/technology/hydraulic-fracturing/.

Arnauld, CH. "Figuring out fracking wastewater: Scientist are analyzing the liquid that comes from fracking wells to determine the best ways to treat it," *Chemical & Engineering News*, March 16, 2015, 8-12.

BALLOTPEDIA. "*Public Policy on Energy*," Accessed May 30, 2017. https://ballotpedia.org/Coalbed_methane.

Bolster, D; Barahona, M; Dentz, M; Fernandez-Garcia, D; Sanchez-Vila, X; Trinchero, P; Valhondo, C; Tartakovsky, DM. "Probabilistic risk analysis of groundwater remediation strategies." *Water Resources Research*, 45, no. 6, (2009). doi/full/10.1029/2008WR007551.

Champlin, SD. "Depositional and Diagenetic History and Petroleum Geology of the Jurassic Norphlet Formation of the Alabama Coastal Waters Area and Adjacent Federal Waters Area." *Marine Georesources & Geotechnology*, 17, (1999), 215-232.

Chaney, PL. "*Climate: Encyclopedia of Alabama*." Accessed March 26, 2019. http://www.encyclopediaofalabama.org/article/h-1283.

Cook, T. "Ground-Shaking Research: How Humans Triggers Earthquakes." *Earth*, May/June 2015, 56-63.

Davis, Charles. Fracking and environmental protection: An analysis of U.S. state policies." *The Extractive Industries and Society*, 4 no. 1, (2017), 63-68.

Echchelh, A; Hess, T; Sakrabani, R. "Reusing oil and gas produced water for irrigation of food crops in drylands." *Agricultural Water Management*., 206, (2018), 124 -134.

"Alabama Oil and Gas Regions." In *Encyclopedia of Alabama*. n.d. Accessed July 29, 2017. http://www.encyclopediaofalabama.org/article/m-5184.

Energy Information Administration [EIA]. "*Alabama State Energy Profile*". Accessed July 5, 2017. https://www.eia.gov/state/print.php?sid=AL.

Energy Information Administration [EIA]. "*U.S. Coalbed Methane: Past, Present, and Future. Panel 2 of 2.*" Accessed July 5, 2017. https://www.eia.gov/oil_gas/rpd/cbmusa2.pdf.

Environmental Protection Agency [EPA]. *Coal Mine Methane Country Profiles*. Coalbed Methane Outreach Program. 381. Accessed May 20, 2019. https://www.globalmethane.org/tools-resources/archive/coal_overview.aspx.

EPA. Evaluation of Impacts to Underground Sources of Drinking Water by Hydraulic Fracturing of Coalbed Methane Reservoirs. *Office of Ground Water and Drinking Water*, (4606M). EPA 816-R-04-003. pp. 463. 2004. Accessed March 21, 2018. https://fracfocus.org/sites/default/files/publications/evaluation_of_impacts_to_underground_sources_of_drinking_water_by_hydraulic_fracturing_of_coalbed_methane_reservoirs.pdf.

EPA. *Coalbed Methane Extraction: Detailed Study Report*. U.S. Environmental Protection Agency, Washington, DC, EPA-820-R-10-022. 2004. Accessed May 12, 2019. https://www.epa.gov/sites/production/files/2015-06/documents/cbm_report_2011.pdf.

EPA. *National Underground Injection Control Inventory – Federal Fiscal Year 2015*. U.S. Environmental Protection Agency, Washington, DC. Accessed February 26, 2018. https://www.epa.gov/sites/production/files/2016-10/documents/ underground_injectio n _control_inventory_fy_2015_0.pdf .

EPA. *Assessment of the Potential Impacts of Hydraulic Fracturing for Oil and Gas on Drinking Water Resources (External Review Draft)*. U.S. Environmental Protection Agency, Washington, DC, EPA/600/R-15/047. 2015. Accessed May 12, 2019. https://cfpub.epa.gov/si/si_public_record_report.cfm?Lab=NCEA&dirEntryID=244651.

EPA. *Regulation of Hydraulic Fracturing Under the Safe Drinking Water Act* U.S. Environmental Protection Agency, Washington, DC, 2014. Accessed May 12, 2019. http://water.epa.gov/type/groundwater/uic/class2/hydraulicfracturing/wells_hydroreg.cfm.

EPA. *Exemption of Oil and Gas Exploration and Production Wastes from Federal Hazardous Waste Regulation* U.S. Environmental Protection Agency, Washington, DC, 2002. EPA530-K-01-004. Accessed May 21, 2018. https://www.documentcloud.org/documents/371040-epa-rcra-exemptions-oil-gas.

Gallegos, TJ; Varela, BA. "Trends in Hydraulic Fracturing Distributions and Treatment Fluids, Additives, Proppants, and Water Volumes Applied to Wells Drilled in the United States from 1947 through 2010—Data Analysis and Comparison to the Literature." *U.S. Geological Survey*, Scientific Investigations Report 2014–5131, 2014. doi.org/10.3133/sir 20145131.

Golden, JM; Wiseman, HJ. "The Fracking Revolution: Shale Gas as a Case Study in Innovation Policy." *Emory Law Journal*, 64, no. 4, (2015), 955-1040. Accessed June 13, 2017, http://law.emory.edu/elj/content/volume-64/issue-4/articles/fracking-revolution-study-innovation-policy.html.

Geraci, Matthew, Ali, Syed Jehangeer, Romolt, Courtney, Rossmann, Regina. "The Environmental Risks and Oversight of Enhanced Oil Recovery in the United States: An overview of Class II well trends and regulations in EPA's Underground Injection Control Program." *Clean Water Action/Clean Water Fund.* August 2, 2017, 62. Accessed November 26, 2018. https://www.cleanwateraction.org/publications/EOR-risks.

Government Accounting Office. "*EPA Program to Protect Underground Sources from Injection of Fluids Associated with Oil and Gas Production Needs Improvement*." GAO-14-556. June 27, 2014. Accessed April 1, 2019. https://www.gao.gov/assests/670/ 664499.pdf.

Haas, Benjamin, Polson, Jim, Kuntz, Phil, Elgin, Ben. "Fracking Hazards Obscured in Failure to Disclose Wells." *Bloomberg News* August 14, 2012. Accessed April 27, 2019. http://www.bloomberg.com/news/articles/2012-08-14/fracking-hazards-obscured-in-failure-to-disclos e-wells.

Hackley, PC; Karlsen, AW. "Geologic assessment of undiscovered oil and gas resources in Aptian carbonates, northern Gulf of Mexico Basin, United States." *Cretaceous Research*, Elsevier, 48, 2014, 225-234.

Hall, B; Kerr, ML. *1991-1992 Green index: A state-by-state guide to the nation's environmental health* (Washington, D.C.: Island Press, 1991), 169.

Harrington, JR; Gelfand, MJ. *Tightness-looseness across the United States*. Retrieved September 7, 2014, from www.pnas.org/cgi /doi/10.1073/pnas.13179 37111.

Karacan, CO. "Production history matching to determine reservoir properties of important coal groups in the Upper Pottsville formation, Brookwood and Oak Grove fields, Black Warrior Basin, Alabama," *Journal of Natural Gas Science and Engineering*. 10, November 29, 2013. 51-67. dx.doi.org/10.1016/ j.jingse.2012.10.005.

King, GE. "Hydraulic Fracturing 101: What Every Representative, Environmentalist, Regulator, Investor, University Researcher, Neighbor, and Engineer Should Know About Estimating Frac Risk and Improving Frac Performance in Unconventional Gas and Oil Wells," *Society of Petroleum Engineers, Hydraulic Fracturing Technology Conference*, The Woodlands, TX, February 6-8, 2012, 80.

Kopaska-Merkel, DC; Dean, LS; Moore, JD. Hydrogeology and vulnerability to contamination to contamination of major aquifers in Alabama: Area 4. *Geological Survey of Alabama Circular*, 199 D, 2005.

Lustgarten, A. "Progress and Controversy Arrive with New Rules for Fracking on Public Lands: An initial review of rules issued by the Interior Department shows the federal government has taken important steps to protect drinking water resources, while not adopting the strictest regulations in place in some states," *ProPublica* Accessed April 28, 2015. http://www.propublica.org/article/progress-and-controversy-arrive-with-new-rules-for-fracking-on-public-lands.

Lustgarten, A. "Injection Wells: The Poison Beneath Us. Lax oversight, uncertain science plague program under which industries dump trillions of gallons of waste underground," *ProPublica*, (2012, June 21). Accessed April 11, 2019. https://www.propublica. org/article /injection-wells-the-poison-beneath-us.

McDonald, Norris. "Who Owns the Groundwater?" *Center for Environment, Commerce, & Energy*. (2010). Accessed February 26, 2018. http://cenvironment.Blogspot.com/2010/01/who-owns-groundwater.html.

Melici, RC. "*Coalbed Methane Production in the Appalachian Basin*," U.S. Geological Open-File Report 02-105 (2000). Accessed August 15, 2017. //pubs/usgs/gov/of/2002/02-105/of 02-105.html.

Morrison, J. "Disclosure debate: Uneven state requirements and trade-secret protections hinder release of data on chemicals on fracking fluid," *Chemical & Engineering News*, March 16. 2015, 13.

NASEM (National Academies of Sciences, Engineering, and Medicine). *Flowback and Produced Waters: Opportunities and Challenges for Innovation* (Proceedings of a Workshop. Washington, DC: The National Academies Press, 2017). https://doi.org/10.17226/24620.

National Research Council. *Management and Effects of Coalbed Methane and Produced Water in the Western United States*. (Washington D.C.: The National Academies Press, 2010) 220. Accessed May 16, 2019. https://doi.org/10.17226/12915.

Pashin, JC; Groshong, Jr. RH. "Structural control of coalbed methane production in Alabama," *International Journal of Coal Geology*, 38, (1998), 89-113.

Resources for the Future. "A Review of Shale Gas Regulations by State," *OurEnergyPolicy.org*. July 16, 2012. Accessed November 3, 2014. https://www.ourenergypolicy.org/a-review-of-shale-gas-regulations-by-state/s.

Richardson, N; Gotlieb, M; Krupnick, A; Wiseman, H. "The State of State Shale Gas Regulation." *Center for Energy Economics and Policy*., (2013, March 3). Accessed 4/30/2015. https://www.rff.org/publications/reports/the-state-of-state-shale-gas-regulation/.

Rozell, DJ; Reavan, SJ. "Water pollution risk associated with natural gas extraction from the Marcellus Shale," *Risk Analysis*, 32, (2012), (8), 1382-1393.

Sadasivam, N. "Criticizes EPA Oversight of Injection Wells." *ProPublica*. July 29, 2014. Accessed April 12, 2019. https://www.propublica.org/article/report-criticizes-epa-oversight-of-injection-wells.

Showshack, R. "Interior Department Announces Fracking Standards," *EOS*, April 15, 2015.

Shipman, CR; Lowry, WR. *Congress and the Environment: A Longitudinal Analysis.* Charles R. Shipman, School of Public Health, University of Michigan, and Department of Political Science, University of Iowa. And William R. Lowry, Department of Political Science, Washington University, July 27, 1997. 33 pp.

Smith, MB; Montgomery, CT. *Hydraulic Fracturing* (Boca Raton, FL.: CRC Press, 2015), 777.

Sontag, D; Gebeloff, R. The Downside of the Boom: North Dakota took on the oversight of a multibillion-dollar oil industry with a regulatory system built on trust, warnings and second chances. *New York Time.*, (2014, Nov. 22). Accessed 4/28/2015. http://www.nytimes.com/interactive/2014/11/23/us/north-dakota-oil-boom-downside.html?r=0.

State of Colorado (2015, January 30) General Rules: 205A. *Hydraulic Fracturing Chemical Disclosure.*, Accessed May 21, 2018. https://cogcc.state.co.us/RR_Docs _new/rules/200 Series.pdf.

Tartakovsky, DM. "Probabilistic risk analysis in subsurface hydrology." *Geophys. Res.* Ltt. 34 (2007) L05404, doi:10.1029/2007GL09245.TEDX.

TEDX, "*The Endocrine Disruption Exchange: Partners in Science*," Accessed June 19, 2018 from https://endocrinedisruption.org.

Thakur, Pramod. *Advanced Reservoir and Production Engineering for Coal Bed Methane.* (New York: Elsevier GPP, 2017), 210.

Tew, Jr. Barry H. "Aquifer Recharge Areas of Alabama (map)." *Geological Survey of Alabama.* Accessed 1/17/2018.//www2.gsa.state.al.us/gsa/water/Images/aquifers/AL_aquifers_Big.pdf.

US Energy Information Administration (EIA) (2011) *Table P5. Energy Production Estimates.* Retrieved from http://www.eia.gov/state/seds/sep_prod/pdf/P5.pdf.

USGS "Water Use in Alabama." *Watershed*, (2005). Accessed April 22, 2018. https://www.usgs.gov/centers/lmg-water/science/water-use-alabama-watershed-2005?qt-science_center_objects=1#qt-science_center_objects.

USGS "Water Produced with Coal-Bed Methane." *USGS Fact Sheet FS-150-00* November 2000. Accessed May 10, 2019. file:///C:/Users/David/Desktop/Water%20Management%20and%20Fracking/Water%20Produced%20with%20Coal-Bed%20Methane.pdf.

In: Advances in Hydrogeochemistry Research
Editors: S. Saeid Eslamian and Faezeh Eslamian
ISBN: 978-1-53617-829-6

Chapter 11

The Environmental State of Freshwater Resources in Man-Made Lakes of Serbia

Gordana Dević*
Institute of Chemistry, Technology and Metallurgy, University of Belgrade, Serbia

Abstract

Surface water quality is currently matter of serious concern in developing countries, due to growing populations, rapid industrialization, urbanization and agricultural modernization. Since contaminants are primarily related to anthropogenic activities, pollution by heavy metals in surface water bodies is a worldwide concern because of their great threat to the health and well-being of aquatic organisms. Data concerning water quality and pollution in Serbia is gathered and presented in the National Environmental Protection Programme of the Republic of Serbia (2010 - 2019). In this document, water quality is categorized as low and classification has been made according to the regulations (National Environmental Aproximation Strategy For The Republic Of Serbia). Serbia belongs to the group of countries poor in water resources. Concerning water supply, Serbia mostly relies on surface water resources. The number of natural lakes in total area of Serbia is relatively small, however, when an increasing number of artificial reservoirs are included, the hydrographic network becomes richer and more complex. Hydro reservoirs represent the solution for water supply and hydro energy for Serbia, thus their construction is necessary and of great priority.

A big problem with hydro reservoirs in Serbia is their preservation in time and space, since they are usually irrationally managed. Most reservoirs do not have adequate monitoring and ecological protection. Since, detailed and systematic investigations of environmental state of freshwater sources in man-made lakes of Serbia has not been conducted (data on Serbian lakes and reservoirs, including Palic Lake and the Iron Gate Dam located downstream from several industrial centres, are only available in a global website that handles this issue), and the need to clarify concerns about the quality and safety of potable water, this study will addres these issues. Results of this study are very

* Corresponding Author's Email: gdevic@chem.bg.ac.rs.

important for scientists dealing with these issues, as well as for the people of Serbia and the region.

Keywords: freshwater resources, man-made lakes of Serbia, water quality, pollution, ecological protection

1. Introduction

Lakes and reservoirs are the major resources of surface fresh water. They are larger and deeper than ponds and are not part of the ocean.

For thousands of years, reservoirs and dams have been important tools for administration of extreme hydrological events. Water related disasters, such as floods and droughts, have been successfully mitigated by the intelligent use of storage provided by reservoirs. Up to the last quarter of the 20th century, those responsible for the construction of dams and creation of reservoirs – entrepreneurs, decision makers, engineers, investors – were praised for the acknowledged benefits of their works: water supply, irrigated agriculture, flood control, improved navigation and hydroelectric powerplants (considered clean and unequivocally renewable energy). However, on the other side, alarmist groups and organizations have been exaggeratedly stating that the infrastructure, in general, and dams and reservoirs, in particular, have serious and intolerable environmental impacts. Reservoir construction leads to number of environmental issues, both during building and following completion (Devic, 2015).

Table 1. Ecotechnological methods applied to reservoir watershed management and recovery

Problem to be solved	Methods
Organic pollution	Clean production, Diversion effluents, Purification plants
Excess nutrients and eutrophication	Wetlands, Diversion of wasters, Progressive agriculture practies, Meadow and riparion forest zones on the vegetated banks, Natural and constructed wetlands, Pre-impondments at the inflow
Eutrophication and oxygen depletion of rivers reservoir siltation	River restoration, Re-oxygenation, Erosion control, Rehabilitation of river banks, Reforestation, Groundwater recharge, Pre-impoundment of inflows
Heavy metal contamination	Reduction of polluted effluents, Wetlands
Acidification	Organic matter additions
Salinization	Improved irrigation practies, Decreased fertilizer applications, Decreased road salting
Decreased biodiversity due to reservoir construction	Prohibit introduction of foreign species, Reintroduction of native species, Maintenance of wetlands as nursery grounds, Maintenance of preserved areas for native species

In the last decades of industrialization and rapid population increase, human intervention on the balance of nature has never been so varied and as extensive as it is today. Artificial water bodies are very sensitive to the intense pressure from human activities, which in turn induce many environmental problems, such as water quality deterioration and eutrophication, especially related to incidents of pollution with hazardous and toxic contaminants, such as heavy metals (Bing et al., 2016; Table 1). Some studies also revealed that eutrophication and cyanobacterial bloom, rather than pollution by heavy metals, are the principal challenges to water resource protection of reservoirs, due to the excess input of nutrients (Gao et al., 2016; Sedláček et al., 2016).

2. Pollution of Freshwater Resources

Water storage in reservoirs induces physical, chemical and biological changes to stored water. As a result, the water discharged from a reservoir could have a different composition to that flowing into the reservoir. Temperature changes relate to the thermal mass and surface area of the reservoir for radiant exchange, retention time, thermocline development and whether the released water originates from the surface or a depth (Richardson et al., 2017).

Chemical changes include altered nutrient levels and dynamics, modified oxygen regimes in the water-column and sediment, nitrogen supersaturation in downstream waters, increased mobilization of certain metals and organic contaminants. Nutrients (phosphorous and nitrogen) are released biologically or leached from flooded vegetation and soil. Eutrophication of reservoirs may occur as a consequence of large influxes of organic loading and/or nutrients (Li et al., 2017). In many cases, these are a consequence of anthropogenic influences in the catchments rather than a direct consequence of their presence of the reservoir. Nutrient pulses in conjunction with specific environmental conditions can cause oxygen depletion and increased concentrations of iron and manganese in the bottom layer and increased pH and oxygen in the upper layers of stratified reservoirs. Although the oxygen demand and nutrient levels generally decrease over time as the organic matter decrease, some reservoirs require a period of more than 20 years for the development of stable water-quality regimes. After maturation, reservoirs, like natural lakes, can act as nutrient sinks, particularly for nutrients associated with sediments.

Heavy metal contamination has recently been highlighted as a major reservoir problem in some countries (Xu et al., 2017). Among environmental pollutants, metals are of particular concern, due to their potential toxic effects and ability to bioaccumulate in aquatic ecosystems (Waykar, Petare, 2017). These metals may be introduced into a fluvial environment as a result of natural weathering, erosion and transport processes, as well as from a range of anthropogenic activities. The release of heavy metals can occur both in dissolved and particulate form. When released in particulate form or adsorbed on particles, heavy metals can settle down and be deposited in the river bottom sediments under the favourable hydraulic conditions provided by natural and reservoir lakes. Reservoir sediments usually accumulate at high rates (typically > 2 cm a^{-1}), compared to lacustrine sediments (often < 3 cm a^{-1}; Audry et al., 2010). Due to these rapid sedimentation rates, reservoir sediments are considered as being little affected by early diagenesis processes and provide preserved historical heavy metal inputs. A special problem is the contamination of newly-formed reservoirs with mercury. When the terrestrial landscape is flooded to create a reservoir, the organic carbon within the soil and vegetation begins to decompose. Under the anoxic conditions, prevalent in new reservoirs, sulphur-reducing bacteria thrive causing microbial methylation of inorganic mercury, naturally accumulated through long-term wet and dry atmospheric deposition, into the highly toxic form methyl mercury (MeHg) (Beldowski et al., 2014). Elevated production of MeHg in new reservoirs may last for more than 13 years and be rapidly transferred through the food chain causing increased metal levels in fish, which may persist for up to 30 years after flooding (Beldowski et al., 2014).

A major impact on surface water quality may stem from biological processes, such as bacterial contamination. Water turbidity could be increased by high plankton concentrations, sometimes accompanied with proliferation of mosquitoes and other insects, as well as rats and other pests. In the surrounding area and downstream from the reservoir, groundwater levels generally rise due to damming, increased infiltration and rise in the hydraulic base level (Wildi, 2010).

3. Case Study: *Serbia*

Serbia belongs to the group of countries poor in water potential. In Serbia only 8% of available water resources originate in the country, the other 92% are transit waters that flow into Serbia through the Danube, Sava, Tisza and other waterways (Devic et al., 2016). A significant resource, when considering surface waters, is the Danube–Tisa–Danube system. This system covers an area of 20,000 km^2 and represents a network of channels with a total length of about 700 km. Groundwater potential is estimated to be in range from 60,000 L/s to 90,000 L/s, of which about 21,000 L/s is used as drinking water supply. There are many sources of mineral and thermal waters in the territory of Serbia. More than 1,000 sources of thermal and mineral water are registered. The total yield of thermal mineral water is more than 1,000 L/s. Raw water intended for drinking comes from underground sources of water (59%), surface sources (24%) and springs (17%) (Devic et al., 2014a).

Hydro reservoirs represent the solution for water supply and hydro energy for Serbia, thus their construction is necessary and of great priority. (Stankovic 2005). According to Bogdanović and Pavić, 2003, “There are 28 large reservoirs with the volume of 10 millions m^3 (Table 2) and more reservoirs have a volume of less than 10 million m^3, and there are about 100 so-called small reservoirs made by the damming of small water courses." Most of these reservoirs have significant problems related to water quality, which are manifested in many different ways: frequently from the lush bloom of toxic blue–green algae, to increased concentrations of certain substances that negatively affect, restrict or in extreme cases completely eliminate the use of these water resources for planned and possible uses (e.g., accumulation Vrutci) and often lead to significant advers effect on the aquatic ecosystem.

Most reservoirs do not have adequate ecological protection, irrational management and only few could boast about the presence of proper monitoring (Devic et al., 2014b). A few of them are included in complex hydro-biological research (first biological research match with the construction of first reservoirs, Zlatković, 2010). The artificial lakes of Vojvodina are small in size and known only to local residents. These shallow lakes are generally located in the northern, lowland region of Vojvodina. The Drina, Uvac, Lim and Morava Rivers (Central Serbia) have the highest number of artificial lakes. In contrast to these areas, Sumadija is a part of Serbia with a large number of small reservoirs, the use of which is limited. Some lakes are smaller than 2 km^2, while only two are larger than 40 km^2. Only six lakes reach depths exceeding 50 m. Agricultural, municipal and industrial effluents have altered the composition of surface-water and restricted its use. The Serbian surface-water resources serve drinking and irrigation purposes, hydroelectric power production, industrial uses, flood defence, fisheries and recreational activities (Table 2).

Current State in Man-Made Water and Sediment Pollution in Serbia

Water supply in various areas of Serbia is different in terms of infrastructure, number of connected households and reliability of water supply systems is in continuous increase, due to migrations from villages to cities. Water quality in watercourses in Serbia is generally low and it is further deteriorating. Examples of very clean water, Class I and I/II, are very rare, and are found in mountainous regions in central Serbia. The most polluted watercourses are in Vojvodina. There are significant regional differences in the water quality between central Serbia and Vojvodina. The main problem in central Serbia is that more than 40% of the samples were bacteriologically contaminated and did not satisfy the quality criteria. In Vojvodina, the primary problems with the physical and chemical water quality parameters are turbidity, iron, arsenic, nitrates and naturally elevated content of manganese. In many areas, the groundwater can not be used for drinking purposes without previous treatment. Most drinking water sources are not sufficiently protected from point and non-point pollution, and hence, there is a significant risk of epidemic outbreaks.

The water quality suffers especially from eutrophication caused by nutrients and organic pollutants. The deterioration of water quality is partially attributed to transboundary pollution of the waters entering Serbia. Transboundary rivers are contaminated with nutrients, oil, heavy metals and organic components. Despite a large pollution load, the quality of the Danube remains in class II–III mainly due to its large dilution capacity. Construction of the Hydroelectric Plant System Djerdap, particularly the dam and the reservoirs, resulted in a number of negative impacts to the environment. Each year, the Danube River tributaries deposit about 20 million m^3 of sediment into these reservoirs. Toxic pollutants discharged from large industrial centres (Novi Sad, Pancevo, Smederevo, and Belgrade) and wastewaters from the upstream countries loaded with both organic pollution and heavy metals are trapped in the man-made Lake Djerdap.

Table 2. Large reservoirs in Serbia

Reservoir	Watercourse	The height of the dam (m)	Length (m)	Volume ($10^6 m^3$)	Purpose
Vlasina	Vlasina	34	239	176	W, E
Medjuvrsje	Zapadna Morava	31	190	18	E
Zvornik	Drina	42	269	89	E
Borsko Lake	Brestovaca	54	350	12	WI
Kokin Brod	Uvac	82	1227	273	E
Gracanka	Gracanica	54	270	32	W
Bajina Basta	Drina	90	461	340	E
Batlava	Batlava	46	302	39	W
Potpec	Lim	46	212	44	E
Djerdap	Dunav	61	1278	2550	E
Gazivode	Ibar	108	520	370	E, I, F, IW
Tisa	Tisa	25	341	160	I, F
Lisina	Bozicka reka 53	53	244	10	E
Celije	Rasina	52	220	60	W, F
Bovan	Moravica	52	151	59	W, F
Uvac	Uvac	110	307	213	E
Paljuvi Vis	Kladnica	16	760	14	WI
Vrutci	Djetinja	77	241	54	W
Lazici	Beli Rzav	131	218	170	E
Gruza	Gruza	52	288	65	W, F
Brestovac	Pusta reka	31	330	10	W
Djerdap 2	Dunav	52	899	868	E
Grliste	Grliska reka	32	101	12	W
Zavoj	Visocica	86	262	170	E
Barje	Veternica	75	326	41	W, F
Prvonek	Banjska	88	250	20	W
Rovni	Jablanica	74	430	52	W
Selova	Toplica	73	210	70	W

E-energy, W-water supply, WI-supply of water industry, F-flood defense, I-irrigation

Heavy Metals Pollution

Waters in the lakes are of the Ca and HCO_3 type. The mechanism that controls the water chemistry of the studied lakes is the rock dominance process according to Gibbs (Devic et al., 2014b).

The statistical summaries of the heavy metals in Serbian reservoirs analysed using Box and Whisker plots are shown in Figure 1. Nickel, lead and zinc showed the highest variance in the distribution of their concentrations - large spatial variations. Figure 1 also shows that the order of the relative abundance of the major metals in samples was Fe>Mn (mg/kg) while that of heavy metals was Zn > Ni > Pb > Cr > Cu > Co > Cd (mg/kg). It is obvious that the major heavy metals displayed wide differences in their variance. However, these metals may in higher concentrations render the water unfit for living organisms and human use.

Comparison of metal contamination data of the manmade lake sediments of Serbia with the published data of other reservoirs (Table 3) revealed that the sediments of the manmade lakes are severely polluted with some heavy metals, while the sediments of other sites are moderate to slightly polluted. The extent of metal pollution in this study wasn'tt more serious than that in Lake Rawal, Pakistan (Iqbal, Sadam., 2014), Lake Daye China (Mao et al., 2014) and Lake Vembanad, India (Priju, Narayana., 2007), but much worse than Lake Volvi Greece (Gantidis et al., 2007), Lake Texoma, USA (An, Kampbel, 2003) and Lake Sapanca, Turkey (Arman et al., 2009), (Table 3).

High concentrations of heavy metals is possibly caused by anthropogenic inputs from surface runoff and atmospheric deposition. Heavy metals in the atmosphere could originate from fossil fuel combustion, traffic, and waste inceneration, which are widespread throughout the study area. Cu containing agrochemicals are commonly used for disease control in livestock and poultry. Urban surface runoff is an additional source of Cu and Zn.

Domestic wastes and agricultural products were the major anthropogenic sources of Mn in the study area (Bovan Lake), mainly due to use of manganese-containing products, such as fertilizers and fungicides (Sakan et al., 2014). Diagenetic mobilization, biodegradation and recycling of heavy metals in bed sediments might also contribute to the heavy metals in water column. Moreover, the fine particles of surface sediments have a larger surface area for adsorption of heavy metals as compared to the relatively coarse particle of bed sediments. Unfortunately, organic matter could not be estimated in suspended particles, however, it is speculated that input of organic waste through sewage drainage and surface runoff also partly contributes to increased concentration of heavy metals in the suspended sediments.

The region Vrutci Lake, a specially contaminated area, was a rapidly developing area with serious non-point sources pollution problems (Blic, 2018). The sediment quality was also influenced by heavy metals *via* industrial sewage from the processing of agricultural products and chemical and machinery industries. Due to outdated and faulty waste water treatment technology and lower scale of production, a significant amount of untreated waste water was discharged into the reservoir system that deteriorated sediment quality. Ni, Cr and Co concentrations were most enriched in the Vrutci reservoir sediments. Vrutci Lake is an ecological bomb that represents a severe the health threat to tens of thousands of people. The pollution is mainly caused by permanent and accidental pollution from industrial plants located in this area.

Table 3. Maximum values of heavy metals in sediments (mg/kg) in the Serbian reservoirs and content of these elements in different world lake sediments

Location	Cd	Cu	Cr	Pb	Ni
Serbia	5.10	66.42	625	124.15	521
China	0.76	39.09	76.98	34.08	40.54
Switzerland	18.4	727	337	620	87
USA	2	38	30	10	17
Turkey	0.76	29.98	14.48	2.44	29.99
Pakistan	0.745	1.293	2.389	8.796	-
India	11	40	200.9	-	92

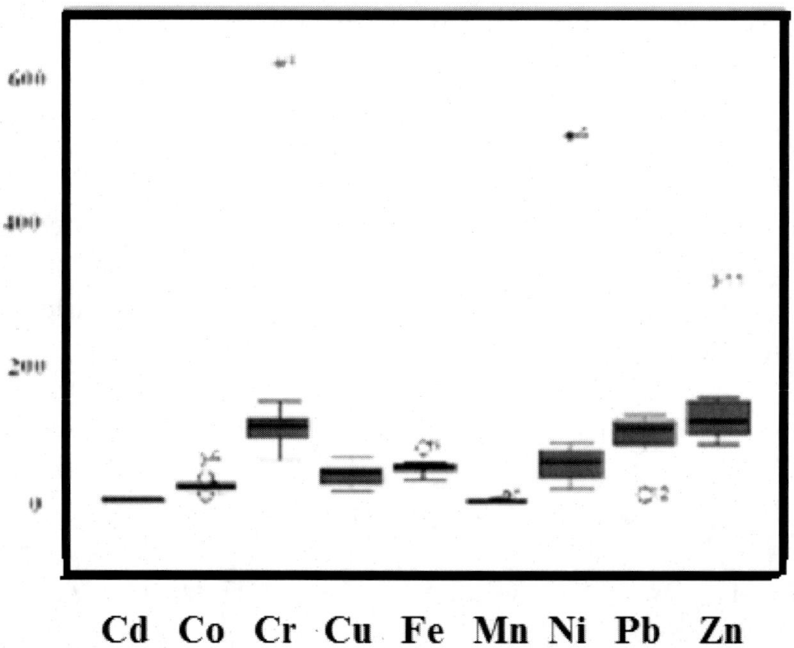

Figure 1. Box and whisker plot showing the variation of analyzed heavy metals in the reservoirs samples from Serbia.

The sites (Celije Lake, Krusevac and Barje Lake, Leskovac) are moderately polluted from the surrounding urban and semi urban areas surrounding town of Krusevac, and communal and industrial wastewater from the town Leskovac. Around the Celije Reservoir, main industrial sites are textile and chemical factories, while in Leskovac, the main industrial sites include food, pharmaceutical, machinery and leather production facilities. The overall level of economic development is low and the rate of wastewater treatment is insuficient.

Lake sediments were evaluated with respect to their pollution status by employing enrichment factors and comparison with sediment quality guidelines. Zn, Cu, Pb and As were found to have significant contributions from anthropogenic sources, particularly in the inner part. Most metals were found at concentrations that could adversely affect freshwater biota and sites could be considered of medium-low and medium-high priority with respect to the likelihood of toxicity (Šakan et al., 2017).

Cyanobacteria Pollution

Data on cyanobacterial occurrence, cyanotoxin production and the possible impact on humans and aquatic ecosystems in the Republic of Serbia were evaluated in over 70 data sources (Svircev et al., 2017). The microalgal and cyanobacterial research consisted of over 250 analyses, of which the majority (about 150) were conducted since the year 2000. Qualitative determination of cyanobacteria showed the presence of more than 70 species in investigated aquatic ecosystems, with 24 species from 13 genera being recorded in blooms (Svirčev et al., 2014).

Research from 2005 to 2007 showed the presence of common cyanotoxin, microcystin (MC) in Lake Palić and Lake Ludos. Furthermore, in 2012, the invasive *Cyl. raciborskii* was also found in these lakes (Institute of Public Health of Serbia, 2013).

Unlike the Vojvodina region where groundwater is used for water supply, central Serbia has a large number of surface reservoirs used for drinking water supply. There are more than 20 reservoirs used as sources of drinking water, and constant mass occurrences of cyanobacteria have been observed in nine of them.

Emergence of bloomforming cyanobacteria, as well as presence of cyanotoxins, has been documented in following Serbian drinking water supply reservoirs: Bovan, Bresnica, Garaši, Grlište, Grošnica, Gruža, Krajkovac and Pridvorica, the. (Svirčev et al., 2014). In Ćelije, the reservoir used for drinking water supply for the city of Kruševac, a bloom of *A. circinalis*, *Aph. flos-aquae* and *M. aeruginosa* was observed in 2004. MC was found in water samples from the reservoir (650 µg MC-LR L^{-1}) and in the tap water (2.5 µg L^{-1}). Recently, in December 2013, cyanobacterial blooms of *P. rubescens* has occurred in the Vrutci reservoir that is used for water supply of city of Užice, where 70.000 inhabitants were potentially exposed to cyanotoxins. Based on the number of cells per mL and the concentration of MCs, according to World Health Organization (WHO, 1999), water from reservoir Vrutci could be classified as high-risk water for recreation and drinking water extraction purposes. Furthermore, a questionnaire and epidemiological results showed that health problems possibly related to cyanotoxins (diseases of digestive system, skin and subcutaneous tissue) already occurred at least two years prior to the incident. This might be a sign that the population of city of Užice could have been exposed to the cyanobacteria and cyanotoxins even two years before the observed bloom in 2013 (Svirčev et al., 2016).

Epidemiological studies conducted in Serbia have revealed a significant correlation between an increased incidence of several cancers (brain, heart, mediastinum and pleural, ovarian, testicular, gastric, colorectal, retroperitoneal and peritoneal, leukaemia, malignant skin melanoma, and primary liver cancer) and cyanobacterial blooms in reservoirs used for drinking water supply (Svirčev et al., 2015). Reservoirs used for hydropower generation have been poorly investigated.

It could be concluded that cyanobacteria and cyanotoxins are omnipresent in different water bodies throughout the Republic of Serbia. The Serbian Cyanobacterial Database represents a treasure of information on cyanobacteria and their toxins, and serves as a model for other *countries in the region and beyond* (Svircev et al., 2017).

Legislation

In 2004, after the adoption, of general Law on the Protection of the Environment, the ecological legislation in Serbia has undergone important and profound changes. Two major characteristics of this transformation were, on the one hand, the adoption of numerous special laws and, on the other, the progressive harmonization of national legislations with EU law. However, in many areas, the EU law is only partially transposed, especially when the application of legal provisions depends on governmental decrees and/or ministerial decisions. Given the technicity of numerous ecological standards, national legislation often remains practically inapplicable without a series of acts implementing standards globally defined by provisions of general and special legislation.

The Water Law from 2010 defines legal water status, integrative management of water facilities and water land management, sources and mode of financing in the water sector, implementation of this law, as well as other relevant water issues. The lack of a national water protection strategy hampers strategic investment planning in the water sector. Completion of the surface and groundwater monitoring network is pending. Water pollution is another major problem, mainly due to the outdated technology, lack of pollution abatement installations, inadequate storage and disposal of byproducts, untreated industrial and municipal waste water, drainage water from agriculture, leachate from landfills, and pollution related to river navigation. Serbia will need to comply with the Water Framework, Urban Waste Water, Drinking Water, Groundwater, Water Quality Standards directives requiring Serbia to invest in the relevant water management and wastewater treatment facilities in the coming years.

While the European Commission has noted in its 2016 assessment report, that Serbia is still at the early stage of aligning policies and legislation with the environment and acquisition, even though some progress in further alignment has been made, the Serbian Government has, according to Post-Screening Documents, undertaken to fully harmonize its national legal and regulatory framework with respect to environmental issues. In

December 2016, the Government of Republic of Serbia adopted a Water Management Strategy in the Republic of Serbia until 2034, which defines long-term policy of water governance in the Republic of Serbia, sustainable activities in areas of water use, water protection, water courses regulation and protection of harmful water effects.

4. CONCLUSION

Water resources are under increased pressure in almost all parts of the world. In such circumstances, it is also common to have conflicts between different water sectors (for instance, tourism *vs.* environmental use; municipal and industrial supply *vs.* agricultural water use, etc.), and interest groups. In most cases, related problems could be efficiently solved through public participation and the involvement of stakeholders. Traditional public participation in water management is mostly focused on problem-solving, rather than on other important contexts, such as: stakeholders' understanding of the problem, motivation (willingness) to participate, preferences, understanding the solving methodology, and expectations that the participatory process will lead to the desired solution(s).

A healthy environment is rapidly becoming a very important issue in human life and a criterion for the quality of life. To achieve this target, many people consider "returning to nature" and respecting the ancient laws of nature as the ultimate solution.

Management techniques used for improving the water quality in reservoirs include: artificial mixing and oxygenation, sediment removal, sediment aeration, sediment covering, phosphorus inactivation, fish management, hydraulic regulation, algicides, light reduction, macrophyte control and ecorecommendation (ERM), and exactly ERM represents the "return to nature" aiming to preserve or re-establish the natural balance of the ecosystems.

REFERENCES

An, Y. J. & Kampbell, D. H. (2003). "Total, dissolved, and bioavailable metals at Lake Texoma marinas". *Environmental Pollution*, *122*, 2, 253-259.

Audry, S., Cécile Grosbois, Hubert Bril, Jörg Schäfer & Jakub Kierczak. (2010). "Postdepositional redistribution of trace metals in reservoir sediments of a mining/smelting-impacted watershed (the Lot River, SW France)." *Applied Geochemistry*, *25* (6), 778-794.

Beldowski, J., Beldowski, M., Miotk, M. & Penkowiak, J. (2014). "Total, methyl and organic mercury in sediments of the Southern Baltic Sea". *Marine Pollution Bulletin* 87(1-2) doi: 10.1016/j.marpolbul.2014.07.001.

Bing, H. & Wu, Y. (2016). "Current state, sources, and potential risk of heavy metals in sediments of Three Gorges Reservoir China". *Environmental Pollution*, *214*, 485-496. doi: 10.1016/j.envpol.2016.04.062.

Bogdanović, Ž. & Pavić, D. (2003). Hydroaccumulations in Serbia. In: Ivanc A., Miljanović B. (Eds). Hydroaccumulation, a multidisciplinary approach to sustainable development. *Monograph, Novi Sad.*, 3-7.

Devic, G. (2015). "*Environmental impacts of reservoirs, in Environmental Indicators*", eds. Robert H. Armon, Osmo Hänninen, Springer, p. 561-575.

Devic, G., Đorđevic, D. & Sakan, S. (2014). "Natural and anthropogenic factors affecting the groundwater quality in Serbia". *Science of the Total Environment*, 469-469, 933-942. doi: 10.1016/j.scitotenv.2013.09.011.

Devic, G., Đorđevic, D. & Sakan, S. (2014b). "Freshwater environmental quality parameters of man-made lakes of Serbia". *Environ Monit Assess.*, *186*, 5221–5234 doi: 10.1007/s10661-014-3771-0.

Devic, Gordana., Sakan, Sanja. & Đorđevic, Dragana. (2016). "Assessment of the environmental significance of nutrients and heavy metal pollution in the river network of Serbia". *Environmental Science and Pollution Research*, *23*, 1, 282-297, doi: 10.1007/s11356-015-5808-5.

Environmental Protection Agency, Republic of Serbia (SEPA). (2017). "*The state of environment in the Republic of Serbia.*" Ministry of Environment and Spatial Planning, Republic of Serbia, Belgrade, Serbia. URL: http://www.sepa.gov.rs/download// Zivotna sredina2017.pdf.

Gao, Q., Cheng, Q., Li, Y. & Yu, M. (2016). "Analysis and assessment of the nutrients, biochemical indexes and heavy metals in the Three Gorges Reservoir, China, from 2008 to 2013." *Water Research*, *92*. doi: 10.1016/j.watres.2015.12.055.

Invest-export. Brussels. *Water Sector in Serbia Wastewaters-overview*, Economic and Trade Office Embassy of Belgium Belgrade, May 2017.

Iqbal, J. & Shah, M. H. (2014). "Study of seasonal variations and health risk assessment of heavy metals in Cyprinus carpio from Rawal Lake, Pakistan". *Environ Monit Assess.*, *186*(4), 2025-37. doi: 10.1007/s10661-013-3515-6.

Li, T., Chu, C., Zhang, Y., Ju, M. & Wang, Y. (2017). "Contrasting Eutrophication Risks and Countermeasures in Different Water Bodies: Assessments to Support Targeted Watershed Management". *Int J Environ Res Public Health.*, *14*, 7, 695. doi: 10.3390/ ijerph14070695.

Öztürk, G., Özözen, O., Minareci, E. & Minareci, M. (2009). "Determination of heavy metals in fish, water and sediments of Avsar dam lake in Turkey". *Iranian Journal of Environmental Health Science & Engineering.*, *6*, 2, 73-80.

Priju, P. & Narayana, A. (2007). "Heavy and Trace Metals in Vembanad Lake Sediments". *International Journal of Environmental Research*, *1*(4), 280-289. ISSN: 1735-6865.

Richardson, D. C, Melles, S. J., Pilla, R. M., Hetherington, A. L., Knoll, L. B., Williamson, C. E., Kraemer, B. M., Jackson, J. R., Long, E. C. & Moore, K. (2017). "Transparency, Geomorphology and Mixing Regime Explain Variability in Trends in Lake Temperature and Stratification across Northeastern North America (1975–2014)". *Water*, *9*, 442.

Sakan, S., Đevic, G., Relic, D., Andjelkovic, I., Sakan, N. & Đordević, D. (2015). "Evaluation of sediment contamination with heavy metals: the importance of determing appropriate background content and suitable element for normaliyation." *Environ Geochem Health*, *37*, 97-113. doi: 10.1007/s10653-014-9633-4. https://www.blic.vesti.rs>vesti April, 26, 2018.

Sakan, S., Sakan, N., Andjelkovic, I., Trifunovic, S. & Djordjevic, D. (2017). "Study of potential harmful elements (arsenic, mercury and selenium) in surface sediments from Serbian rivers and artificial lakes". *Journal of Geochemical Exploration*, *180*, 24-34.

Sedlacek, J., Novakova, T. & Babel, O. (2016). "Sedimentary record and anthropogenic pollution of a complex, multiple source fed dam reservoirs: An example from the Nové Mlýny reservoir, Czech Republic." *Science of The Total Environment*, *574*. doi: 10. 1016/j.scitotenv.2016.08.127.

Stankovic, S. (2005). *Lakes of Serbia*. ISBN 10: 8617129905 ISBN 13: 9788617129901 p. 223. Publisher: Institute for Textbooks, Belgrade.

Svirčev, Y., Drobac, D., Tokodi, N., Lužanin, Y., Munjas, A. M., Nikolin, B., Vuleta, D. & Meriluoto, J. (2014). "Epidemiology of Cancers in Serbia and Possible Connection with Cyanobacterial Blooms". *Journal of Environmental Science and Health*, *Part C*, *32*, 4, 319-337, doi: 10.1080/10590501.2014.967053.

Svirčev, Z., Lujić, J., Marinović, Z., Drobac, D., Tokodi, N., Stojiljković, B. & Meriluoto, J. (2015). "Toxicopathology induced by microcystins and nodularin: A histopathological review". *J Environ Sci Health C Environ Carcinog Ecotoxicol Rev.*, *33*(2), 125-167.

Svirčev, Z., Obradović, V., Codd, G. A.., Marjanović, P., Spoof, L., Drobac, D., Tokodi, N., Petković, A., Nenin, T., Simeunović, J., Važić, T. & Meriluoto, J. (2016). "Massive fish mortality and Cylindrospermopsis raciborskii bloom in Aleksandrovac Lake". *Ecotoxicology*, *25*, 1353. https://doi.org/10.1007/s10646-016-1687-x.

Svirčev, Z., Tokodi, N. & Drobac, D. (2017). "Review of 130 years of research on cyanobacteria in aquatic ecosystems in Serbia presented in a Serbian Cyanobacterial Database". *Advances in Oceanography and Limnology*, *8*(1), doi: 10.4081/aiol.2017. 6360.

The Institute of Public Health of Serbia, "*Dr Milan Jovanović Batut, 2013, Results of the National Health Survey of the Republic of Serbia*."

Thevenon, F. & Poté, J. (2012). "Water Pollution History of Switzerland Recorded by Sediments of the Large and Deep Perialpine Lakes Lucerne and Geneva". *Water Air and Soil Pollution.*, *223*, 6157–6169. 10.1007/s11270-012-1347-6.

Waykar, B. B. & Petare, R. K. (2017). "Comparative variationsin protein contents of three freshwater snails bellamya bengalensis, mellanoides tuberculata and lymnaea acuminata collected from dedargaon reservoir of dhule district (M. S.) India, on exposure to heavy metals". *International Journal of Current Advanced Research*, *6* (12), 7914-7918. doi: http://dx.doi.org/10.24327/ijcar.2017.7918.1252.

WHO (World Health Organization). (1999). *Toxic Cyanobacteria in Water: A Guide to their Public Health Consequences, Monitoring, and Management, I.* Chorus and J. Bartram, (Eds.). E&FN Spon, London, UK.

Wildi, W. Environmental hazards of dams and reservoirs. In: *NEAR curriculum in Natural Environmental Science*. Genève: Section des sciences de la Terre et de l'environnement, 2010. (Terre & Environnement; 88) https://archive-ouverte.unige.ch/unige:90831.

Xu, Y., Wu, Y., Han, J. & Li, P. (20107). "The current status of heavy metal in lake sediments from China: Pollution and ecological risk assessment". *Ecol Evol.*, *7*, 14, 5454-5466. doi: 10.1002/ece3.312.

Zlatkovic, S., Šabić, M.. Milinčić, M., Knežević-Vukčević, J. & Stanković, S. (2010). "Geographical and biological analysis of the water quality of Bovan lake, Serbia". *Arch. Biol. Sci. Belgrade*, *62* (4), 1083-1087.

In: Advances in Hydrogeochemistry Research
Editors: S. Saeid Eslamian and Faezeh Eslamian
ISBN: 978-1-53617-829-6

Chapter 12

HYDROGEOCHEMISTRY OF CIRCUM-NEUTRAL MINE WATERS IN THE ANGLÈS OLD MINING AREA, NORTH EASTERN SPAIN

Andrés Navarro*

Department of Fluid Mechanics,
Universitat Politècnica de Catalunya, Terrassa, Spain

ABSTRACT

Circum-Neutral mine drainage may constitute a significant environmental problem due to the mobilization of metalloids such as As, Sb, Se, and metals such as Cd, Pb, and Zn. The neutral pH conditions of these waters may be attributed to the dissolution of acid-consuming carbonate minerals associated with carbonate gangue or host rocks.

In the abandoned Anglès mining district, mine water from Coral adit (Osor area) and mine shafts (Bonmatí area), generally, had neutral pH from carbonate and silicate dissolution, and showed high concentrations of Fe, Mn, Ni, Pb and Zn. Mine water samples showed high concentrations of Ca (116-189 mg/L), HCO_3^- (290-377 mg/L), SO_4^{-2} (156-288 mg/L) and an elevated pH (7.08-7.68), which may indicate calcite dissolution through the Coral adit. In addition, the elevated concentrations of Mn (0.19-1.8 mg/L) and Zn (>0.25-2.8 mg/L) suggest the dissolution of sphalerite and Mn oxides. Moreover, mine water showed elevated concentrations of some cations, such as Na (31.4-42.5 mg/L), Mg (18.0-28.0 mg/L), and K (4.5-6.2 mg/L), above their concentrations in surface water.

Geochemical modeling showed that albite and other aluminosilicates are not stable in these waters and likely dissolve; equilibrium with kaolinite also is a main processes controlling water chemistry.

Keywords: mine water, circum-neutral, drainage, hydrogeochemistry, modeling

* Corresponding Author's Email: andres.f.navarro@upc.edu.

INTRODUCTION

Although less well documented, neutral mine drainage may also constitute a significant environmental problem due to the mobilization of metalloids such as As, Sb, Se, and metals such as Cd, Pb, and Zn (Heikkinen et al. 2009; Plante et al. 2011; Jang and Kwon 2011). The geochemistry of the mineral-water interaction is dependent on the ore and gangue minerals, local water chemistry, and geologic setting, and some authors have shown the relationship between the types of mineral deposits and their environmental signature (Plumlee et al. 1999; Seal and Foley 2002; Seal and Hammarstrom 2003; Seal et al. 2008; Da Pelo et al. 2009). Thus, the drainage-water compositions of polymetallic vein and related deposits that contain significant carbonate minerals in their gangue or wallrock alteration tend to have near-neutral pH, and elevated concentrations of Zn and Cu (Plumlee et al. 1999; Navarro and Martínez 2010; Hidalgo et al. 2010; Papp et al. 2017; Majzlan et al. 2018; Navarro and Carulla 2018). In the same way, some epithermal deposits may be associated to mine water and/or leachates of circum-neutral pH (Carrillo-Chávez et al. 2003; Navarro et al. 2006, 2009; Bori et al. 2016).

Besides, polymetallic replacement deposits hosted by carbonate sedimentary rocks tend to have mine-drainage water compositions with near-neutral pH; elevated concentrations of dissolved SO_4^{2-}, Fe, Zn, As, Sb, and Tl may be generated under these conditions (Plumlee et al. 1999; Al et al. 2000; Navarro et al. 2004; Lee et al. 2005; Bucher et al. 2009; Navarro and Cardellach 2009; Lindsay et al. 2009). Associated with Skarn and Mississippi Valley deposits have been detected waters with high pH and elevated concentrations of Zn, As and other metalloids (Irîbar et al. 1999; Pfeifer et al. 2007; Frau et al. 2009; Cidu et al. 2007, 2009, 2012; Daniele et al. 2013). Also, elevated concentrations of Mn, Ni, and Fe in pore water from low sulfide tailings from Ni-Cu sulfide and chromite ores were documented by Heikkinen et al. (2009), Heikkinen and Räisänen (2009) and Kumari et al. (2017).

The neutral pH conditions of these waters may be attributed to the dissolution of acid-consuming carbonate minerals associated with carbonate gangue or host rocks. In some situations, precipitation of jarosite, gypsum, goethite, Fe oxy-hydroxides, siderite, and other secondary phases control Ca, Fe, Na, K, SO_4^{2-} and trace element concentrations (Blowes et al. 1994, 1998; Jurjovec et al. 2002; Navarro and Martínez 2010). Even when carbonate rocks host the affected aquifers, the near-neutral groundwater may show high dissolved concentrations of As, Fe, Ni, SO_4^{2-}, Zn, Cd, and Pb (Younger 2000; Blowes et al. 1998; Ezekwe et al. 2013).

In old mining areas, such as the abandoned mining district of Anglés (Girona, NE Spain), the exploitation of Pb-Fe-Ba (Bonmatí area) and F-Ba-Pb-Zn (Osor area) ores until 1980 generated an important source of contamination by copper, lead, zinc and other metals in surface waters, groundwater and soils located in the vicinity of the main mine-waste impoundments. The main consequences of unremediated mine sites containing sulfide minerals, mainly pyrite and/or pyrrhotite, are the generation of mine drainage waters and the possible mobilization of metals and metalloids (Plumlee et al., 1999; Lottermoser 2010). Depending on the neutralization capacity of the mining wastes, tailings and ore host rocks, the production of neutral mine waters may be of environmental concern.

The main objectives of this study were to evaluate metal mobilization and release of toxic elements from mine wastes and mine waters, in pH-Eh conditions associated to circum-neutral mine waters. In addition, geochemical modeling was used in order to evaluate the geochemical processes that may explain the geochemistry of mine water, and the relationship with the mineralogical characteristics of mine wastes and ore mineralogy.

ANGLES OLD MINING AREA

The abandoned Anglés mining area (Figure 1) lies some 35 km SE of Girona, in the La Selva basin and Montseny-Guilleries massif, which forms part of the Catalonian Coastal Range (CCR) in the NE section of the Iberian Peninsula (Navarro et al. 2015). This old mining area comprised, approximately, 100 km^2 located between 1120-1130 and 830-840 U.T.M. coordinates. The Anglés mineral deposits are located in the Paleozoic materials of the Montseny-Guilleries massif and occur as stratabound and vein deposits associated with igneous and metasedimentary rocks (Figure 1). These Paleozoic materials comprise the Lower Infra-Caradocian medium to high-grade metamorphic sequence, which can be divided into the high-grade Osor complex and the medium to low-grade Susqueda complex that are two pelitic series separated by orthogneiss.

Stratigraphically above lies the St. Martí Sacalm Caradocian volcano-sedimentary formation, the Silurian black slates and the carbonatic Devonian St. Martí de Llémana formation. The Paleozoic sequence shows a low and high-grade metamorphism (amphibolite facies) associated with the Variscan orogeny. The Paleozoic materials were intruded by post-tectonic granitoids, mainly biotite granodiorites of calc-alkaline composition.

The main deposits exploited in this area were the stratabound polymetallic deposits of Pb-Fe-Ba in the Bonmatí sector (Figure 1), associated with Susqueda Complex marbles and the F-Ba-Pb-Zn vein deposits of possible Jurassic age in the Osor area (Piqué et al. 2008). Thus, several mineralized veins formed at low temperatures are found enclosed within the Paleozoic materials (Canals and Cardellach 1997; Piqué et al. 2008); these veins include the following:

- Pb-rich veins of predominantly carbonatic gangue (dolomite and ankerite) with significant vertical development are enclosed exclusively within the Hercynian basement (Bellmunt area, in the south of the CCR);

- Metal-poor and fluorite-rich veins are also enclosed within the basement (Osor veins);
- Barite-rich veins are associated with normal faults and minor sulfides (mainly galena) (Serrat Blanc mines in the Bonmatí sector).

The polymetallic stratabound deposits in the Bonmatí sector are hosted in marbles associated with Susqueda Complex metasedimentary materials, comprising disseminated galena and chalcopyrite in a carbonate-rich matrix, with calcite and ankerite as the dominant gangue minerals. These deposits were mined by underground mining until 1936 (Carlota mine) and by small open pits until 1972. In the second half of the 20th century, between 1962 and 1972, mine wastes, including tailings, were accumulated at the Bonmatí gravimetric plant (Figure 1). Ore production during this period was in the Bonmatí area of 8000 t/y, producing 3000 t of barite concentrates and 500 t of Pb concentrates per year.

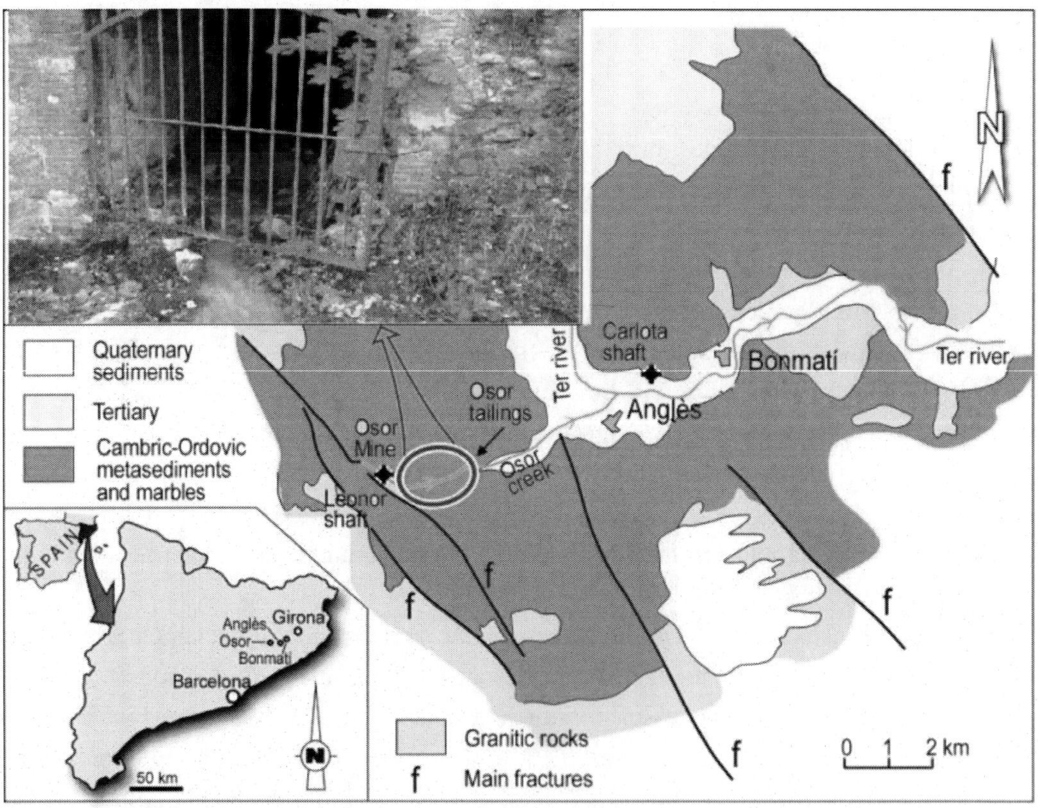

Figure 1. Situation map of the study area. Synthetic geologic map and detail of Coral adit.

The Osor vein deposit is located 4 kmSE of the Anglés town, and includes several geologically similar and thick (1-4m) fluorite-barite-sphalerite-galena veins, exploited to a depth of 300m. Gangue minerals included quartz, barite, calcite, pyrite and silicates (mainly muscovite, albite and biotite). Exploitation of these veins concluded in 1980, after reaching yearly productions of 20000-30000 t of fluorite, 2000 t of Pb concentrates and 3000 t of Zn concentrates. The Osor flotation tailings were homogeneous in grain size and composition, occupying an area of $3150m^2$ with a mean thickness of 15 m, while the gravimetric tailings from Bonmatí extended over approximately $22500m^2$, and comprised a total affected volume of $8000m^3$ of sediments and contaminated soils. Mine drainage was mainly through the Coral adit (Figure 1), which drains the Osor vein system, with an estimated discharge into the Osor Creek of between 800 and $1100m^3$ /day of near neutral contaminated mine water. On an annual basis, mine water discharge was highest in the autumn or spring, and generally lowest in the late summer and early winter.

The area studied is characterized by a typically Mediterranean climate, with a mean yearly rainfall of 736mm and a mean annual temperature of 15° C, with mean maximum and minimum temperatures of 21.1 and 9.5° C, respectively. The only important water course in the area is the Ter River, which flows near the Bonmatí tailings dumping area, and its tributary, the Osor Creek, which receives contaminated sediment and episodic draining waters from the Osor tailings area and contaminated mine water from the Coral adit. Estimated mean discharge of the Ter River varied widely throughout the year, ranging from $3\text{-}15m^3/s$ in the summer to $17\text{-}52m^3$ /s in the wet period, mainly during early spring and fall (Navarro et al. 2015).

MATERIALS

Geochemistry and Mineralogy of Mineralization and Mine Wastes

Identification and analysis of the mineral phases in selected samples were performed in the laboratories of the University of Barcelona (UB) through X-ray diffraction (Table 1). Once the materials had been dried and ground, their geochemical composition was analyzed using instrumental neutron activation analysis (INAA) in Actlabs (Ontario, Canada). The following elements were quantified: Au, Ag, As, Ba, Br, Ca, Co, Cr, Cs, Fe, Hf, Hg, Ir, Mo, Na, Ni, Rb, Sb, Sc, Se, Sn, Sr, Ta, Th, U, W, Zn, La, Ce, Nd, Sm, Eu, Tb, Yb, and Lu. In addition, the concentrations of the following elements were determined by acid digestion and subsequent analysis by inductively coupled plasma atomic emission spectrometry (ICP-AES): Ag, Cd, Cu, Mn, Mo, Ni, Pb, Zn, Al, Be, Bi, Ca, K, Mg, P, Sr, Ti, V, Y, and S (Table 2).

Mine Waters

Mine waters were collected from the Coral adit, which drained the abandoned mine of Osor (Girona, NE Spain) and the Grevolosa creek which recharges the old mining works. Water sampling was carried out between 2009 and 2014 and comprised six campaigns (Tables 3 and 4). Flow-rate quantification of the Coral adit was performed using a broad-crested weir, which is a structure in an open channel that has a horizontal crest above which the fluid pressure may be considered hydrostatic (Munson et al. 2009). Also, groundwater samples from old mine shafts and two piezometer wells (PZ-5, PZ-12) excavated into the contaminated area of Bonmatí (Figure 1), were collected between 2009 and 2010.

The pH, redox potential (Eh; mV), temperature and electrical conductivity (EC; μS/cm) measurements were calibrated using standard solutions and measured in situ with portable devices (HACH model sensION TM378). The samples were filtered using a cellulose nitrate membrane with a pore size of 0.45 μm. Samples for cation analysis were later acidified to pH < 2.0 by adding ultra-pure HNO3.

The samples were collected in 110 mL high-density polypropylene bottles, sealed with a double cap, and stored in a refrigerator until analysis. Metal concentrations were measured by inductively coupled plasma mass spectrometry (ICP-MS) at Actlabs (Ontario, Canada). Chloride, nitrate and sulfate concentration were analyzed on unfiltered samples by ion chromatography (IC). Water alkalinity was analyzed by titration. Standard reference material NIST 1640 (ICPMS) was used to confirm accuracy. In order to preserve the carbonate equilibrium and CO2 (g) levels in water samples, the samples were quickly closed and rapidly taken to the laboratory after sampling.

Hydrogeochemical analyses of leachates were performed using the PHREEQC[1] (Parkhurst et al. 1999) numerical code (version 3.0.6-7757) to evaluate the speciation of dissolved constituents and calculate the saturation state of the effluents. The MINTEQ thermodynamic database was used for chemical equilibrium calculations. In the reaction-path modeling and Eh-pH diagrams was used The Geochemist´s Workbench (Bethke and Yeakel 2011).

The mineralogical composition of solid samples is given in Table 1. XRD analyses showed that quartz, cerussite, calcite, barite, muscovite and kaolinite were the most abundant phases detected in mine wastes and mineralizations from the Bonmatí sector. In the Osor sector, quartz, fluorite, barite, calcite, galena, sphalerite, albite, biotite, muscovite and kaolinite were detected by XRD and visual inspection. Pyrite was present in both mining sectors, although it was more abundant in the Osor vein deposit.

[1] The PHREEQC geochemical model use solution equilibrium models to calculate aqueous speciation and saturation state of mineral phases, among other possibilities, and have been applied to evaluate the mobility of heavy metals.

Table 1. Identified main mineral phases in ore and mine wastes

Mineral phase	Formula
Mine wastes (Bonmatí mining area)	
Quartz	SiO2
Cerrusite	PbCO3
Mineral phase	Formula
Calcite	CaCO3
Barite	BaSO4
Jarosite	KFe3 (SO4)2 (OH)6
Hydrozincite	FZn5 (CO3)2 (OH)6
Moscovite-Kaolinite	KAl2(Al,Si)4 O10 (OH,F)2
Rustumite	Ca10(Si2O7)2 (SiO4) Cl2 (OH)2
Pyrite	FeS2
Mine wastes (Osor mining area)	
Quartz	SiO2
Galena	PbS
Sphalerite	ZnS
Fluorite	CaF2
Calcite	CaCO3
Albite	Na0.986(Al1.005Si2.995O8)
Barite	BaSO4
Moscovite-Kaolinite	KAl2(Al,Si)4 O10 (OH,F)2 Al4(Si4O10) (OH)8
Biotite	KMg1.5Fe1.5Al Si3 O10 (OH)2
Pyrite	FeS2

Table 2 shows the element concentrations in ore and mine wastes from the Bonmatí and Osor areas. Table 2 also includes analyses of Bonmatí ore samples from nearby mines (Carlota, Serrat Blanc and Ermita mines) that are typical of the ore processed in the area. Bonmatí tailings showed high amounts of Cd, Cu, Pb, Zn, As, Ba, Se and Sb, above the regulatory levels in Catalonia for soils of industrial use (level CAL in Table 2). The Bonmati tailings (Table 2) had a higher As content (1850 mg·kg-1) and an elevated Zn content, which reached 10.0% in the Bonmati tailings, which may be due to the absence of Zn recuperation in the old gravimetric plant.

Mine waste sample from the Osor sector (Table 2) contained high amounts of Pb (>5000 mg kg-1) due to the presence of argentiferous galena and high concentrations of Na, K and Mg from silicates and possibly jarosite. The high content of Zn (11300 mg kg-1) and Cd (24.9 mg kg-1) was associated with sphalerite, whereas the elevated concentration of Sb may be linked to galena or undetected sulfosalts. Osor flotation tailings sample (Table 2) contains high amounts of Al, Ba and Sr, due to concentration of gangue material in the flotation processes.

Table 2. Concentrations of metals and metalloids in ore samples and mine wastes. Values in mg·kg^{-1} except Fe and Sr (%). *: ore samples of the Bonmatí sector. CAL(**): The Catalonian soil intervention values (industrial use) in ppm. ND: non determined

Element	Ag	Cu	Cd	Pb	Ni	Zn	As	Ba	Bi	Cr	Fe	Hg
Unit	ppm	ppm	ppm	ppm	ppm	ppm	ppm	ppm	ppm	ppm	%	ppm
Detection limit	0.3	1	0.3	1	1	1	0.01	50	2	0.3	0.01	0.05
Bonmatí tailings	325	>10000	97.8	>5000	453	>100000	1850	1800	<2	198	2.63	13
Carlota mine*	1.7	42	0.5	>5000	5	26	9	1600	< 2	30	5.14	< 1
Ermita mine*	2.5	22	0.8	>5000	19	52	2.8	<50	<2	<2	5.44	<1
Serrat Blanc mine*	2.0	21	<0.3	39	9	21	11	330000	<2	71	0.27	<1
Osor mining wastes	29.9	11	24.9	>5000	3	11300	15.2	250	0.4	<2	0.35	3
Osor tailings	0.6	47	7.6	940	18	2370	12.5	5110	0.3	39	1.78	<1
Osor vein	66.2	88	68.5	> 5000	31	34000	164	2190	< 2	119	0.96	5
CAL**	---	1000	55	550	1000	1000	30	1000	---	1000	---	30
Element	Mn	Na	P	Sb	Se	Sr	Th	U	W	Nd	Sm	Sn
Unit	ppm	%	%	ppm	ppm	ppm	ppm	ppm	ppm	ppm	ppm	%
Detection limit	1	0.01	0.001	0.005	0.1	1	0.1	0.01	0.05	0.3	0.001	0.01
Bonmatí tailings	2260	0.18	0.087	761	77	160	<0.2	9.7	22	<5	1.6	<0.01
Carlota mine*	6170	0.03	0.003	4	< 3	58	<0.2	2.5	< 1	53	8.4	<0.01
Ermita mine*	7690	0.03	0.004	0.7	<3	19	0.6	<0.5	<1	43	10.8	<0.01
Serrat Blanc mine*	32	0.01	0.008	2.8	<3	79.3	1.5	<0.05	9	<5	0.3	<0.01
Osor mining wastes	89	0.07	0.008	56.5	<0.1	17	1	<0.5	<1	<5	2	<0.01
Osor tailings	684	1.0	0.041	1	<0.1	105	5.2	2.4	<1	24	5	0.02
Osor vein	109	0.14	0.008	208	<3	74	1.5	<0.5	<1	<5	1.4	<0.01
CAL**	---	-----	---	30	70	---	---	---	---	---	---	0.1

Waters

Dissolved concentrations of major and minor components are given in Tables 3 and 4, showing a considerable difference between the samples obtained during the summer and winter sampling campaigns. Besides, groundwater from Bonmatí area (PZ-5 and PZ-12 in Tables 3 and 4) showed the most high values of dissolved metals, specially Zn, Cu and Pb, which reached some mg/L of these metals dissolved. The hydrogeochemistry of Osor water samples and mine waster showed that superficial waters are a dominant HCO_3-Ca character, while mine water samples of the Coral adit (OS-8, OS-11, GC-1, GC) showed a dominant HCO_3, SO_4-Ca composition.

The composition of these waters, also shows a possible close relationship between the Coral mine water adit (OS-8 and OS-11 samples) and Grevolosa creek (OS-6 samples) (Tables 3 and 4), suggesting that superficial water recharge with rock-vein interaction may be the origin of the hydrogeochemical composition of mine water. The variations in dissolved metal concentrations with pH shows that base metals such as Cd and Cu tended to increase with decreasing pH, indicating higher concentrations in the groundwater from the Bonmatí sector.

Pb and Zn were the predominant dissolved base metals, typically ranging between 110-7800 μg/L and 3800-38000 μg/L, respectively, in contaminated groundwater. In mine water samples, Zn reached 2800 μg/L, showing high concentrations under neutral pH conditions (7.08-7.68). Mine water samples also showed elevated concentrations of Fe and Mn (Table 3).

Oxyanion-forming elements such as As showed significant concentrations under neutral-alkaline pH conditions, whereas the lower concentrations were associated with water samples with low pH (groundwater) and some samples with an elevated pH (superficial water samples). In general, the low concentration of As in waters may be associated with lower percentages of arsenical pyrite and arsenopyrite in the minerals exploited, and/or the possible adsorption onto Fe oxyhydroxides (Navarro and Domènech 2010).

Mine water samples (OS-8 and OS-11) showed high concentrations of Ca (116-189 mg/L), HCO_3 (290-377 mg/L), SO_4 (156-288 mg/L) and an elevated pH (7.08-7.68), which may indicate the calcite dissolution through the Coral adit. Also, the elevated concentrations of Mn (0.19-1.8 mg/L) and Zn (up to 2.8 mg/L) suggest the dissolution of sphalerite and Mn oxides. Moreover, mine water showed elevated concentrations of some cations, such as Na (31.4-42.5 mg/L), Mg (18.0-28.0 mg/L) and K (4.5-6.2 mg/L), above their concentrations in surface water and springs (Table 3). The weathering of some detected silicates (albite and muscovite) may, also, explain the high pH detected. The main hydrochemical reactions associated to mine water in the Osor sector may be the following:

(i) Pyrite and sphalerite oxidation.

$$FeS_2 + 7/2\ O_2 + H_2O \rightarrow Fe^{2+} + 2\ SO_4^{2-} + 2\ H^+ \qquad (1)$$

$$ZnS + 2\ O_2 \rightarrow Zn^{2+} + SO_4^{2-} \qquad (2)$$

(ii) Calcite and silicate (albite and muscovite) dissolution.

Table 3. Main metal contents of groundwater, surface water and mine water samples. Values in µg/L. *: ICP-MS and ICP-OES, ND: non determined. OS-6(S,W,A): Grevolosa creek (summer/winter/autumn), OS-8(S,W,A): Coral mine drainage (summer/winter/autumn), OS-11: Coral mine drainage (summer 2009), GC-1 and GC: Coral mine drainage (summer 2011 and 2014), Carlota: mine shaft in Bonmatí sector, PZ-5(S,W): piezometer in Bonmatí sector (summer/winter), PZ-12(S,W): piezometer in Bonmatí sector (summer/winter). Modified from Navarro et al. (2015)

Analyte Symbol	Na	Mg	Al	K	Ca	Mn	Fe	Co
Unit Symbol	µg/L	µg/L	µg/L	µg/L	µg/L	µg/L	µg/L	µg/L
Detection Limit	5	1	2	30	700	0.1	10	0.005
OS-6 (S)	10300	4160	6	990	19800	2.4	20	0.026
OS-6 (W)	7720	2580	37	750	11900	2.0	30	0.05
OS-6 (A)	8740	3760	5	1020	19600	6	70	0.15
OS-8 (S)	34900	19800	4	5100	134000	879	< 10	21.8
OS-8 (W)	31400	18000	6	4590	116000	193	<10	3.66
OS-8 (A)	28700	16400	3	4380	135000	749	240	20.1
OS-11 (S)	42500	28000	49	6250	189000	1860	440	53.2
GC-1	35000	20000	13	6080	>20000	1980	270	53.7
GC	38400	24300	120	5440	167000	841	120	19.1
Carlota	19400	14800	5	1740	104000	7.3	100	0.05
PZ-5 (S)	ND	ND	ND	ND	ND	ND	ND	ND
PZ-12 (S)	ND	ND	ND	ND	ND	ND	ND	ND
PZ-5(W)	ND	ND	ND	ND	ND	ND	ND	ND
PZ-12 (W)	ND	ND	ND	ND	ND	ND	ND	ND

Analyte Symbol	Ni	Cu	Zn	As	Se	Cd	Ba	Pb	U	
Unit Symbol	µg/L	µg/L	µg/L	µg/L	µg/L	µg/L	µg/L	µg/L	µg/L	
Detection Limit	0.3	0.2	0.5	0.03	0.2	0.01	0.1	0.01	0.001	
OS-6 (S)	0.6	0.8	192	0.17	<0.2	0.57	32.4	13.6	0.36	
OS-6 (W)	0.5	1.5	55.6	0.15	<0.2	0.14	20.2	3.63	0.21	
OS-6 (A)	0.7	2.5	443	0.17	<0.2	1.19	36.8	36.7	0.52	
OS-8 (S)	19.4	0.3	2800	0.6	0.3	2.47	39.9	2.61	8.7	
OS-8 (W)	15.1	0.6	>250	0.47	0.3	1.67	36.4	1.14	6.94	
OS-8 (A)	17.9	2.8	2810	0.42	0.4	1.73	29.8	1.92	7.9	
OS-11 (S)	38.7	10.2	2800	2.2	<0.2	3.05	29.4	19.4	13.0	
GC-1	34.6	1.6	>250	1.67	0.7	3.65	31.2	3.16	11.2	
GC	17.9	7.3	2900	1.59	4.4	1.98	34.1	2.11	9.16	
Carlota	4.9	3.0	204	0.24	0.2	0.19	163	10.2	3.1	
PZ-5 (S)	140	1600	13000	<0.5	ND	52	52	120	ND	
PZ-12 (S)	140	3400	19000	8.3	ND	260	280	7800	ND	
PZ-5(W)	370	4400	38000	<0.5	ND	140	<5	380	ND	
PZ-12 (W)	32	54	3800	<0.5	ND	14	91	110	ND	

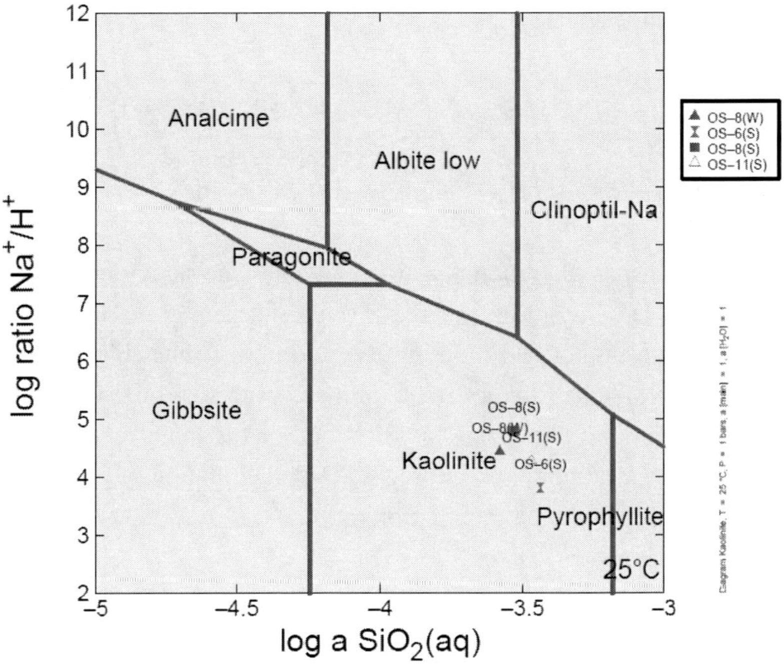

Figure 2. Activity diagram for the feldspar system using selected samples.

Table 4. Main anion contents of groundwater, surface water and mine water samples. Values in mg/L. OS-6 (S,W): Grevolosa creek (summer/winter), OS-8 (S,W,A): Coral mine drainage (summer/winter/autumn), OS-11: Coral mine drainage (summer 2009), GC-1 and GC: Coral mine drainage (2011 and 2014). Carlota: mine shaft in Bonmatí sector. DL: detection limit. Modified from Navarro et al. (2015)

Analyte	pH	Eh	C	F	Cl	NO3 (as N)	PO4 (as P)	SO4	HCO3
Unit	pH unit	mV	µS/cm	mg/L	mg/L	mg/L	mg/L	mg/L	mg/L
DL	---	---	---	0.01	0.03	0.01	0.02	0.03	1
Method	---	---	---	IC	IC	IC	IC	IC	TITR
OS-6 (S)	7.2	200	145	0.2	10	0.1	< 0.02	8.8	72
OS-6 (W)	7.4	103	78.7	0.1	6.1	0.45	<0.02	8.7	45
OS-8 (S)	7.68	197	950	1.9	28	< 0.03	< 0.06	180	324
OS-11 (S)	7.08	21	1180	2.1	29	< 0.03	< 0.06	288	377
OS-8 (W)	7.38	-2	860	2.1	26	<0.02	<0.04	156	290
OS-8 (A)	7.35	-2	563	2.31	26.9	<0.03	<0.06	171	293
Carlota	7.13	108	446	0.07	39	6.11	< 0.04	44	271
GC-1	7.0	-420	380	ND	ND	ND	ND	ND	ND
GC	7.46	98.6	678	2.73	22.4	0.4	ND	210.7	408

$$CaCO_3 + H^+ = Ca^{2+} + HCO_3^- \text{ (pH > 6.4)} \quad (3)$$

$CaCO_3 + 2H^+ = Ca^{2+} + H_2CO_3^-$ (pH < 6.4) (4)

$2\ NaAlSi_3O_8 + 2H^+ + 9\ H_2O \rightarrow 2\ Na^+ + Al_2Si_2O_5(OH)_4 + 4H_4SiO_4$ (5)

$2\ KAl_3Si_3O_{10}(OH)_2 + 2H^+ + 3\ H_2O \rightarrow 2\ K^+ + 3Al_2Si_2O_5(OH)_4$ (6)

Progress of the feldspar alteration reactions changes the activities of the envolved solutes: Ca^{2+}, Na^+, K^+, H^+, $SiO_2(aq)$, thus the activity diagram Na^+/ H^+ versus $SiO_2(aq)$, showed that all the samples plot into the stability field of kaolinite (Figure 2). Therefore, albite and other aluminosilicates are not stable in these waters and must dissolve, also the equilibrium with this mineral phase (kaolinite) is one of the main processes controlling water chemistry. At the Coral portal was observed the precipitation of Fe oxy-hydroxides, because the possible oxidation and precipitation of Fe:

(i) Fe2+ oxidation and oxy-hydroxides precipitation.

Fe2+ + 1/4 O2 + H+ → Fe3+ +1/2 H2O (7)

FeS2 + 14 Fe3+ + 8 H2O → 15 Fe2+ + 2 SO42- + 16 H+ (8)

Fe3+ + 3 H2O → Fe (OH)3 + 3 H+ (9)

These neutralization processes and the scarce quantity of pyrite, may explain the elevated pH of waters and to a lesser extent, the mobility of metals such as Zn, and the possible immobilization of Fe in mine waters, resulting in low concentrations of Fe at mine waters (samples OS-8 and OS-11 in Table 3).

Geochemical Modeling

Calculations using PHREEQC for representative water samples (Table 5) showed that the most abundant species of Fe are Fe^{2+} and $FeSO_4$. The aqueous speciation of Pb was dominated by $PbCO_3^0$ (Table 5) with other species (Pb^{2+}, $PbHCO_3^+$, $PbOH^+$) lower by one and two orders of magnitude, in a similar proportion to other Pb contaminated areas.

The aqueous speciation of Zn showed that Zn^{2+} is the most abundant species. The remaining species ($ZnSO_4^0$, $ZnCO_3^0$, $Zn(CO_3)_2^{-2}$) are, generally, lower by one and two orders of magnitude.

Table 5. Distribution of species for water samples and leachates. Data calculated using PHREEQC and database MINTEQ. Values in mol·kg^{-1}

Species	OS-6(S)	OS-8(S)	OS-11(S)	Carlota shaft
Cu				
$Cu(OH)_2$	$5.92 \cdot 10^{-9}$	$2.26 \cdot 10^{-9}$	$2.17 \cdot 10^{-8}$	$6.16 \cdot 10^{-9}$
$CuCO_3$	$4.27 \cdot 10^{-9}$	$2.20 \cdot 10^{-9}$	$9.75 \cdot 10^{-8}$	$1.81 \cdot 10^{-8}$
Cu^{+2}	$1.40 \cdot 10^{-9}$	$7.84 \cdot 10^{-11}$	$1.28 \cdot 10^{-8}$	$2.54 \cdot 10^{-9}$
$CuHCO_3^+$	$5.29 \cdot 10^{-10}$	$9.67 \cdot 10^{-11}$	$1.73 \cdot 10^{-8}$	$2.79 \cdot 10^{-9}$
Fe				
Fe^{2+}	$3.27 \cdot 10^{-7}$	$9.34 \cdot 10^{-8}$	$6.92 \cdot 10^{-6}$	$1.74 \cdot 10^{-6}$
$FeSO_4$	$2.82 \cdot 10^{-9}$	$8.43 \cdot 10^{-9}$	$8.46 \cdot 10^{-7}$	$4.27 \cdot 10^{-8}$
$Fe(OH)^{2+}$	$2.23 \cdot 10^{-8}$	$4.33 \cdot 10^{-8}$	$9.09 \cdot 10^{-8}$	$2.01 \cdot 10^{-9}$
$Fe(OH)_3$	$3.93 \cdot 10^{-9}$	$2.16 \cdot 10^{-8}$	$1.12 \cdot 10^{-8}$	$2.87 \cdot 10^{-10}$
Pb				
$PbCO_3$	$8.07 \cdot 10^{-9}$	$1.16 \cdot 10^{-8}$	$8.07 \cdot 10^{-8}$	$4.32 \cdot 10^{-9}$
Pb^{2+}	$8.19 \cdot 10^{-10}$	$1.24 \cdot 10^{-10}$	$3.16 \cdot 10^{-9}$	$1.82 \cdot 10^{-10}$
$PbHCO_3^+$	$4.90 \cdot 10^{-10}$	$2.50 \cdot 10^{-10}$	$7.03 \cdot 10^{-9}$	$3.25 \cdot 10^{-10}$
$PbOH^+$	$2.15 \cdot 10^{-10}$	$8.14 \cdot 10^{-11}$	$4.95 \cdot 10^{-10}$	$3.49 \cdot 10^{-11}$
Zn				
Zn^{2+}	$2.37 \cdot 10^{-6}$	$1.46 \cdot 10^{-5}$	$2.27 \cdot 10^{-5}$	$1.91 \cdot 10^{-6}$
$ZnSO_4$	$2.91 \cdot 10^{-8}$	$1.87 \cdot 10^{-6}$	$3.96 \cdot 10^{-6}$	$6.73 \cdot 10^{-8}$
$ZnCO_3$	$2.70 \cdot 10^{-7}$	$1.61 \cdot 10^{-5}$	$6.90 \cdot 10^{-6}$	$5.31 \cdot 10^{-7}$
$Zn(CO_3)_2^{-2}$	$5.06 \cdot 10^{-9}$	$4.81 \cdot 10^{-6}$	$6.40 \cdot 10^{-7}$	$3.63 \cdot 10^{-8}$

The calculated saturation index, showed that most waters were saturated with respect to ferrihydrite and goethite, with the exception of the Carlota shaft sample, which showed the most elevated concentrations of Fe, and some mine water samples (samples OS-8 and OS-11 of Coral adit), which were near to equilibrium. On the other hand, waters were undersaturated as regards Fe sulfate minerals and gypsum, with the exception of water sample OS-11 (Coral adit drainage) which showed a value close to equilibrium for jarosite.

As regards Pb and Zn minerals, all the water samples were undersaturated with anglesite, cerussite, smithsonite, goslarite and zincosite, which may explain the mobilization of Pb and Zn in aqueous phase. Some samples showed SI values near to equilibrium with respect to cerussite, which suggests the possible control of dissolved Pb by $PbCO_3$. In addition, mine drainage waters (OS-8 and OS-11) were close to equilibrium as regards smithsonite, which may indicate control of dissolved Zn by this mineral phase.

Furthermore, carbonate minerals, calcite, dolomite and siderite were undersaturated in almost all the samples, showing values near to equilibrium in mine drainage waters. Eh-pH conditions were congruent with these data, showing that under the Eh-pH conditions of waters, the most stable species for Fe, Pb and Zn were oxy-hydroxides of Fe, $PbCO_3^0$ and $ZnSO_4^0$ (Figures 3 to 5).

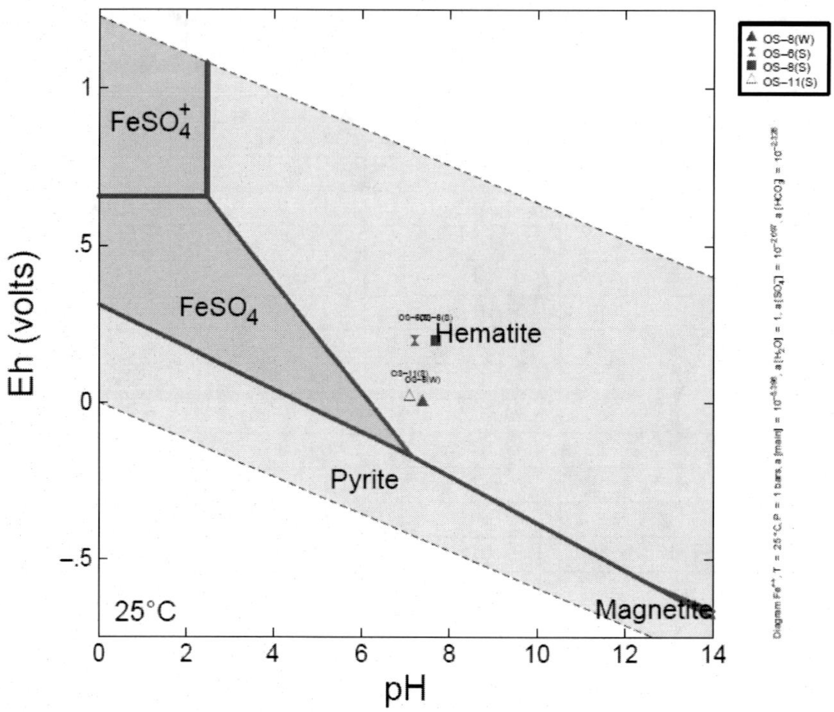

Figure 3. Eh-pH diagram for Fe using selected samples.

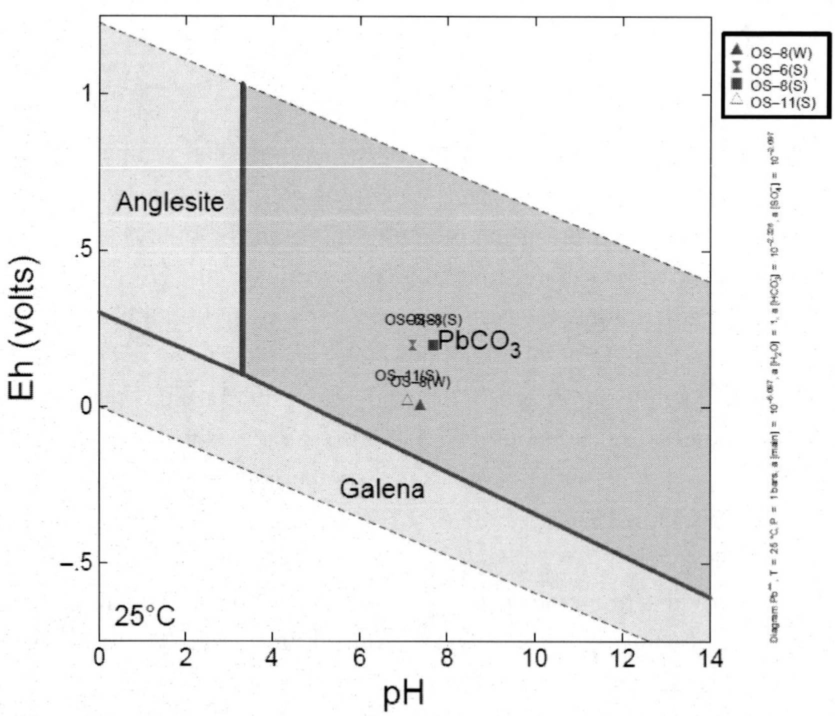

Figure 4. Eh-pH diagram for Pb using selected samples.

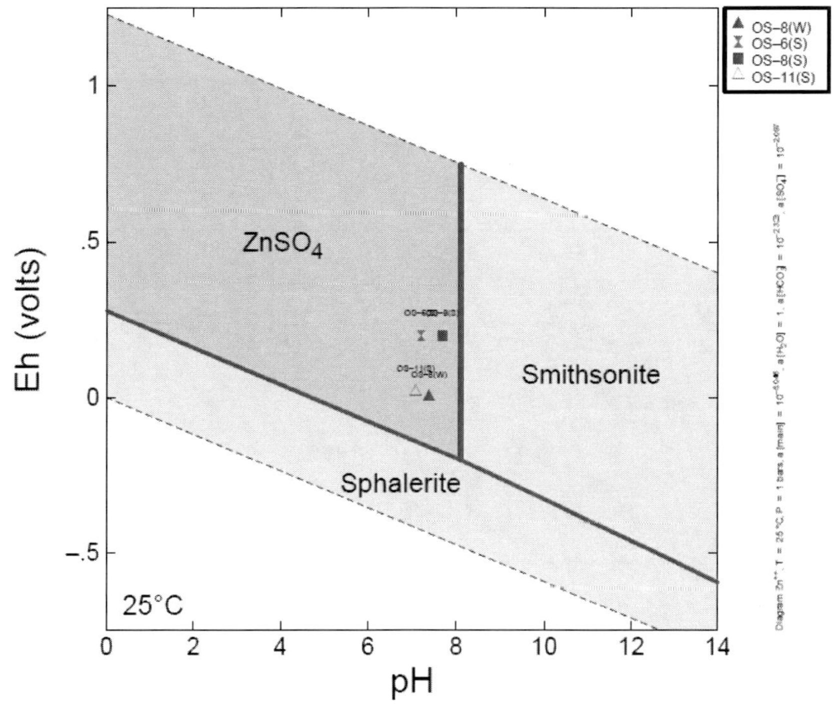

Figure 5. Eh-pH diagram for Zn using selected samples.

Table 6. Rate constants for sulfide and silicate weathering used in reaction-path modeling

Mineral	Mass (g)	Rate constant ($mol \cdot cm^{-2} \cdot s^{-1}$)	Reference
Albite	14	$3.1 \cdot 10^{-16}$	Blum and Stillings (1995).
Muscovite	10	$1.0 \cdot 10^{-17}$	Nagy (1995)
Pyrite	0.35	$3.1 \cdot 10^{-14}$	McKibben and Barnes (1986)
Sphalerite	2	$4.3 \cdot 10^{-14}$	Seal et al. (2008)
Galena	1	$5.0 \cdot 10^{-14}$	Kruse et al. (2009)

A reaction-path model was carried out assuming either chemical equilibrium among minerals and waters or kinetic control of weathering reactions, with the main data summarized in Table 6.

The starting water composition was surface water of the Grevolosa Creek (samples OS-6 in Tables 3 and 4), showing the results of reaction-path modeling values of pH near 6.5-7.0, and a chemical composition similar to Coral mine water (Figure 6). Thus, rate controlled simulations may provide a realistic description of the weathering reactions which led to the production of mine water from the possible infiltration of surface water and their interaction with the identified minerals of the Osor vein deposit and host-rock.

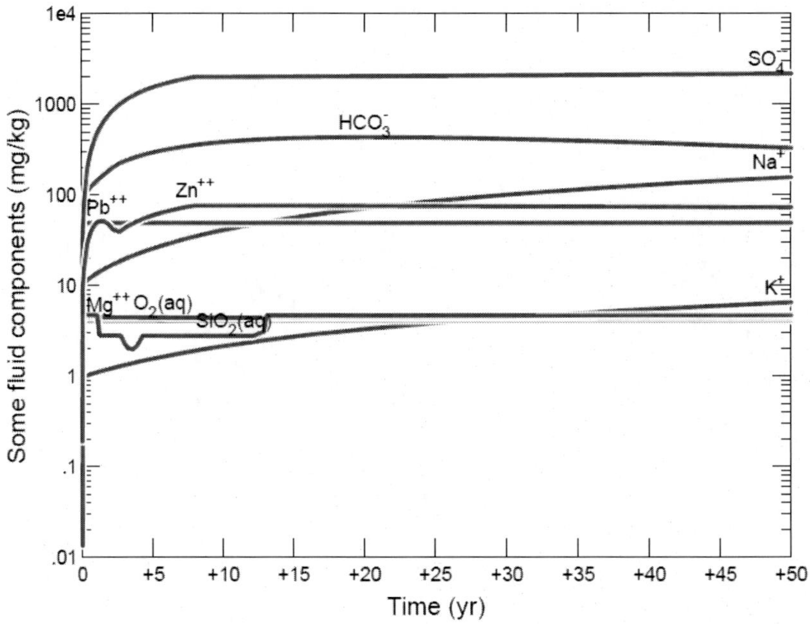

Figure 6. Evolution of water chemistry with time for simulations using sulfide-oxidation rate constants based on molecular O_2 as the oxidant.

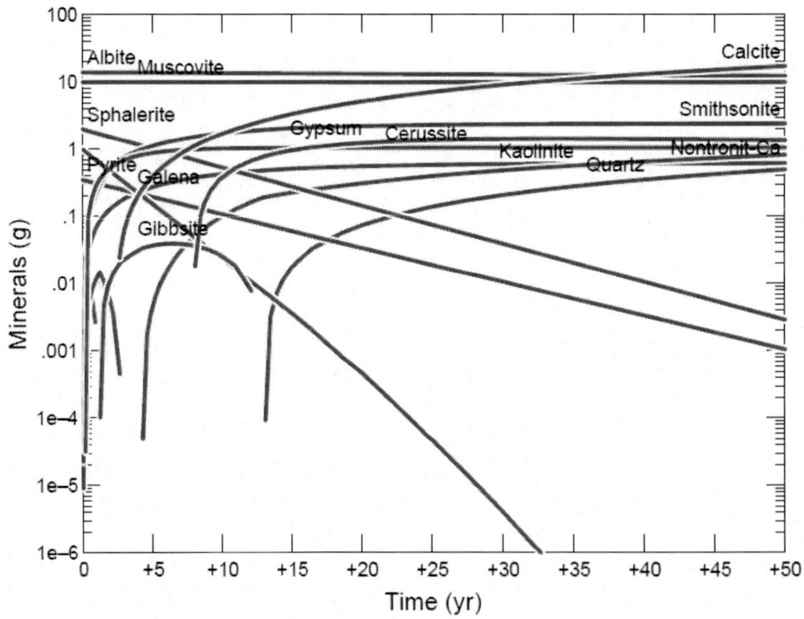

Figure 7. Dissolution and precipitation of various phases with time for simulations using sulfide-oxidation rate constants.

The simulations of the progressive reaction of host-rock with superficial water showed that the most prominent mineralogical changes over the course of the modeling experiments were the possible precipitation of smithsonite and kaolinite, and the removal of limited amounts of starting phases (albite, muscovite), resulting in the sub-horizontal lines for these phases (Figure 7). However, significant amounts of major sulphides (galena, pyrite, sphalerite) were consumed during the simulations covering 30 years.

CONCLUSION

The old mining activities in the Anglés area have contaminated groundwater and surface water in extensive zones, in spite of the neutral character of leachates and mine water generation. In Bonmatí sector, tailings showed high amounts of Cd, Cu, Pb, Zn, As, Ba, Se and Sb, above the regulatory levels in Catalonia for soils of industrial use.

Mining wastes from the Osor sector contained high amounts of Pb (>5000 mg kg-1), Zn (11300 mg kg-1) and Cd (24.9 mg kg-1), possibly associated with sphalerite. The Osor tailings sample had high concentrations of Al, Ba and Sr, due to the composition of gangue and soil, and sediment samples revealed high concentrations of As, Ba, Cd. Co, Ni, Pb and Zn, which are above the metal concentration of the Catalonian soil intervention values.

Mine water samples showed high concentrations of Ca (116-189 mg/L), HCO_3^- (290-377 mg/L), SO_4^{-2} (156-288 mg/L) and an elevated pH (7.08-7.68), which may indicate calcite dissolution through the Coral adit. In addition, the elevated concentrations of Mn (0.19-1.8 mg/L) and Zn (>0.25-2.8 mg/L) suggest the dissolution of sphalerite and Mn oxides. Moreover, mine water showed elevated concentrations of some cations, such as Na (31.4-42.5 mg/L), Mg (18.0-28.0 mg/L), and K (4.5-6.2 mg/L), above their concentrations in surface water.

The reaction-path modeling simulated the progressive reaction of host-rock with superficial water, resulting in mine water of Coral adit. The results showed that the most prominent mineralogical changes over the course of the modeling experiments were the possible removal of limited amounts of starting phases (albite, muscovite). Also, significant amounts of major sulphides (galena, pyrite, sphalerite) were consumed during the simulations covering 30 years.

REFERENCES

Al TA, Martin CJ, Blowes DW (2000) Carbonate-mineral/water interactions in sulfide-rich mine tailings. *Geochim Cosmochim Acta*, 64 (23): 3933–3948.

Bethke CM, Yeakel S (2011) *The Geochemist´s Workbench, Release 9.0, Essential Guide*. Aqueous Solutions LLC, Champaing, Illinois.

Bori J, Vallès B, Navarro, A, Riva, MC (2016) Geochemistry and environmental threats of soils surrounding an abandoned mercury mine. *Environmental Science Pollution Research*, 23: 12941-12953.

Blowes DW, Ptacek CJ (1994) Acid-neutralization mechanisms in inactive mine tailings. In: Jambor JL, Blowes DW (eds), The Environmental Geochemistry of Sulfide Mine-Wastes. *Mineralogical Assoc of Canada Short Course Handbook* 22, p 271–292.

Blowes DW, Jambor JL, Hanton-Fong CJ (1998) Geochemical, mineralogical and microbiological characterization of a sulphide-bearing carbonate-rich gold-mine tailings impoundment, Jouetl, Québec. *Appl Geochem*, 13: 687–705.

Blum AE, Stilligs LL (1995) Feldspar dissolution kinetics. In: White AF, Brantley SL (eds). Chemical Weathering Rates of Silicate Minerals. *Reviews in Mineralogy*, vol. 31, 291–351.

Bucher K, Zhu Y, Stober I (2009) Groundwater in fractured crystalline rocks, the Clara mine, Black Forest (Germany). *Int. J. Earth Sci*, 98: 1727-1739.

Canals A, Cardellach E (1997) Ore lead and sulphur isotope pattern from the low-temperature veins of the Catalonian Coastal Ranges (NE Spain). *Miner Depos*, 32: 243–249.

Carrillo-Chávez A, Morton-Bermea O, González-Partida E, Rivas-Solórzano H, Oesler G, García-Meza V, Hernández E, Morales P, Cienfuegos E (2003). Environmental Geochemistry of the Guanajuato Mining District, Mexico. *Ore Geology Reviews*, 23: 277-279.

Cidu R, Biddau R, Nieddu G (2007) Rebound at Pb-Zn Mines Hosted in Carbonate Aquifers: Influence on the Chemistry of Ground Water. *Mine Water Environ*, 26: 88-101.

Cidu R, Biddau R, Fanfani L (2009) Impact of past mining activity on the quality of groundwater in SW Sardinia (Italy). *Journal of Geochemical Exploration,* 100: 125-132.

Cidu R, Dadea C, Desogus P, Fanfani L, Manca PP, Orrù G (2012) Assessment of environmental hazards at abandoned mine sites: A case study in Sardinia, Italy. *Applied Geochemistry*, 27: 1795-1806.

Da Pelo S, Musu E, Cidu R, Frau F, Lattanzi P (2009) Release of toxic elements from rocks and mine wastes at the Furtei gold mine (Sardinia, Italy). *Journal of Geochemical Exploration,* 100: 142-152.

Daniele L, Corbella M, Vallejos A, Díaz-Puga M, Pulido-Bosch A (2013) Geochemical simulations to assess the fluorine origin in Sierra de Gador groundwater (SE Spain). *Geofluids*, 13: 194-203.

Ezekwe IC, Ezekwe AS, Chima GN (2013) Metal Loadings and Alkaline Mine Drainage from Active and Abandoned Mines in the Ivo River Basin Area of Southeastern Nigeria. *Mine Water Environ*, 32: 97–107.

Frau F, Ardau C, Fanfani L (2009) Environmental geochemistry and mineralogy of lead at the old mine area of Baccu Locci (south-east Sardinia, Italy). *Journal of Geochemical Exploration,* 100: 105-115.

Heikkinen PM, Räisänen ML, Johnson RH (2009) Geochemical Characterization of Seepage and Drainage Water Quality from Two Sulphide Mine Tailings Impoundments: Acid Mine Drainage versus Neutral Mine Drainage. *Mine Water Environ*, 28: 30–49.

Heikkinen PM, Räisänen ML (2009) Trace metal and As solid-phase speciation in sulphide mine tailings-Indicators of spatial distribution of sulphide oxidation in active tailings impoundments. *Appl Geochem*, 24: 1224–1237.

Hidalgo MC, Rey J, Benavente J, Martínez J (2010). Hydrogeochemistry of abandoned Pb sulphide mines: the mining district of La Carolina (southern Spain). *Environ Earth Sci*, 61: 37-46.

Iríbar V, Tames P, Antigüedad I, da Silva A (1999) Water contamination and remedial measures at the Troya abandoned Pb-Zn mine (Tha Basque Country, Northern Spain). *Environmental Geology*, 39(7): 800-806.

Jang M, Kwon H (2011) Pilot-scale tests to optimize the treatment of net-alkaline mine drainage. *Environ Geochem Health*, 33: 91–101.

Jurjovec J, Ptacek CJ, Blowes DW (2002) Acid neutralization mechanisms and metal release in mine tailings: A laboratory column experiment. *Geochim Cosmochim Acta*, 66(9): 1511–1523.

Kruse NA, Younger PL (2009) Development of thermodynamically-based models for simulation of hydrogeochemical processes coupled to channel flow processes in abandoned underground mines. *Appl Geochem*, 24: 1301–1311.

Kumari B, Tiwary RK, Srivastava KK (2017) Physico-Chemical Analysis and Correlation Study of Water Resources of the Sukinda Chromite Mining Area, Odisha, India. *Mine Water Environ*, 36: 356-362.

Lee JY, Choi JC, Lee KK (2005) Variations in heavy metal contamination of stream water and groundwater affected by an abandoned lead-zinc mine in Korea. *Environ Geochem Health*, 27: 237–257.

Lindsay MBJ, Condon PD, Jambor JL, Lear KG, Blowes DW, Ptacek CJ (2009) Mineralogical, geochemical, and microbiological investigation of a sulfide-rich tailings deposit characterized by neutral drainage. *Appl Geochem*, 24: 2212–2221.

Lottermoser B (2010). *Mine Wastes*. 2 Ed. Springer, Berlín.

McKibben MA, Barnes HL (1986) Oxidation of pyrite in low temperature acidic solutions – Rate laws and surface textures. *Geochim Cosmochim Acta*, 50: 1509–1520.

Majzlan J, Stevko M, Chovan M, Luptáková J, Miloovská S, Milovský R, Jelen S, Sýcorová M, Pollok K, Göttlicher J, Kupka D (2018). Mineralogy and geochemistry of the copper-dominated neutral mine drainage at the Cu deposit Lúbietová-Podlipa (Slovakia). *Applied Geochemistry*, 92: 59-70.

Munson BR, Okiishi TH, Huebsch WW (2009) *Fundamentals of Fluid Mechanics*. Hoboken, NJ, USA.

Nagy KL (1995) Dissolution and precipitation kinetics of sheet silicates. In: White AF, Brantley SL (eds). *Chemical Weathering Rates of Silicate Minerals. Reviews in Mineralogy* vol. 31, 173–233.

Navarro A, Collado D, Carbonell M, Sánchez, JA (2004) Impact of mining activities on soils in a semi-arid environment: Sierra Almagrera district, SE Spain. *Environmental Geochemistry & Health,* 26: 383–393.

Navarro A, Biester H, Mendoza JL, Cardellach E (2006) Mercury speciation and mobilization in contaminated soils of the Valle del Azogue Hg mine (SE, Spain). *Environmental Geology*, 49: 1089-1101.

Navarro A, Carbonell M (2008) Assessment of groundwater contamination caused by uncontrolled dumping in old gravel quarries in the Besós aquifers (Barcelona, Spain). *Environ Geochem Health*, 30: 273–289.

Navarro A, Cardellach E (2009) Mobilization of Ag, Heavy metals and Eu from the waste deposit of Las Herrerías mine (Almería, SE Spain). *Environ Geol*, 56 (7): 1389–1404.

Navarro A, Cardellach E, Corbella M (2009) Mercury mobility in mine waste from Hg-mining areas in Almería, Andalusia (SE, Spain). *Journal of Geochemical Exploration*, 101: 236-246.

Navarro A, Martínez F (2010) Evaluation of Metal Attenuation from Mine Tailings in SE Spain (Sierra Almagrera): A Soil-Leaching Column Study. *Mine Water Environ*, 29: 53–67.

Navarro A, Domènech LM (2010) Arsenic and metal mobility from Au mine tailings in Rodalquilar (Almería, SE Spain). *Environ Earth Sci*, 60: 121–138.

Navarro A, Font X, Viladevall M (2015) Metal Mobilization and Zinc-Rich Circumneutral Mine Drainage from the Abandoned Mining Area of Osor (Girona, NE Spain). *Mine Water Environ*, 34 (3), 329-342.

Navarro A, Carulla N (2018) Evaluation of Geothermal Potential in the Vicinity of the Flooded Sierra Almagrera Mines (Almería, SE Spain). *Mine Water Environ*, 37 (1): 137-150.

Papp DC, Cociuba I, Baciu C, Cozma A (2017) Origin and Geochemistry of Mine Water and its Impact on the Groundwater and Surface Running Water in Post-mining Environments: Zlatna Gold Mining Area (Romania). *Aquat Geochem*, 23: 247-270.

Parkhurst DL, Appelo CAJ (1999) *User´s Guide to PHREEQC (version 2)-a computer program for speciation, batch-reaction, one-dimensional transport, and inverse geochemical calculations*, USGS Water-Resources Investigations Report 99–4259, 312 pp.

Pfeifer HR, Häussermann A, Lavanchy JC, Halter W (2007) Distribution and behavior of arsenic in soils and waters in the vicinity of the former gold-arsenic mine of Salanfe, Western Switzerland. *Journal of Geochemical Exploration*, 93: 121-134.

Piqué A, Canals A, Grandia F, Banks DA (2008) Mesozoic fluorite veins in NE Spain record regional base metal-rich brine circulation through basin and basement during extensional events. *Chem Geol*, 257: 139–152.

Plante B, Benzaazoua M, Bussière B (2011) Predicting Geochemical Behaviour of Waste Rock with Low Acid Generating potential Using Laboratory Kinetic Tests. *Mine Water Environ*, 30: 2–21.

Plumlee GS, Smith KS, Montour MR, Ficklin WH, Mosier EL (1999) Geologic controls on the composition of natural waters and mine waters. In: Filipek LH, Plumlee GS (eds) *The Environmental Geochemistry of Mineral Deposits. Part B: Case Studies and Research Topics. Reviews in Economic Geology*, vol 6B, Chelsea, MI, USA, p 373–432.

Saatsaz M, Azmin-Sulaiman WN, Eslamian S, Mohammadi K (2013) Hydrogeochemistry and groundwater quality assessment of Astaneh-Kouchesfahan Plain, Nothern Iran. *International Journal of Water*, 7: 44-65.

Seal RR, Foley NK (2002) *Progress on Geoenvironmental Models for Selected Mineral Deposit Types*. U.S. Geological Survey Open-File Report 02–195, USA.

Seal RR, Hammarstrom JM (2003) Geoenvironmental models of mineral deposits: examples from massive sulfide and gold deposits. In: Jambor JL, Blowes DW, Ritchie AIM (eds) *Environmental Aspects of Mine Wastes*. Mineralogical Association of Canada, Short Course Series 31.

Seal RR, Hammarstrom JM, Johnson AN, Piatak NM, Wandless GA (2008) Environmental geochemistry of a Kuroko-type massive sulfide deposit at the abandoned Valzinco mine, Virginia, USA. *Appl Geochem*, 23: 320–342.

Younger PL (2000) Nature and practical implications of heterogeneities in the geochemistry of zinc-rich, alkaline mine waters in an underground F-Pb mine in the UK. *Appl Geochem*, 15: 1383–1397.

In: Advances in Hydrogeochemistry Research
Editors: S. Saeid Eslamian and Faezeh Eslamian
ISBN: 978-1-53617-829-6

Chapter 13

EVALUATION OF NATURAL WATER QUALITY IN THE JUNGAR BASIN IN CENTRAL ASIA AND ITS IMPLICATIONS ON REGIONAL WATER RESOURCE MANAGEMENT

***Bing-Qi Zhu*[1,*] *and Saeid Eslamian*[2,3]**
[1]Key Laboratory of Water Cycle and Related Land Surface Processes, Institute of Geographic Sciences and Natural Resources Research, Chinese Academy of Sciences, Beijing, China
[2]Department of Water Engineering, Collage of Agriculture, Isfahan University of Technology, Isfahan, Iran
[3]Center of Excellence in Risk Management and Natural Hazards, Isfahan University of Technology, Isfahan, Iran

ABSTRACT

The use of water resources in arid lands is strongly limited by their quantity. To add to such knowledge, this study evaluates the natural water quality and its suitability for drinking, agricultural and industrial purposes in the Jungar Basin (Central Asia), using chemical-physical indicators. The waters are neutral to alkaline and most of them are soft-fresh waters. The total dissolved solid (TDS) varies over two orders of magnitude. Much of the solutes and physicochemical parameters in water are under the highest desirable limits of the World Health Organization (WHO) for drinking purpose and most waters are of good water quality for irrigation. The effects of local pollution are minimal in the montane and piedmont areas of these watersheds but are significant in the oases and central areas of the drainage basins. Although the headwaters of the Jungar Basin represent natural background conditions (soft-fresh water in salinity and hardness) and population densities within the catchment are relatively low, the river basin is facing relatively high

[*] Corresponding Author's Email: zhubingqi@igsnrr.ac.cn.

anthropogenic pressure on water quality in the low reaches. The main contributors to the nutrient emissions are agricultural land use and, to a lesser extent, unban settlements with a high proportion of households without connection to wastewater treatment plants. Proposals for regional water resources management are advised, i.g. the geographic data and information should be detailedly included in the assessment and monitoring procedure, a water quality model should be built, and information technology such as visualization technology and the internet should be used.

Keyworlds: natural water; water quality; drinking water; irrigation water; Industrial quality; Central Asia

1. INTRODUCTION

Environmental pollution has become of public interest long time ago. Not only the developed countries have been affected by this environmental problem, but also the developing nations suffer the impact of pollution (Listori and World-wide BanK, 1990; Niemi et al., 1990; Tomas et al., 2017; Erickson et al., 2017), due to disordered economic growth associated with the exploration of virgin natural resources.

Among the environmental issues, water pollution affects human health and agricultural production to a high extent. Knowledge of water pollution and its environmental characteristics is thus vital to understand water suitability for various purposes (Cicchella et al., 2010; Erickson et al., 2017; Misaghi et al., 2017). Water quality is an indicator of water pollution. Natural water (surface water and groundwater) quality in an area is a function of physical, chemical, biological, and radiological characteristics of water (Diersing, 2009) that are greatly influenced by geological formations, climate, and topography (Helena et al., 2000; Subramani et al., 2005; Zhu and Yang, 2007; Zhu et al., 2011, 2012, 2013; Balasubramanian et al., 2015; Tomas et al., 2017; Erickson et al., 2017). In addition, human activities such as industrialization, agriculture, mining, and urbanization produce effluents affect natural water quality (Dinelli et al., 2010; Zhu et al., 2011; Singh et al., 2012; Oyarzun et al., 2015; Tomas et al., 2017; Erickson et al., 2017). Normally, natural water contamination occurs gradually with little impact in the initial period of deterioration, but if it is not controlled at the right time, this water may not be suitable for any purpose for a long time (Rao et al., 1997; Tomas et al., 2017; Misaghi et al., 2017). Thus the importance of regular monitoring of natural water quality should be underlined, especially when there are significant changes in industrial, agricultural, and urbanization activities.

Over the past century, the world has experienced an undeniable temperature increase and thereby inevitably is faced with water environmental problems, which accelerate greatly in global water cycle processes (Menzel and Burger, 2002). This is especially significant in arid lands of Central Asia, because most of natural waters in Central Asia originate from peripheral mountain glaciers (Zhu et al., 2012, 2013, 2015, 2017) and are fragile to global warming. Both of the quantity and quality of water resources in this context are the most important issues threatening the economic and social development of the Central Asian countries (Ososkova et al., 2000; Food and Agriculture Organization, 2003; Crosa et al., 2006; Bezborodov et al., 2010), such as Nepal, India, Bangladesh and China, the pollution of rivers is severe and critical near urban stretches due to huge amounts of pollution caused by urban activities (Crosa et al., 2006; Yadav and Rajesh, 2011). An estimated 580 people in India die of water pollution related illnesses every day (Rahman et al., 2003). With the rapid economic growth in China, urban water quality is now facing serious challenges (Yan et al., 2016). Approximately 90% of the water in the cities of China is polluted to varying degrees (Huang et al., 2010). According to the 2009 statistics from the Ministry of Environmental Protection (MEP) of China, there were 80 water pollution accidents that year. In China's key cities, 27% of the centralized drinking water sources did not meet official standards, and half a billion Chinese had no access to safe drinking water (Moe and Rheingans, 2006). Not only in Asia, water pollution is the leading cause of deaths and diseases worldwide. Across the globe, 6000 children die of diseases and 30 million people die of cancers caused by drinking polluted water each year (WHO, 2008). In developing countries, 80% of human diseases is associated with polluted drinking water. Therefore, the quality of natural water is a powerful environmental determinant of health. The assurance of water use safety is a foundation for the prevention and control of waterborne diseases. Safe and clean water is a vital resource for everyone.

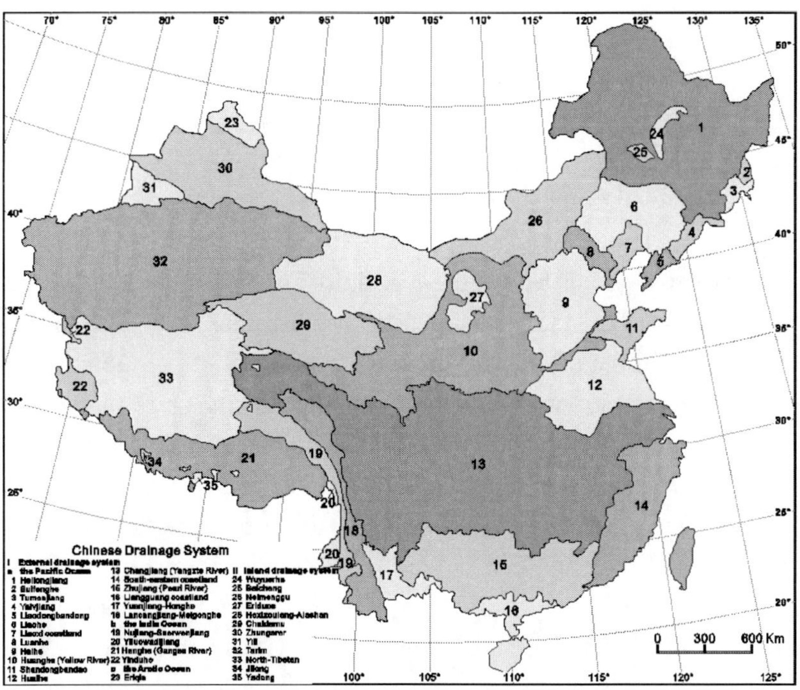

Figure 1. Geographical distribution of drainage systems in China. The 3 watersheds in northwestern China of Central Asia, numbered 23, 30 and 31, are investigated in this study.

The Jungar Basin, including the Erlqis, Jungar and Yili watersheds in northern Xinjiang, northwestern China (No. 23, 30 and 31, respectively, in Figure 1), are large watersheds in Central Asia. These watersheds are extremely arid in climate and together represent one of the least-studied areas in China. At present, the Jungar Basin is facing a tremendous change in climate and land-use intensification due to expansions in the agricultural sector, an increase of cattle and livestock and a growth of urban settlements due to migration of the rural population to the cities. It is expected that these may lead to unfavourable changes in natural water quality. In future, a possible strong need of water coming from other than mountain glaciers would result in a very expensive water collection system to supply the local population and economy. Therefore, it is essential to diagnose the quality of natural water for the regional water resource management. A key challenge is the identification of the main anthropogenic pollution sources and the assessment of their spatial distribution. Until now, however, despite continuing speculation about the evolution of water resource quantity in the Jungar Basin, very little is known about the water quality of natural waters in these watersheds, because little information and chemical data are available within the international scientific literature with regard to this drainage system.

In order to add to such knowledge, this study investigates the geochemical solutes from the river water, groundwater, lakes and reservoirs in the Jungar Basin and identify the main anthropogenic pollution sources and their spatial distribution. The water quality and its suitability for drinking, irrigation and industrial purposes in this region are emphasisly evaluated.

2. Materials and Methods

This study focuses on the Jungar Basin in the northwestern China in Central Asia. The drainage basin includes three watersheds, namely the Yili, Jungar and Erlqis watersheds. The study areas investigated lies between 78° and 90°E and 42° and 50°N, covers an area of approximately 603,000 km^2 and is bounded by the Tianshan Mountains to the south and by the Altai Mountains to the north. The Jungar watershed is the central part of the study area. It is formed by a structural depression filled with Quaternary and Tertiary unconsolidated sediments as much as 500-1000 m thick (Zhu et al., 2011). The large geographic distance from the surrounding oceans and the presence of the rain-shadow effect due to the surrounding orographic conditions are responsible for the arid climate of these watersheds (Zhu et al., 2012, 2017).

These watersheds in Central Asia are controlled by an arid temperate continental climate. The mean annual air temperature is about 5°C, with a minimum of -10°C to -20°C in January and a maximum of 28°C to 33°C in August (Zhu et al., 2012). The regional precipitation is derived mainly from the westerly winds, with a mean annual precipitation rate of 60-150 mm in the center desert and 200-500 in the surrounding mountainous areas

(Zhu et al., 2012, 2013). The potential evapotranspiration is approximately 1000-3500 mm/a (Zhu et al., 2012). The Yili and Jungar watersheds are inland watersheds and the Erlqis watershed is a tributary of the Ob River which drains into the Arctic Ocean (Zhu et al., 2011).

Water sample locations were selected from various parts of the Erlqis, Jungar and Yili watersheds (Figure 2), including river stems, stream channels, hill slopes, wells, lakes, ponds, man-made trenches and reservoirs. The sampling sites were chosen carefully in order to maximise representation of the diverse ecohydrological environments within the river systems. Taking into account that the tributaries reflect a much broader variety of geologic, biologic, and population patterns than do main stem rivers (Pawellek et al., 2002; Zhu et al., 2013), it was of interest to sample tributary water to look for common features reflected in their hydrochemistry.

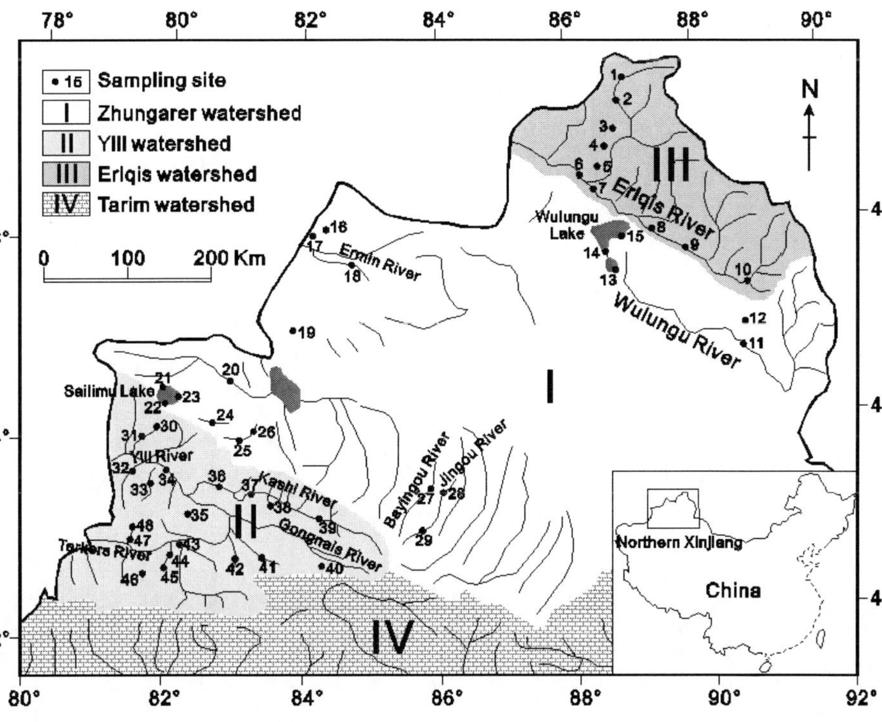

Figure 2. Map of the river distributions and sampling site locations of the 3 studied watersheds in Central Asia.

The physicochemical parameters such as temperature (T), pH, electrical conductivity (EC), electrical potential (EP/Eh) and total dissolved solid (TDS) were measured in situ with a calibrated portable Multi-Parameter Analyzer (Eijkelkamp 18.28). The bias of the analytical procedures is <2% for pH and EP, <1% for EC, TDS and T. Alkalinity (as HCO_3^-) was determined with a Hach digital titrator using the Gran method (Wetzel and Likens, 2000) The concentrations of major cations (Li^+, Na^+, NH_4^+, K^+, Mg^{2+}, Ca^{2+}) and anions (F^-

, Cl^-, NO_2^-, Br^-, NO_3^-, $H_2PO_4^-$, SO_4^{2-}) were determined by ion chromatography (Dionex 600 IC) with deionized water (EC < 2.1 μS/cm) as the dilute base. A balance of ionic charge and a conservation of mass occur, respectively, indicating a high reliability of the analytical data.

3. Results

The pH values of the selected water samples vary from 7.0 to 9.81 with a median value of 7.85 indicating that the waters are generally neutral to slightly alkaline. The TDS range between 24.6 mg/L and 6200 mg/L, varying over two orders of magnitude from fresh (TDS < 1,000 mg/l) to brackish (1,000 mg/l < TDS < 10,000 mg/l).

The variation of major cation (Na^+, K^+, Ca^{2+} and Mg^{2+}) and anion (Cl^-, HCO_3^-, SO_4^{2-} and NO_3^-) concentrations measured in the studied water samples is illustrated in the Box and Whisker plot (Figure 3), where Ca^{2+} and HCO_3^- are the dominant cation and anion, respectively. Nevertheless, many of the water samples contained high SO_4^{2-} concentrations. The order of relative abundance of major cations in these waters is $Ca^{2+} > Na^+ > Mg^{2+} > K^+$ (on mg/L basis) while that of anions is $HCO_3^- > SO_4^{2-} > Cl^- > NO_3^-$. Large percentages of contribution to the TDS of water samples are from HCO_3^-, Ca^{2+} and SO_4^{2-}.

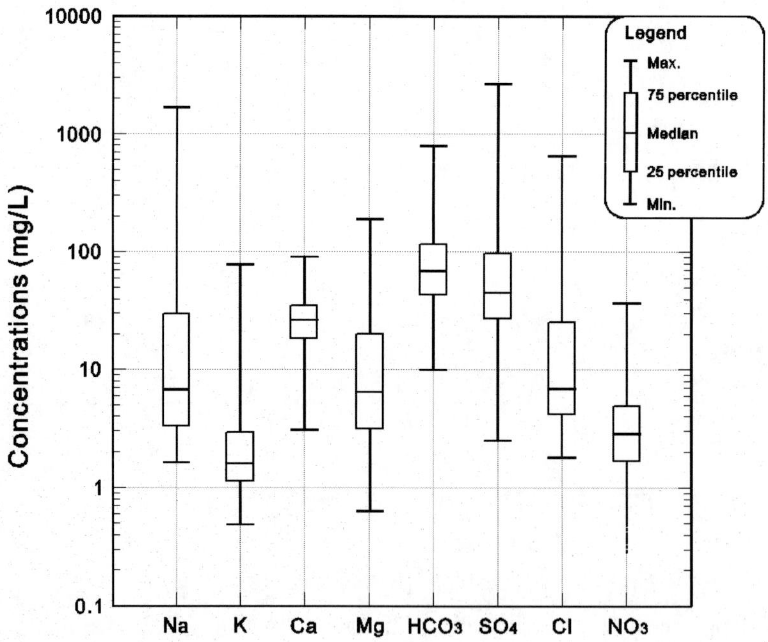

Figure 3. Box and Whisker plots showing the variation of major ion concentrations in the studied water samples.

The concentrations of major ions measured in the studied water samples are presented in the Piper Trilinear Plot (Figure 4). This figure shows that the natural waters in the study area can be broadly divided into 4 major types: type-1, Ca-HCO_3; type-2, Ca-Mg-HCO_3-SO_4; type-3, Ca-Mg-SO_4 and type-4, Na-SO_4. Major waters are of Ca-HCO_3, Ca-Mg-HCO_3-SO_4 and Ca-Mg-SO_4 types and their chemical properties are dominated by alkaline earths and the HCO_3^- and SO_4^{2-} anions. The Na-SO_4 type water occupies the section near the right corner of the central diamond. These waters are of the saline water type. Water types distributed in the Erlqis and Yili watersheds are mainly type-1 and type-2, while those in the Jungar watershed are type-1 and type-2 in the mountain areas and type-3 and type-4 in desert plain areas. In general, water types of the 3 watersheds display a kind of zonal distribution from mountainous area to pediment, to oases and to desert plain, namely a variational trend in rough order of type-1, type-2, type-3 and type-4.

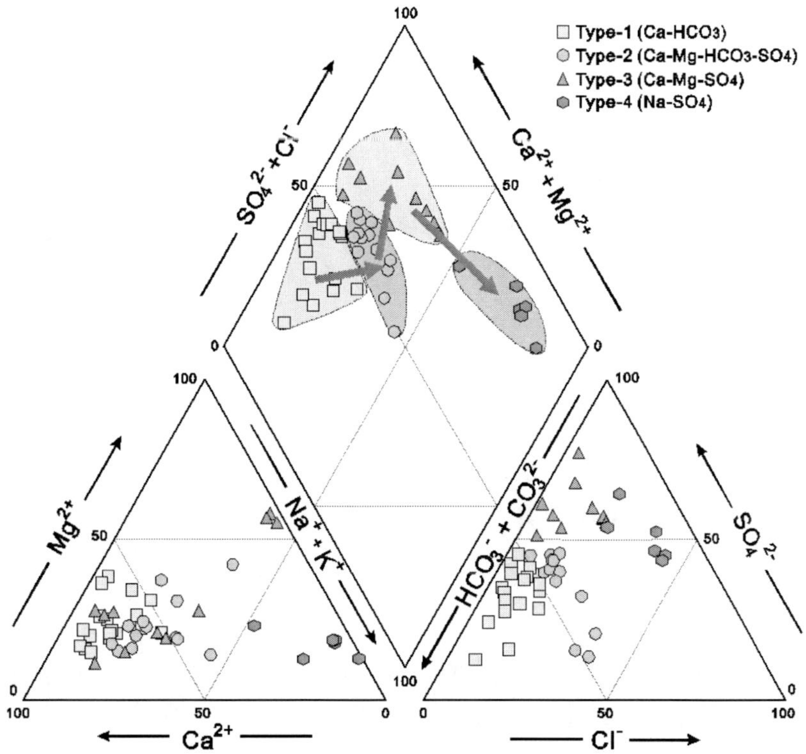

Figure 4. Piper diagram showing major ion composition of the natural waters in the three watersheds in Central Asia.

4. DISCUSSION

4.1. Water Quality and Its Suitability for Drinking, Irrigation and Industrial Purposes

Water quality refers to the chemical, physical, biological, and radiological characteristics of water (Diersing, 2009). It is a measure of the condition of water relative to the requirements of one or more biotic species and or to any human need or purpose (Johnson et al., 1997). It is most frequently used by reference to a set of standards against which compliance can be assessed. The most common standards used to assess water quality relate to health of ecosystems, safety of human contact and drinking water. The World Health Organization (WHO) has produced international norms on water quality in the form of guidelines, which are used as the basic regulations and standards in developing and developed countries.

In this study, the natural water quality for drinking, irrigation and industrial purposes in the Jungar Basin is assessed based on the recommended standards of the World Health Organization (WHO, 2008), the National Standard of the People's Republic of China (NSPRC, 2002) and the water quality criteria adopted by Johnson (1983) and Subba Rao et al., (2012).\

4.1.1. Drinking Water Quality

The total dissolved solids (TDS) versus the total hardness (TH) plot (Figure 5a) shows that most surface and groundwater samples in these watersheds are soft-fresh and hard-fresh water. Several water samples with high TDS (>1000 mg/L) and TH (>300 mg/L) values, classed as hard-brackish water in Figure 5a, are from the Jungar watershed. TDS, TH, pH and the concentrations of Ca^{2+}, Mg^{2+}, Na^{+}, Cl^{-}, SO_4^{2-}, NO_3^{-} and F^{-} in about 0~20% of the studied water samples are higher than the highest desirable limits of 1000, 450, 8.5, 75, 30, 200, 250, 250, 50, and 1.5 mg/L respectively, prescribed for drinking purpose by WHO (2008) and NSPRC (2002). All groundwater samples in this study are below the limits of the WHO (2008) and NSPRC (2002). This means that most of the surface waters (80%) and all of groundwaters in the study area are suitable for drinking and domestic purposes.

The classification of the studied natural waters based on total hardness (TH) shows that 50% of the water samples fall in the soft category (Figure 5a, TH < 100 mg/L) and 50% in the moderately to very hard (Figure 5a, TH > 100 mg/L). In most of the water samples (> 87%), the alkaline earth metals exceed the alkalis metals (Ca + Mg > Na + K) (Figure 4) and also in a large part of water samples (40%), the weak acid exceeds strong acids (HCO3 > SO4 + Cl) (Figure 4). Therefore, such water has only temporary hardness according to Handa's classification (Handa, 1964). The maximum allowable limit of TH for drinking purpose by WHO is 500 mg/L and the most desirable limit is 100 mg/L (WHO, 2008). According to this standard, 50% of the water samples in the Jungar Basin is under the desirable limit and 90% of the water samples is under the maximum allowable limit, respectively.

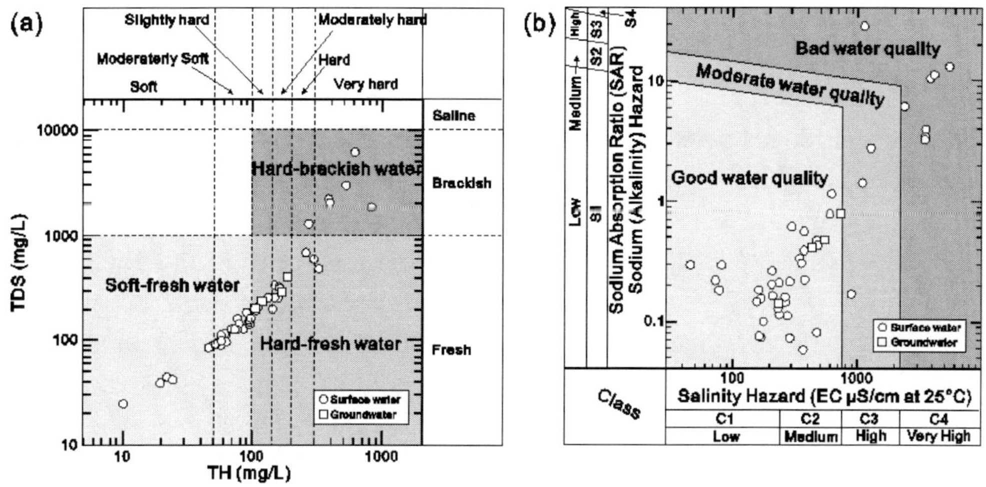

Figure 5. Natural water quality for drinking and irrigation purposes in the Erlqis, Jungar and Yili watersheds in Central Asia.

4.1.2. Irrigation Water

To assess the water quality for irrigation purposes, the Na or alkali-hazard expressed in terms of sodium adsorption ratio (SAR) is widely used. If waters used for irrigation are high in Na^+ and low in Ca^{2+}, the ion-exchange sites may become saturated with Na^+, which destroys the soil structure, because of dispersion of the clay particles. Such soils reduce plant growth. The SAR for the studied waters is computed, using the formula (Hem, 1991): $SAR = Na^+/((Ca^{2+} + Mg^{2+})/2)^{0.5}$, where the ion concentrations are expressed in meq/L. The computed SAR values for water samples in the 3 watersheds range from 0.06 to 28.3, with an average value of 1.97. The SAR versus EC plot (Figure 5b) shows that most surface water samples are of good water quality (75%). A small number of surface water samples are of moderate quality (7%) and poor quality (18%) types. The 4 groundwater samples analyzed are all of good water quality. The moderate to bad quality waters have higher Na^+ concentrations.

It is generally known that agricultural practices can pose a serious threat to natural water, particularly groundwater quality. It is especially true in arid areas as the growing season with its high water demands coincides with the dry period. In addition, the use of fertilizers with a strong polluting effect on groundwater is a widespread agricultural practice in northwestern China. The use of N fertilizers in particular is well established throughout the world and increased use has led to extremely high NO_3^- concentrations in groundwaters (Stigter et al., 1998). Compared with the NO_3^- contents (0.5~7.7 mg/L) in the studied surface waters, groundwater samples (such as samples 16, 25, 26) collected from different irrigated oases in the Jungar watershed have evidently higher concentrations of NO_3^- (22-37 mg/L), indicating the strong influence of local agricultural practices in these zones.

The three inorganic N species (NO_3^-, NO_2^-, NH_4^+) have different measured distributions in the studied waters. Nitrite was not detected in any sample, and NH_4^+ only occurred in a few samples in this study. Nitrate was detected in most of the studied samples. Its concentrations do not show any systematic variations with any ion concentrations and parameters or along the river courses, suggesting that point sources, presumably of anthropogenic origin, play an important role. In comparison with the mean value (2.46 mg/L) of NO_3^- in this study, somewhat higher concentrations of NO_3^- (7~37 mg/L) were observed in several studied samples collected from oases with a dense population or farming, particularly groundwater and river water taken from agricultural and residential regions. This indicates an evident anthropogenic input of NO_3^- in these places, probably from agricultural activities (fertilizers) or municipal and industrial discharges. These inputs are possibly modified by biological processes in the river itself, with oxidation of NH_4^+ to NO_2^- and subsequently to NO_3^- (Pawellek et al., 2002), which could explain the distribution patterns of the 3 inorganic N species in the studied waters.

4.1.3. Water Quality for Industrial Use

Industries frequently suffer from the common undesirable effects of incrustation and corrosion, which are the chemical reactions caused by inferior water quality. The incrustation involves a deposition of undesired material of $CaCO_3$ on surface of metals, while the corrosion is a chemical action on metals that results in the metal being eaten away.

The quality requirements for industrial water supplies range widely. In this study, the following water quality criteria adopted by Johnson (1983) and Subba Rao et al., (2012) are used to evaluate the incrusting and corrosive properties of the natural waters in the Jungar Basin:

(1) Water, with HCO_3 more than 400 mg/l or SO_4 more than 100 mg/l, may cause incrustation, and
(2) Water, with pH < 7 or TDS more than 1,000 mg/l or Cl more than 500 mg/l, may cause corrosion.

According to the above two standards, the content of HCO_3 exceeds the limit of 400 mg/l in approximately 4.17% of the natural water samples in the Jungar Basin. The concentration of SO_4 is more than 100 mg/l in approximately 25.0% of the natural water samples in the Jungar Basin. Such natural water quality can develop incrustation on metal surfaces and hence is not recommend for industrial use.

The natural waters studied are free from corrosion, as the pH is more than 7 for all of the water samples. However, highly mineralized water, with TDS more than 1,000 mg/l, is observed from approximately 16.67% of the total water samples in the Jungar Basin. Therefore, water sample S12-S15, S19 and S21-S23 can cause corrosion on metal surfaces.

In addition, except for the water sample S13-14 and S19, the concentration of Cl does not exceed the limit of 500 mg/l in the other water samples in the Jungar Basin.

4.2. Implication for Regional Water Resources Management

The chemical speciation and fractionation of ions and nutrients fluxes in the natural waters in this study demonstrate the wide variety of point and diffuse sources impacting upon the water quality of the Jungar Basin. The analytical dataset indicates that chemical solutes and nutrient inputs to the oasis areas in the central basin are significantly impacted by diffuse agricultural sources, which suggesting the mobilization of N-rich pore water from the surrounding agricultural catchment via throughflow and groundwater flow pathways. Such relationships are well documented in the literature in agricultural catchments (Johnes and Burt, 1991, 1993; Heathwaite and Johnes, 1996; Jarvie et al., 2002, 2005). However N flux from major point source discharges is also important in the Jungar Basin, as is shown in the data for groundwater in the central Jungar watershed. At these sites the impact of this discharge on background hydrochemistry is evident with all N species concentrations.

In general, the basin-scale investigation for natural water quality in the Jungar Basin shows that even though most rivers in the basin are in relatively pristine condition, at least 20% water samples in forty-eight rivers/lakes are potentially changed and/or polluted due to natural and anthropogenic factors, including the central Jungar oasis and surrounding tributaries. Physical-chemical analyses of the water samples taken in the Yili, Jungar and Erqis watersheds show a relative enrichment in total hardness (Figure 5) in comparison to natural background conditions and maximal permissible concentrations of the international standard of WHO (2008). The remarkable increase of Nitrate-N concentration from 2.46 to 37 mg/L in the natural waters is an indication for the increasing nutrient release into oasis areas by diffuse sources, mainly urban areas without connection to treatment plants in the Jungar Basin. For nutrients concentrations and loads, such as nitrogen and orthophosphate-P in arid land rivers, a similar trend between them has been observed (Hofmann et al., 2013). As revealed by a nutrient emission modelling, urban settlements are generally the main sources for nitrogen and phosphorus emissions contributing about 55% (nitrogen) and 52% (phosphorus) of the total emissions (Hofmann et al., 2011). The proportion of point sources (PS) was much higher for nitrogen (30% of total N emissions) than for phosphorus (15% of total Permissions). Since only about 35% of the total population in a river basin are connected to PSs, thus unconnected urban areas represent an important proportion of the total emissions (38% of phosphorus and 25% of nitrogen emissions) (Hofmann et al., 2011). However, this unconnected part is often disturbed by river bank erosion, which is another significant source of nutrient release (Hartwig et al., 2012; Theuring et al., 2013). This process is triggered by an increasing degradation of riparian vegetation due to high livestock densities with free access to the running waters. Once the trends occur upstream, it help to explain the nutrient enrichment further downstream in the main stem, such as in the Selenga river (Hofmann et al., 2011).

The Jungar catchment is an international drainage system in terms of water resources, which is shared between China (upstream) and Kazakstan (downstream) for the Yili watershed and between China (upstream) and Russia (downstream) for the Elqis watershed, respectively. By international standards, the Jungar Basin is characterized by a relatively low mean population density of around 13.8 people/km^2 (<83.6 people/km^2 in Asia) (Deng and Duan, 2009). However, localized concentrations of population (such as 177.3 people/km^2 in the Urumqi City in the Jungar Basin), an often poor state of urban waste water infrastructures, high livestock densities in the riverine floodplains and large-scale mining activities are potential threats to the aquatic ecosystems in the Jungar. In the recent past, water quality monitoring in the China parts and the Kazakstan parts of the river basin was not harmonized, which is maybe best reflected by the fact that the two countries are located in the upstream and downstream sections of the Yili river watersheds, and thus faced different water resources problems. This followed by discrepancies in water policy objectives and understanding of monitoring purposes.

Therefore, the water quality analysis in this study indicates that water management in the Jungar Basin faces several challenges. Such as in the Yili watershed, one of the three watersheds in the Jungar Basin, the Yili River and its tributaries do not only constitute a transboundary river basin with a lack of harmonized monitoring; the large size, low population densities and challenges related to the political and economic transformation also result in a scarcity of environmental data. In such a situation, a comprehensive monitoring of water-resources is almost prohibitive. Therefore, one promising approach is to combine intensive monitoring in a representative model region with a more synoptic monitoring in the larger basin. This strategy seems plausible for the other two watersheds in the Jungar Basin, which are comparable with regard to the (bio-) physical and socio-economic environment (Zhu et al., 2011).

The current situation of natural water quality in the Jungar Basin has potential implications for regional water resources management. It can be summarized briefly below.

(1) Natural water quality in the Jungar Basin is characterized by geography, which depends on the location, and the conditions of the water source from which it is treated and transferred. The geographic data and information, which are strongly correlated with natural water quality, should be included in the assessment and monitoring of drinking, agricultural and industrial water quality in the regional and local management of water resources.
(2) With the development of infrastructure, natural water data are abundant in the Jungar Basin, but the value of the information about water quality is poor. The comprehensive assessment and monitoring based on a water quality model should be built, as which will play an important role in ensuring safe and clean drinking,

agricultural and industrial water in the future. For example, we can build a model based on combinational weighted aggregative index methods, which include two processes – the weight distributions of drinking, agricultural and industrial water indexes and comprehensive scores of these water qualities. In this model, the distribution of proper weights is a crucial procedure that greatly affects the comprehensive evaluation results. A small change of weight may cause a significant variation of the evaluation results. Subjective and objective weighting methods should be employed to obtain a new and robust weight distribution.

(3) The openness and transparency of water quality information to the public is beneficial to the improvement of water quality. For example, the United States has one of the safest water supplies in the world (Diersing, 2009). However, the national statistics in China do not specifically inform citizens on the quality and safety of the water coming out of the tap.

(4) Information technology, such as visualization technology and the internet, will make the assessment and monitoring of water quality more intuitive and convenient. The comprehensive assessment and monitoring of the drinking, agricultural and industrial water qualities in the Jungar Basin will provide a powerful tool for government and health supervision agencies that need to make informed decisions.

CONCLUSION

This paper focused on the major ion chemistry of natural waters in the Jungar Basin in northwestern China (Central Asia) to characterize their hydrogeochemistries and water qualities. Results show that the natural waters in the study areas are neutral to slightly alkaline in nature and most of them are soft-fresh water. The water salinities vary from fresh to brackish. Total dissolved solids (TDS), total hardness (TH), pH and the concentrations of Ca^{2+}, Mg^{2+}, Na^{+}, Cl^{-}, SO_4^{2-}, NO_3^{-} and F^{-} in about 0~20% of the studied water samples are higher than the highest desirable limits of drinking water prescribed for drinking purpose by WHO (2008) and NSPRC (2002). Most groundwater samples are below the limits of the WHO (2008) and NSPRC (2002). Regards to the water quality for irrigation purposes, most surface water samples are of good water quality (75%). A small number of surface water samples are of moderate quality (7%) and poor quality (18%). The groundwater samples are all of good water quality for irrigation purposes. The moderate to bad quality waters have higher Na^{+} concentrations. Inorganic N demonstrates no statistically significant change in the study areas, but an increase in the lower reaches downstream from the Jungar watersheds, which may be due to urban expansion in the lower catchment. Point sources, presumably of anthropogenic origin, play an important role in oasis areas, but the effects of local pollution are minimal in the montane and piedmont areas. The current situation of natural water quality in the basin has potential implications for regional water resources management, i.g. the geographic data and information should be included in the assessment and monitoring of natural water quality in the regional and local management of water resources, a comprehensive assessment and monitoring based on a water quality model should be built, the water quality information should be open and transparent to the public and information technology such as visualization technology and the internet should be used for the assessment and monitoring of water quality in the basin.

ACKNOWLEDGMENTS

This work study was financially supported by the National Natural Science Foundation of China (41771014) and the National Key Research and Development Program of China (2016YFA0601900). The author is grateful to Profs. Xiaoping Yang and Xiaoguang Qin for their generous help in the research work.

REFERENCES

Balasubramanian N, Sivasubramanian P, Soundranayagam J, Chandrasekar N, Gowtham B (2015) Groundwater classification and its suitability in Kadaladi, Ramanathapuram, India using GIS techniques. *Environmental Earth Sciences* 57:1–23.

Bezborodov GA, Shadmanov DK, Mirhashimov RT, Yuldashev T, Qureshi AS (2010) Mulching and water quality effects on soil salinity and sodicity dynamics and cotton productivity in Central Asia. *Agriculture Ecosystems & Environment* 138(1-2):95-102.

Cicchella D, Albanese S, De Vivo B, Dinelli E, Giaccio L, Lima A, Valera P (2010) Trace elements and ions in Italian bottled mineral waters: identification of anomalous values and human health related effects. *Journal of Geochemical Exploration* 107(3):336–349.

Crosa G, Froebrich J, Nikolayenko V, Stefani F, Galli P, Calamari D (2006) Spatial and seasonal variations in the water quality of the Amu Darya River (Central Asia). *Water Research* 40:2237-2245.

Deng D, Duan H (2009). The problem of the population development and the characteristics of the population space distriubtion in urban regions of northern Xinjiang. *Journal of Arid Land Resources and Environment* 23(8):53-60.

Diersing N (2009) *Water Quality: Frequently Asked Questions.* Florida Keys National Marine Sanctuary, Key West, FL.

Dinelli E, Lima A, De Vivo B, Albanese S, Cicchella D, Valera P (2010) Hydrogeochemical analysis on Italian bottled mineral waters: effects of geology. *Journal of Geochemical Exploration* 107(3):317–335.

Erickson JJ, Smith CD, Goodridge A, Nelson KL (2017) Water quality effects of intermittent water supply in Arraiján, Panama. *Water Research* 114:338-350.

Food and Agriculture Organization (2003) *Land Degradation Assessment in Drylands (LADA). FAO AGLL Technical Report.* State Design and Research. Uzgipromeliovodkhoz Institute, Ministry of Agriculture and Water Resources; Tashkent Republic of Uzbekistan. 43 pp.

Handa BK (1964) Modified classification procedure for rating irrigation waters. *Soil Science* 68(4):264–269.

Hartwig M, Theuring P, Rode M, Borchardt D (2012) Suspended sediments in the Kharaa River catchment (Mongolia) and its impact on hyporheic zone functions. *Environmental Earth Sciences* 65:1535–1546.

Heathwaite AL, Johnes PJ (1996) Contribution of nitrogen species and phosphorus fractions to stream water quality in agricultural catchments. *Hydrological Processes* 10:971-83.

Helena B, Pardo R, Vega M, Barrado E, Fernandez J (2000) Temporal evolution of groundwater composition in an alluvial aquifer (Pisuerga River, Spain) by principal component analysis. *Water Research* 34(3):807–816.

Hofmann J, Hurdler J, Ibisch R, Schaffer M, Borchardt D (2011) Analysis of recent nutrient emission pathways, resulting surface water quality and ecological impacts under extreme continental climate: The Kharaa River Basin (Mongolia). *International Review of Hydrobiology* 96 (5):484–519.

Hofmann J, Rode M, Theuring P (2013) Recent developments in river water quality in a typical Mongolian river basin, the Kharaa case study. *Proceedings of the IAHS-IAPSO-IASPEI Assembly*, Gothenburg, Sweden.

Huang F, Wang XQ, Lou LP, Zhou ZQ, Wu JP (2010) Spatial variation and source apportionment of water pollution in Qiantang River (China) using statistical techniques. *Water Research* 44(5):1562–1572.

Jarvie HP, Neal C, Williams RJ, Neal M, Wickham HD, Hill LK (2002) Phosphorus sources, speciation and dynamics in the lowland eutrophic River Kennet, UK. *Science of the Total Environment* 282:175-203.

Jarvie HP, Neal C, Withers PJA, Wescott C, Acornley RA (2005) Nutrient hydrochemistry for a groundwater-dominated catchment: the Hampshire Avon, UK. *Science of the Total Environment* 344:143-58.

Johnes PJ, Burt TP (1991) Water quality trends in theWindrush catchment: nitrogen speciation and sediment interactions. *IAHS Publ* 203:349-57.

Johnes PJ, Burt TP (1993) Nitrate in surface waters. In: Burt TP, Heathwaite AL, Trudgill ST (eds) *Nitrate: processes, patterns and control*. Chichester, Wiley. p.269-320.

Johnson DL, Ambrose SH, Bassett TJ, Bowen ML, Crummey DE, Isaacson JS, Johnson DN, Lamb P, Saul M, Winter-Nelson AE (1997) Meanings of environmental terms. *Journal of Environmental Quality* 26:581-589.

Johnson EE (1983) *Groundwater and wells*. Ist Indian edition, Jain Brothers, 440 p.

Listori JJ, World-wide Bank (1990) *Environmental Health Components for Water Supply, Sanitation and Urban Projects*. World-wide Bank, Washington, DC.

Menzel L, Burger G (2002) Climate change scenarios and runoff response in the Mulde catchment (southern Elbe, Germany). *Journal of Hydrology* 267:53–64.

Misaghi F, Delgosha F, Razzaghmanesh M, Myers B (2017) Introducing a water quality index for assessing water for irrigation purposes: A case study of the Ghezel Ozan River. *Science of The Total Environment* 589:107-116.

Moe CL, Rheingans RD (2006) Global challenges in water, sanitation and health. *Journal of Water and Health* 4 (Suppl 1):41–57.

Niemi GJ, Devore P, Detenbeck N, Taylor D, Lima A (1990) Overview of case studies on recovery of aquatic systems form disturbance. *Environmental Management* 14: 571–587.

NSPRC (2002) *Standards for Drinking Water Quality (GB3828-2002)*. National Standard of People's Republic of China.

Ososkova T, Gorelkin N, Chub V (2000) Water Resources of Central Asia and Adaptation Measures for Climate Change. *Environmental Monitoring and Assessment* 61(1):161-166.

Oyarzun R, Jofre E, Morales P, Maturana H, Oyarzun J, Kretschmer N, Aravena R (2015) A hydrogeochemistry and isotopic approach for the assessment of surface water–groundwater dynamics in an arid basin: the Limarı′ watershed, North-Central Chile. *Environmental Earth Sciences* 73(1):39–55.

Pawellek F, Frauenstein F, Veizer J (2002) Hydrochemistry and isotope geochemistry of the upper Danube River. *Geochimica & Cosmochimica Acta* 66:3839-3854.

Rahman MM, Mandal BK, Chowdhury TR, Sengupta MK, Chowdhury UK, Lodh D, Chanda CR, Basu GK, Mukherjee SC, Saha KC, Chakraborti D (2003) Arsenic groundwater contamination and sufferings of people in North 24-Parganas, one of the nine arsenic affected districts of West Bengal, India. *Journal of Environmental Science and Health Part A: Toxic/Hazardous Substances and Environmental Engineering* 38 (1):25-59.

Rao YS, Reddy TV, Nayudu PT (1997) Groundwater quality in the Niva River basin, Chittoor district, Andhra Pradesh, India. *Environmental Geology* 32(1):56–63.

Singh A, Mondal GC, Singh TB, Singh S, Tewary BK, Sinha A (2012) Hydrogeochemical processes and quality assessment of groundwater in Dumka and Jamtara districts, Jharkhand, India. *Environmental Earth Sciences* 67(8):2175–2191.

Stigter TY, Van Ooijen SPJ, Post VEA, Appelo CAJ, Dill AMMC (1998) A hydrogeological and hydrochemical explanation of the groundwater composition under irrigated land in a Mediterranean environment, Algarve, Portugal. *Journal of Hydrology* 208:262-279.

Subba Rao N, Surya Rao P, Venktram Reddy G, Nagamani M, Vidyasagar G, Satyanarayana NLVV (2012) Chemical characteristics of groundwater and assessment of groundwater quality in Varaha River Basin, Visakhapatnam District, Andhra Pradesh, India. *Environ Monitoring and Assessment* 184:5189–5214.

Subramani T, Elango L, Damodarasamy SR (2005) Groundwater quality and its suitability for drinking and agricultural use in Chithar River Basin, Tamil Nadu, India. *Environmental Geology* 47(8):1099–1110.

Theuring P, Rode M, Behrens S, Kirchner G, Jha A (2013) Identification of fluvial sediment sources in a meso-scale catchment, Northern Mongolia. *Hydrological Processes* 27:845-856.

Tomas D, Curlin M, Maric AS (2017) Assessing the surface water status in Pannonian ecoregion by the water quality index model. *Ecological Indicators* 79:182-190.

WHO (2008) *Guidelines for Drinking Water Quality*. World Health Organization, Geneva.

Yadav SS, Rajesh K (2011) Monitoring water quality of Kosi River in Rampur District, Uttar Pradesh, India. *Advances in Applied Science Research* 2 (2):197–201.

Yan W, Li J, Bai X (2016) Comprehensive assessment and visualized monitoring of urban drinking water quality. *Chemometrics and Intelligent Laboratory Systems* 155:26-35.

Zhu B, Yang X (2007) The ion chemistry of surface and ground waters in the Taklimakan Desert of Tarim Basin, western China. *Chinese Science Bulletin* 52: 2123-2129.

Zhu B, Yang X, Rioual P, Qin X, Liu Z, Xiong H, Yu J (2011) Hydrogeochemistry of three watersheds (the Erlqis, Zhungarer and Yili) in northern Xinjiang, NW China. *Applied Geochemistry* 26: 1535–1548.

Zhu B, Yu J, Qin X, Rioual P, Xiong H (2012) Climatic and geological factors contributing to the natural water chemistry in an arid environment from watersheds in northern Xinjiang, China. *Geomorphology* 153–154:102–114.

Zhu B, Yu J, Qin X, Rioual P, Liu Z, Zhang Y, Jiang F, Mu Y, Li H, Ren X, Xiong H (2013) The Significance of mid-latitude rivers for weathering rates and chemical fluxes: evidence from northern Xinjiang rivers. *Journal of Hydrology* 486:151–174.

Zhu B, Yu J, Rioual P, Gao Y, Zhang Y, Xiong H (2015) Climate Effects on Recharge and Evolution of Natural Water Resources in middle-latitude Watersheds Under Arid Climate. In: Ramkumar M et al., (Eds) *Environmental Management of River Basin Ecosystems*, *Springer Earth System Sciences*. Pp. 91-109.

Zhu B, Wang X, Rioual P (2017) Multivariate Indications between environment and ground water recharge in a sedimentary drainage basin in northwestern China. *Journal of Hydrology* 549:92-113.

In: Advances in Hydrogeochemistry Research
Editors: S. Saeid Eslamian and Faezeh Eslamian
ISBN: 978-1-53617-829-6

Chapter 14

EVOLUTION AND CHARACTERISTICS OF NATURAL WATERS UNDER ARID CLIMATE IN NORTHERN TIANSHAN MOUNTAINS, CENTRAL ASIA

***Bing-Qi Zhu*[1,*] *and Saeid Eslamian*[2,3]**
[1]Key Laboratory of Water Cycle and Related Land Surface Processes, Institute of Geographic Sciences and Natural Resources Research, Chinese Academy of Sciences, Beijing 100101, China
[2]Department of Water Engineering, Collage of Agriculture, Isfahan University of Technology, 8415683111, Isfahan, Iran
[3]Center of Excellence in Risk Management and Natural Hazards, Isfahan University of Technology, 8415683111, Isfahan, Iran

ABSTRACT

This paper analyzed the physico-chemical characteristics of natural waters in a drainage system of northern Xinjiang (Central Asia) for an identification of the geological evolution and recharge mechanism of natural waters in an arid environment. The studied waters are different in mineralization but are typically carbonate rivers and alkaline in nature. No Cl-dominated water type occurs, indicating an early stage of water evolution. Regolith and geomorphological parameters controlling ground-surface temperature may play a large role in the water geological evolution. Three main morphological and hydrological units are reflected in water physico-chemistry. Climate influences the salinization of natural waters substantially. Direct recharge from seasonal snow and ice-melt water and infiltration of rain into the ground are significant recharge processes for natural waters, but recharge from potential deep groundwater may be less important. The enrichment of ions in lakes has been mainly caused by evaporation rather than through the quality of the recharged water.

* Corresponding Author's Email: zhubingqi@igsnrr.ac.cn.

Keywords: hydrogeochemistry, arid environment, recharge mechanism, climatic effect, Central Asia

1. INTRODUCTION

Water in the Earth system is not only one of the key components but also a major agent of the biogeochemical cycles that link the lithosphere, biosphere and atmosphere. In the central Asia, high mountains, particularly the Tibetan Plateau and Tianshan Ranges, function as barriers for atmospheric circulation and keep moisture from reaching an extensive region in western China, causing arid and hyperarid climatic conditions (Domros and Peng 1988; Sun et al. 2010). Fresh water in these regions is definitely a non-substitutable resource upon which humans and ecosystems depend. The arid land of northwestern China is significant repositories of information relating to hydrological evolution and climatic changes in central and eastern Asia. Northern Xinjiang in northwestern China, the geographical centre of the Asian Continent, is an extremely arid and ecologically weak region. An understanding of the source of natural waters in the northern Xinjiang watersheds is significant not only to policy makers for regional planning, but also to scientists interested in the hydrological cycles in an arid environment, and the climate change in Northern China and other arid regions of Asia.

Most of the rivers in the northern Xinjiang drainage system originate from peripheral mountain glaciers and snow and drain areas of diverse geology and climatic conditions, with lower reaches in the desert environment. Geochemical studies of these waters are of considerable importance for a better understanding of the sources of the natural water resources and the influencing factors. At present, the Xinjiang province has become the focus of attention in China due to its rapidly increasing population and an evident climatic variation. Both of these factors consequently increase stress on regional water resources and environmental protection. Understanding the evolution and recharge mechanisms of natural water is a fundamental issue for both aquatic ecologists and water resource managers (Meyer et al. 1988; Kimbadi et al. 1999). Until now, except for a few studies about the chemical weathering and weathering flux in these regions (Zhu et al. 2011, 2012, 2013a, 2013b, 2015, 2016a, 2016b), little works were carried out in the northern Xinjiang drainage systems. Using the area of northern Xinjiang as an example, this paper provides insight into the physicochemistry of natural waters in arid environment. The geochemical evolution and recharge mechanisms of natural water and the climate effect in this region are emphatically discussed.

2. Regional Setting

The geographical location of the northern Xinjiang (China) ranges between 78° and 90°E and 42° and 50°N. It is approximately 603,000 km^2 in area and is bounded by the Tianshan Mountains to the south and the Altai Mountains to the north. Elevations increase from < 500 m above sea level (asl) in the centre of the Zhungarer Basin to >3000 m asl in the south and north ranges (Figure 1c). The topography is generally flat in the centre plain and is rather cragged in the peripheral mountainous areas. The wide piedmonts and pediment plains are marked by the Gobi desert, grassland for herding and oasis areas with intensive agricultural activities. The land is controlled by an arid temperate continental climate. The mean annual air temperature is about 5°C, with a minimum of - 10°C to - 20°C in January and a maximum of 28°C to 33°C in August. The regional precipitation is derived mainly from westerlies, with a mean annual precipitation rate of 60-150 mm in the centre plain and 200-500 mm in the surrounding mountainous areas (Figure 1b). While the potential evapotranspiration is approximately 1000-3500 mm per year, it varies in aridity both seasonally and along the elevation gradient across northern Xinjiang, due to the distribution patterns of seasonal precipitation, temperature and relative humidity and the orientation of delivery of moisture by the westerlies.

The northern Xinjiang drainage system includes three watersheds: the Zhungarer, the Yili and the Erlqis (Ma 2002) (Figure 1a). The Zhungarer and Yili watersheds are inland continental watersheds and the Erlqis watershed drains into the Arctic Ocean. Water resources in northern Xinjiang are mainly distributed in Yili (170.9×10^8 m^3), Arltai (129×10^8 m^3) and Tacheng districts (61.6×10^8 m^3) and the total volume of the surface water in northern Xinjiang is about 435.7×10^8 m^3. The temporal distribution of runoff in a year is extremely heterogeneous with 50-70% in summer, 10-20% both in spring and autumn, and less than 10% in winter. The uneven spatial-temporal distribution of water resources and dry climate environment determine the complete dependence of oasis agricultural activities on water. The Zhungarer Basin (379,000 km^2) is located at the central part of northern Xinjiang and is the geographic centre of the Asian Continent. The basin is a structural depression filled with unconsolidated Quaternary and Tertiary sediments as much as 500-1000 m in thickness (XETCAS et al. 1978), and is an extension of the Palaeozoic Kazakhstan block surrounded by Palaeozoic folded mountains. Aeolian deposits are widespread in this basin. The Gulbantonggute Sand Sea (48,800 km^2) is located in the central part of the basin. A large geographic distance from the surrounding oceans and the presence of the rain-shadow effect due to the surrounding orographic conditions is responsible for the arid climate of the Zhungarer Basin, which is necessary for the creation of the sand desert. Most rivers in the Yili watershed converged into the Yili River at their lower reaches. The Yili River is a large international river running from its sources in the northeast Borohoro Mountains and southeast Halik Mountains, both branches of the eastern Tianshan Mountains, and flowing eastward through the Yili basin into Kazakstan (Figure 1). It drains igneous and metamorphic Precambrian (An€) and Variscan (γ_4) granite, Carboniferous (C_1, C_{2+3}) carbonatite and limestone and Quaternary sediments (Ma 2002) (Figure 1d). The Yili River is about 430 km long inside the Yili watershed and contains three major tributaries in the catchment: the Kashi River in the northeast, the Gongnais River in the east and the Terkers River in the south and southeast. Much of the drainage areas of the tributaries are dominated by the sedimentary, mostly carbonate, lithologies (Figure 1d). The Erlqis watershed has headwaters in the southern slopes of the Arltai Mountains (Figure 1a). Rivers converge into the Erlqis River at their lower reaches, and the Erlqis River is also a large international river, flowing through Kazakstan northward through Russia and finally into the Arctic Ocean. The stem channel of the Erlqis River is about 500 km long inside the Erlqis watershed. The Buerjin River is one of the most important tributaries of the Erlqis River and is about 250 km long and originates in the glaciers of the Youyi Mountains (4373 m asl). The lithologic outcrops in the catchment are mainly the Devonian marine carbonate, clastite (D2, D2 + 3), and the igneous Variscan (γ_4) granite and Quaternary sediments (Ma 2002) (Figure 1d).

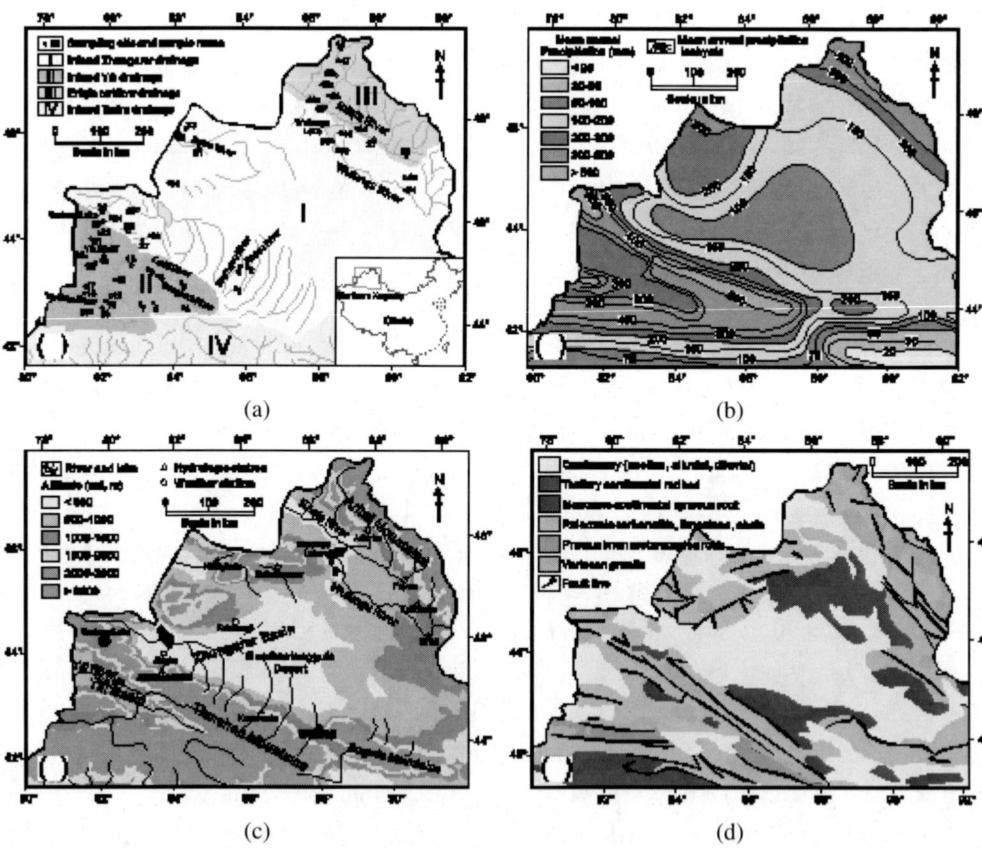

Figure 1. (a) Hydrological setting and sampling locations map, (b) mean annual precipitation isohyets distribution map, (c) topographical map and (d) lithological-distribution map of the northern Xinjiang (Central Asian) in China.

3. METHODS

Field investigations were carried out on the whole region of northern Xinjiang during summer-autumn season in 2008. Chinese 1:50,000 scale topographic maps, a prismatic compass, a Thrommen altimeter and a Garmin GPS were used for orientation. All locations and elevations were recorded using GPS and topographic maps. Water samples were collected under natural flow conditions in one-liter polyethylene bottles from various parts of the Yili, Zhungarer, and Erlqis watersheds (Figure 1a), including river stems, stream channels, hill slopes, wells, lakes, ponds, man-made trenches and reservoirs. Taking into account that the tributaries reflect a much broader variety of geological, biological, and population patterns than do main stem rivers (Pawellek et al. 2002), it was of interest to sample tributary water to look for common features reflected in their hydrochemistry. Some pristine streams draining forested catchments were chosen in the Tianshan and the Arltai Mountains. All waters were surface-sampled except for well water extracted from pump shafts. Most water samples were colorless, but a few were yellow, gray or turbid in colour due to dissolved iron content or suspended solid particles. The colored river water samples primarily drained through the south and north piedmont areas of the Tianshan Mountains, with deep coves of loess and loess-like soil. Because no river and spring exist in the hinterland of the Gulbantonggute Desert, no surface water samples was collected for comparison from others in this work.

Water samples were filtered through a 0.45 μm Millipore membrane filters in the field and analyzed in the institute of Geology and Geophysics, Chinese Academy of Sciences (IGGCAS). Analysis of water samples included total dissolved solids (TDS), pH, and ion concentrations. Ions were mainly measured with a DX600 Dionex ion-chromatography. Temperature (T), pH, and TDS were measured *in situ* using a portable Multi-Parameter Analyzer (Eijkelkamp 18.28). The water samples for cation analysis were acidified with hyper-pure HNO_3 below pH 4.5. Cations and anions were determined by ion chromatography (IC, Dionex 600) with deionized water (EC < 2.1 μS/cm) as the dilute base. The eluents used were 20 mmol/L MSA (methanesulfonic acid) at pH 4.5 and conductivity of approximately 400 μS/cm for cation analysis and 3.5 mmol/L Na_2CO_3, 1.0 mmol/L $NaHCO_3$ at pH 8.5 for anion analysis, respectively. One replicate for each sample was taken for ion analysis, and the instrument was recalibrated after every 5 samples. The alkalinity was measured with a Hach digital titrator using the Gran method (Wetzel and Likens 2000) within three days after sampling.

4. RESULTS

Major physio-chemical and chemical properties of the water samples studied are statically listed in Table 1. The data show that the natural water at various sites is different in terms of mineralization, with TDS ranging from 24.6 to 6200 mg/L, the mean being 580 mg/L. More than half of the investigated samples belong to fresh water (TDS < 1000 mg/L), and the remainder to brackish water (TDS 1-10 g/L). All natural water is alkaline, and pH values range from 7.0 to 9.81, with an average value of 7.87. In the Piper's diagram the natural waters are distributed in fields where alkaline earths exceed alkalis and strong acids and weak acids dominate partially in composition (Figure 2). There are four main water types: Ca-HCO_3, Ca-NDA (non-dominant anion), Ca-SO_4 and NDC-NDA (non-dominant cation) or Na-NDA. Ca-type water is most widely distributed in the study area. The Ca-HCO_3 and Ca-NDA types mainly occur in the montane areas and the low-pitched piedmont zones. The Ca-SO_4 and Na-SO_4 types are mostly distributed in the transition band between oasis and desert plain. Mg-type water merely occurs in the Sailimu Lake. There is no Cl-dominated water type that occurs in the study area (Figure 2).

5. DISCUSSION

5.1. Geochemical Evolution and Distribution

The natural waters in northern Xinjiang are similar to the majority of the large rivers on earth, which are typically carbonate rivers and alkaline in nature (Meybeck 1987). Compared with the upper ranges of typical values for major world rivers and the lower ranges for major pristine rivers (most common natural concentrations - MCNC) (Figure 3), such as the Amazon and Mackenzie (Meybeck and Helmer 1989; Meybeck 1996), the mean major-ion concentrations of the study waters are almost in excess of typical natural values of the major world rivers (Figure 3), except for Ca^{2+} and HCO_3^-. This suggests that the natural waters in northern Xinjiang fall close to the upper limit of concentrations in major world rivers, classified as "salted" by Meybeck (1996). Furthermore, the study waters are notably characterized by an enormous range over three orders of magnitude for concentrations and yields than that of global rivers (Figure 3). This indicates a definable spatial variation of water chemistry and quality in the study watersheds.

Cl-type water is usually one of the major water types in the desert environment in northern China, such as in the Taklamakan Desert (Zhang et al. 1995; Zhu and Yang 2007) and the Badain Jaran Desert (Yang and Williams 2003; Zhu and Yang 2010; Zhu et al. 2012). Cl-dominated water type does not exist in the study area (Figure 2), indicating that the natural waters in northern Xinjiang are still at the early stage of water evolution.

The large variation in Ca^{2+} and Na^+ concentrations in the natural water samples from the study area (Table 1 and Figure 2) may theoretically reflect local mineralogical changes in the sediments and in the carbon dioxide produced by biological processes in their surface layers. However, biogenic weathering might not be a key factor, because the natural water here is alkaline in general. The CO_2 reactions in the natural water are possibly not very active at many sites because of alkaline conditions and limited vegetation on the surface. The heterogeneity of the ionic compositions (Figure 2) may also be caused by local variations in the mineralogy of the natural water reservoir and geochemical processes occurring in the aquifer. Principally, the amount of sodium rises as the chloride content increases in the investigated samples (Figure 4), supporting the view that geochemical processes play a role in the chemistry of natural water. Referring to the Na/Cl ratios, part of samples (mostly from mountainous areas) with lower TDS lie very close to the 1:1 ratio line (Figure 4), indicating that sulphate and other salts are relatively abundant in the natural water with higher TDS (mainly under fluvial-plains and desert environments). The regolith and geomorphological parameters controlling ground-surface temperature may play a large role in rock weathering regime and so in the water chemistry, because solution of lithological components is controlled to a large extent by the ground temperature in a periglacial landscape and its surrounding areas (Beylich et al. 2004). Such as in areas of shade, cold ground and thin regolith, the TDS of natural water (surface water and shallow groundwater) is low, whereas the TDS is high in areas exposed to intensive radiation. Data from the artic-oceanic drainage basin of northern Sweden (Latnjavagge) show that water chemistry may vary considerably in a small area with very homogeneous lithology but great slope segments. The regolith in northern Xinjiang area varies in thickness from less than a meter to more than 5 m between upper and lower slope segments.

The water samples from the oasis areas in the Zhungarer basin where the community administrations are located show much higher TDS than samples from more pristine areas. This is probably caused by two processes: (1) direct pollution reflecting increased population and agricultural density and (2) re-seepage of wastewater. It appears that the wastewater can infiltrate into the natural water systems quickly, because the irrigation and housing, and therefore waste and dung depot, are very close and easy to the aquifer due to the close hydrological conditions with low topography in the basin. Compared with the nitrate contents (0.5-7.7 mg/L) in the surface waters, groundwater samples in the Zhungarer watershed have clearly higher concentrations of nitrate (22-37 mg/L). The high concentration of nitrate in these samples is probably related to the decomposition process of untreated waste. In fact, the nitrate concentration is quite high in the majority of the groundwater samples from the entire study area. In the arid and desert environment, both humans and their animals depend on the groundwater (wells) to live. Therefore human influence is quite strong around the wells and the dung depot is often located not far from the wells. This may be the reason for the high concentration of nitrate in the water samples. There is no wastewater treatment system in this region yet. In the interests of the health of local residents it is necessary to add sewage treatment.

Table 1. Statistical summary of the physical parameters and major ion concentrations determined in the study water samples (TH, total hardness; SAR, sodium adsorption ratio)

Parameter	Maximum	Minimum	Average	Median	10th	90th	SD
pH	9.81	7.00	-	7.85	7.33	8.24	0.47
Eh	- 21.0	- 187	- 72.6	- 72.5	- 94.6	- 41.8	27.6
EC (μs/cm)	5380	46.3	864	352	162	3426	1258
TDS (mg/L)	6200	24.6	580	186	86.1	1843	1071
Li^+ (mg/L)	0.13	nd	0.02	0.01	nd	0.04	0.02
Na^+ (mg/L)	1673	1.66	105	6.86	2.60	272	281
NH_4^+ (mg/L)	0.17	nd	0.02	nd	nd	0.03	0.04
K^+ (mg/L)	77.7	0.49	8.48	1.63	0.82	25.9	17.4
Mg^{2+} (mg/L)	190	0.63	27.2	6.49	2.44	81.8	49.5
Ca^{2+} (mg/L)	91.9	3.05	29.1	26.4	14.0	48.0	16.7
F^- (mg/L)	3.04	0.02	0.31	0.18	0.06	0.54	0.51
Cl^- (mg/L)	648	1.83	74.5	6.91	2.97	223	160
NO_3^- (mg/L)	36.5	nd	4.96	2.88	0.72	8.89	6.62
SO_4^{2-} (mg/L)	2655	2.52	205	45.3	9.01	682	437
HCO_3^- (mg/L)	799	9.76	117	69.4	29.7	283	142
TH (mg/L)	824	10.2	184	102	47.9	436	209
SAR	28.2	0.058	1.97	0.278	0.094	4.56	4.85

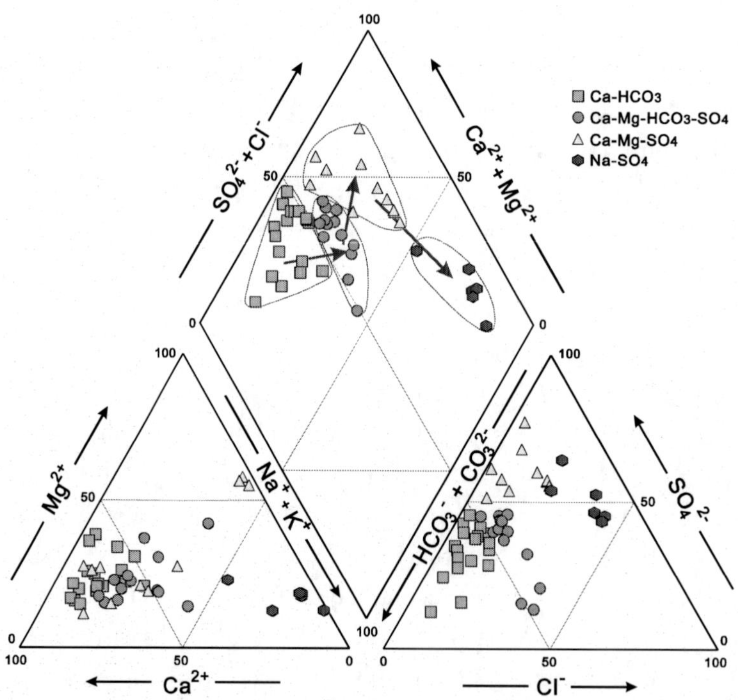

Figure 2. Piper plot of the study water samples.

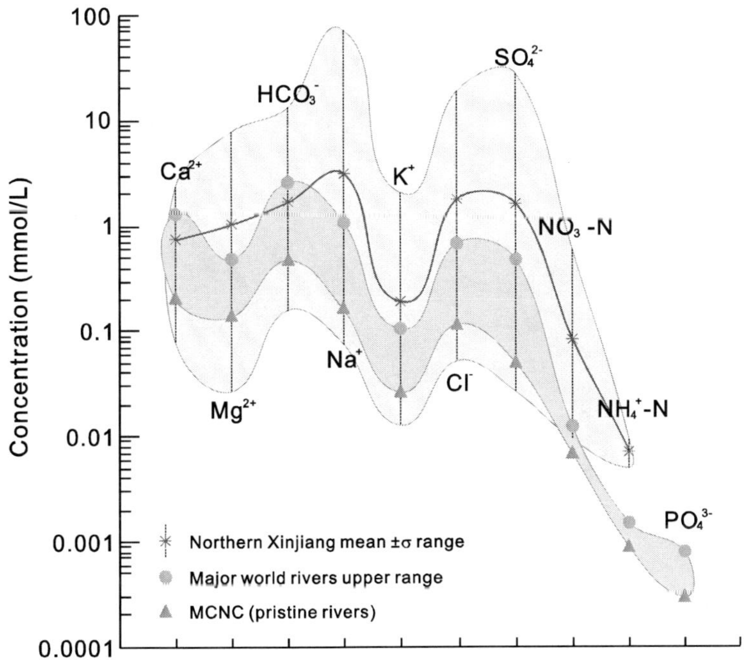

Figure 3. Typical ranges for major ions in the natural waters in northern Xinjiang. Also marked are the upper limits for these components in major world rivers and the median concentrations in pristine rivers (MCNC) (Meybeck and Helmer 1989; Meybeck 1996).

5.2. Recharge Mechanisms of the Natural Waters

The possible mechanisms for natural water (especially river and shallow groundwater) recharge in the areas of the northern Xinjiang are: (a) direct infiltration of rain into the ground; (b) direct recharge from seasonal snow and ice-melt water; (3) recharge from potential deep groundwater. The first two processes are thought to be significant, but the third may be much less important, because the northern Xinjiang watersheds form a various ranges in a desert landscape. Based on the tectonic structure of mountain-basin couples, it is concluded that the recharge of the shallow groundwater is mainly through infiltration of seasonal snow and ice-melt water derived from mountain areas and local precipitation that infiltrates downwards quite quickly owing to the loose nature of the regolith. Due to the large temporal and spatial variability of rainfall, infiltration is related to the amount of precipitation of each event. From dried channels in the fields one can assume that there may be ephemeral streams and ponds in the study area when it rains. As a result, the overall recharge process is quite heterogeneous. The clear difference in ion concentrations suggests that the hydraulic relationship between various sampling sites is weak.

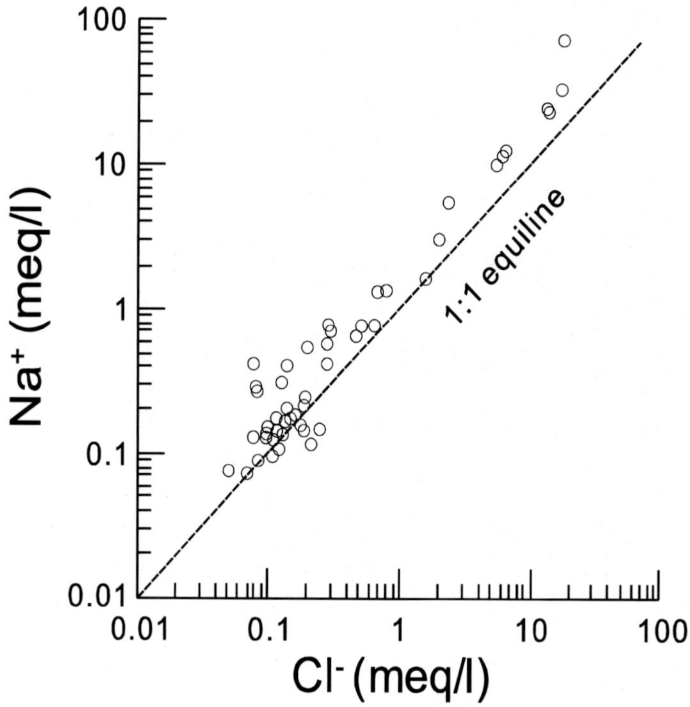

Figure 4. Scatter diagram of the concentrations of Na^+ vs. Cl^- of the study waters.

For the areas of fluvial plain in the Zhungarer, we consider that the precipitation directly onto the rivers, lakes and reservoirs plays a minor role in the recharge today, because the amount of precipitation is much smaller than the evaporation occurring on the water surface. Field observations indicate that all rivers in these areas were recharged mainly by mountain stream water and partially by groundwater at present time. Owing to the isolated geomorphological basin structures, it is unlikely that the recharge mechanisms might have fundamentally changed in the past. The large range in chemical composition of water samples suggests not only the heterogeneity of the recharge conditions in the study areas, but also a possible mixing with enriched waters that could be in different aquifers. However, the input from other regions (via surface runoff or groundwater inflow) seems to be rather insignificant in the local water cycle. Besides, a large proportion of rain is maybe added to the shallow groundwater system because of rapid infiltration in the loose and sandy sediments. Observations and interviews with local residents suggest that the water from the shallow aquifer can meet the water demand of inhabitants with traditional life styles and adequate livestock grazing in the desert environment of northern Xinjiang.

The highest TDS concentrations were observed at some lake depressions (such as the Salimu Lake) around the central Zhungarer watershed. From field observation, it is assumed that the Salimu Lakes are mainly recharged by glacier/snow waters, because it is located in a highland depression surrounded by high mountains and there is surficial inflow but no outflow. The major ions in the lake are Mg^{2+} for cations and SO_4^{2-} for anions. At present, Salimu is one of the largest lakes in northern Xinjiang and has relatively high TDS (1850 mg/L). Compared with the chemistry of other lake samples, the proportions of sodium and chloride are much lower in the Salimu lake water. From a global point of view, the dominant ions in most saline lakes are Na^+ and Cl^-, and occasionally in some less saline lakes, Na^+ and HCO_3^-/CO_3^{2-} (e.g., Day 1993). It is reasonable to believe that the brines of these lakes were formed by long and intensive evaporation, because Na^+ and Cl^- are often associated with evaporites. We can exclude the halite origin for the Salimu Lake. However, Mg^{2+} and SO_4^{2-} are also often associated with evaporites. Dissolution of sulphates (such as gypsum and anhydrite) may be an important source here. In general, major ions in saline waters were derived from the catchment or the substratum (e.g., Kilham 1990). However, atmospheric precipitation should not be ignored as a potentially significant source of ions (Eriksson 1985) in the water of the study area because of frequent dews and rimes. In many large saline lakes of northern China and Mongolia, Cl^-, SO_4^{2-} and HCO_3^- are almost equally predominant anions, while Na^+ and K^+ are strongly dominant cations. The salts of these lakes were thought to have originated from high evaporation and soil salinization (Egorov 1993). The salinity of the Chinese lakes increases westwards, consistent with the trend of increasing aridity. Both natural climatic changes and irrigation activities have caused decadal variations in the salinity in these lakes. From limnological analysis of different world regions, it was concluded that large, deep, saline lakes often contain alkaline, sodium rich waters with considerable chloride, sulphate, and carbonate plus bicarbonate (Melack 1983). The chemistry of lakes in the northern Xinjiang drainage basins is, to a considerable extent, consistent with those of large lakes. Also, from this point of view, it appears that the enrichment of the ions in the lakes has been mainly caused by evaporation, rather than through the quality of the recharged water.

5.3. Response to Climate

The regional distribution pattern of chemical water types (Figure 2) reveals that a transitional trend exists between natural waters in the northern Xinjiang drainage basins. Water types are generally transformed from Ca-HCO_3 to Ca-Mg-HCO_3 to Ca-Mg-SO_4 and to Na-SO_4. This trend is spatially accompanied by the altitude variation of landscapes in the order of mountains, pediments, piedmont plains and desert plains. It indicates a potential control of a vertical zonality of temperature and relevant geological effects on the natural water evolution. Besides this trend, another kind of zonal distribution of hydrogeochemistry also exists in the northern Xinjiang drainage basins. For example, the relatively wet zones, with AMP > 200 mm in three watersheds, consist of the Ca-HCO_3, Ca-NDA and NDC-NDA types. In the relatively dry zones, with AMP < 200 mm, however, the Ca-SO_4 and Na-SO_4 types predominate. The Ca-HCO_3-dominated chemistry of river water in the relatively wet zones couples well with regional carbonate lithological distributions (Figures 1d). It confirms that rock weathering and the regional lithologic distributions have a major effect on the evolution of river chemistry in the wet zones. The precipitation processes can accelerate the rock weathering processes (White and Blum 1995). The water types in dry zones tend to be relatively uniform. Because the lithological distributions are not homogeneous, evaporation processes, which can result in the selective precipitation of solute fraction ($CaCO_3$) and subsequent changes in water chemistry as water moves downstream in the relatively arid zones (Gibbs 1970; Kilham 1990; Feller 2005), could be responsible for the accumulation and evolution of sodium and sulphate salts. It indicates a major effect of climate rather than geology on the evolution of water chemistry in dry zones.

The strong relationship between the annual mean summer precipitations (AMP) in northern Xinjiang and the altitudes of the sampling sites in this study can be well expressed as AMP = 0.14 × altitude + 101.3 ($r = 0.88$, $p < 0.005$). The total dissolved solids in most samples decreased in concentration with the increasing altitude (Figure 5). It indicates a potential influence of precipitation on dissolved materials in the water. The fact that rock types are not distributed according to the altitude (Figure 1c,d) implies that topography and temperature, in addition to rock type and weathering, affects water chemistry indirectly by influencing AMP/evaporation processes.

In the central parts of the study watersheds, evaporation probably influences the salinization of surface and groundwater substantially, because the TDS value of river water clearly varies along their channel courses, such as the Erlqis, Wulungu, Jingou, Gongnais and Yili rivers (Figure 5). The river waters in the middle reaches of the Jingou, Gongnais and Wulungu have TDS values of 90, 95 and 321 mg/L, respectively, and gradually change to higher values of 112, 147 and 2000 mg/L after flowing about 70, 100 and 300 km downward, respectively. The Jingou, Gongnais and Wulungu rivers have few tributaries over their water courses (Figure 1a), so the waters flowing along the lengths of these courses do not mix extensively with water from different tributary sources. In addition, the seasonal changes of discharge and sources cannot be the reasons for the variation in the dissolved solid concentrations for one-day-collected samples from a single river. Accordingly, the increase of TDS from upper reaches toward lower reaches in these rivers can be interpreted in terms of evaporation and dissolution of soil salts along their river courses. Due to the relatively fixed channel course at present, the dissolution of soil salts alone river bed maybe does not vary greatly. So if evaporation processes cause the change of concentration of dissolved material during water flow, given no change in the flow speed of ~ 5 km/h, it is intense enough in these drainage basins that runoff can decrease by one-third to one-fifth during one day.

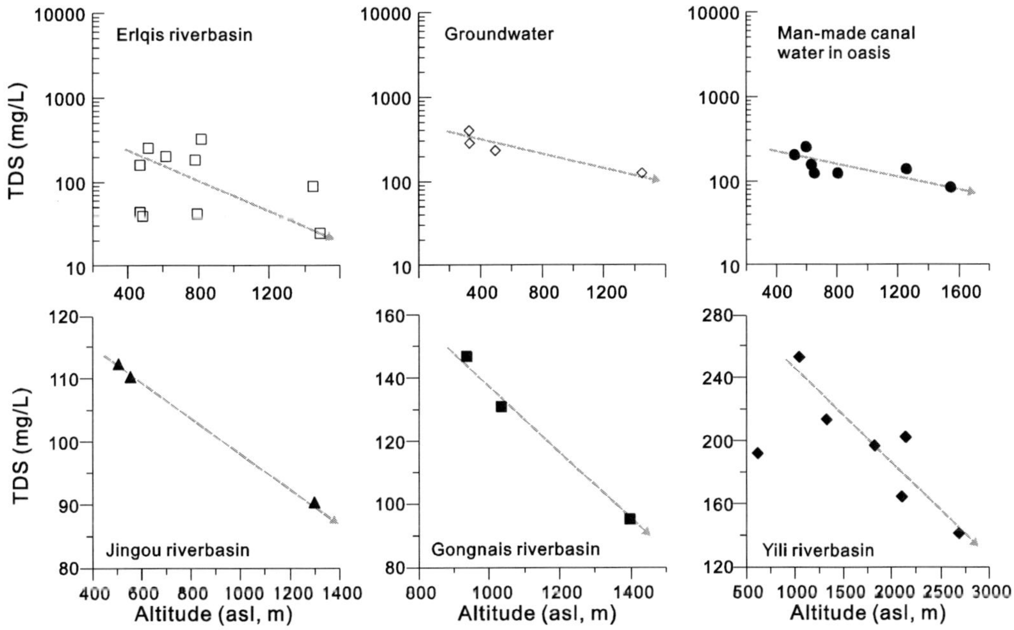

Figure 5. TDS vs. altitude (m, asl) of the study water samples.

6. CONCLUSION

The physico-chemical characteristics of the natural waters in northern Xinjiang drainage system, as well as the climatic, geological and topographical context where they flow and reside, allowed the identification of fluvial geological evolution and recharge mechanism of the natural water resources in arid environment. The natural water at various sites in the study area is different in terms of mineralization and is similar to the majority of the large rivers on Earth in general, which are typically carbonate rivers and alkaline in nature. However, there is no Cl-dominated water type that occurs in the study area, indicating that the natural waters in northern Xinjiang are still at the early stage of water evolution. The regolith and geomorphological parameters controlling ground-surface temperature may play a large role in water chemistry and its evolution. Evaporation influences the salinization of natural water substantially. The high concentration of nitrate in oasis water is probably related to the decomposition process of untreated waste. Direct recharge from seasonal snow and ice-melt water and infiltration of rain into the ground are thought to be significant recharge processes for natural water in the study area, but recharge from potential deep groundwater may be much less important. The chemistry of lakes is to a considerable extent consistent with those of large lakes in the world but the enrichment of the ions in the lakes has been mainly caused by evaporation, rather than through the quality of the recharged water. Two kinds of zonal distribution of water types exist in the northern Xinjiang drainage basins. It confirms that temperature-and-precipitation-dependent geological weathering and lithologic distributions have a major effect on the evolution of water evolution in the wet zones and a major effect of climate (evaporation) rather than geology on the water evolution in dry zones.

ACKNOWLEDGMENTS

This work study was financially supported by the National Natural Science Foundation of China (41771014) and the National Key Research and Development Program of China (2016YFA0601900). The authors are grateful to Profs. Xiaoping Yang and Xiaoguang Qin for their generous help in the research work.

REFERENCES

Beylich, A. A., Kolstrup, E., Thyrsted, T. & Gintz, D. (2004). Water chemistry and its diversity in relation to local factors in the Latnjavagge drainage basin, arctic-oceanic Swedish Lapland. *Geomorphology*, *58*, 125-143.

Day, J. A. (1993). The major ion chemistry of some southern African saline systems. *Hydrobiologia*, *267*, 37-59.

Domros, M. & Peng, G. (1988). *The Climate of China*. Springer, Berlin.

Egorov, A. N. (1993). Mongolian salt lakes: some features of their geography, thermal patterns, chemistry and biology. *Hydrobiologia*, *267*, 13-21.

Eriksson, E. (1985). *Principles and Applications of Hydrochemistry*. Chapman and Hall, London.

Feller, M. C. (2005). Forest harvesting and streamwater inorganic chemistry in western North America: a review. *Journal of the American Water Resources Association*, *41*, 785-811.

Gibbs, R. J. (1970). Mechanisms controlling world water chemistry. *Science*, *170*, 1088-1090.

Kilham, P. (1990). Mechanisms controlling the chemical composition of lakes and rivers: data from Africa. *Limnology and Oceanography*, *35*, 80-83.

Kimbadi, S, Vandelannoote, A., Deelstra, H., Mbemba, M. & Ollevier, F. (1999). Chemical composition of the small rivers of the north-western part of Lake Tanganyika. *Hydrobiologia*, *407*, 75-80.

Ma, L. (2002). *Geological Atlas of China*. Geological Press, Beijing (in Chinese).

Melack, J. M. (1983). Large, deep salt lakes: a comparative limnological analysis. *Hydrobiologia*, *105*, 223-230.

Meybeck, M. (1987). Global chemical weathering of surficial rocks estimated from river dissolved loads. *American Journal of Science*, *287*, 401-428.

Meybeck, M. (1996). River water quality: global ranges, time and space variabilities, proposals for some redefinitions. *Verhandlungen des Internationalen Verein Limnologie*, *26*, 81-96.

Meybeck, M. & Helmer, R. (1989). The quality of rivers: from pristine stage to global pollution. *Palaeogeography Palaeoclimatology Palaeoecology*, *75*, 283-309.

Meyer, J. L., McDowell, W. H., Bott, T. L., Elwood, J., Ishizaki, C., Mclack, J. M., Peckarsky, B., Peterson, B. & Rublee, P. (1988). Elemental dynamics in streams. *Journal of the North American Benthological Society*, *7*, 410-432.

Pawellek, F., Frauenstein, F. & Veizer, J. (2002). Hydrochemistry and isotope geochemistry of the upper Danube River. *Geochimica et Cosmochimica Acta*, *66*, 3839-3854.

Sun, J., Ye, J., Wu, W., Ni, X., Bi, S., Zhang, Z., Liu, W. & Meng, J. (2010). Late Oligocene-Miocene mid-latitude aridification and wind patterns in the Asian interior. *Geology, 38*, 515-518.

Wetzel, R. G. & Likens, G. E. (2000). *Limnological Analyses* (3rd ed.). Springer, New York.

White, A. F. & Blum, A. E. (1995). Effect of climate on chemical weathering in watersheds. *Geochimica et Cosmochimica Acta*, *59*, 1729-1747.

XETCAS (Xinjiang Expedition Team of the Chinese Academy of Sciences), IGCAS (Institute of Geography of the Chinese Academy of Sciences), and DGBNU (Department of Geography of Beijing Normal University). (1978) *Geomorphology in Xinjiang*. Science Press, Beijing (in Chinese).

Yang, X. & Williams, M. A. J. (2003). The ion chemistry of lakes and late Holocene desiccation in the Badain Jaran Desert, Inner Mongolia, China. *Catena*, *51*, 45-60.

Zhang, J., Takahashi, K., Wushiki, H., Yabuki, S., Xiong, J. & Masuda, A. (1995). Water geochemistry of the rivers around the Taklimakan Desert (NW China): crustal weathering and evaporation processes in arid land. *Chemical Geology*, *119*, 225-237.

Zhu, B., Yang, X., Rioual, P., Qin, X., Liu, Z., Xiong, H. & Yu, J. (2011). Hydrogeochemistry of three watersheds (the Erlqis, Zhungarer and Yili) in northern Xinjiang, NW China. *Applied Geochemistry, 26*, 1535-1548.

Zhu, B. & Yang, X. (2007). The ion chemistry of surface and ground waters in the Taklimakan Desert of Tarim Basin, western China. *Chinese Science Bulletin*, *52*, 2123-2129.

Zhu, B. & Yang, X. (2010). The origin and distribution of soluble salts in the sand seas of northern China. *Geomorphology*, *123*, 232-242.

Zhu, B., Yang, X., Liu, Z., Rioual, P., Li, C. & Xiong, H. (2012b). Geochemical compositions of soluble salts in aeolian sands from the Taklamakan and Badanjilin deserts in northern China, and their influencing factors and environmental implications. *Environmental Earth Sciences*, *66*, 337-353.

Zhu, B., Yu, J., Qin, X., Rioual, P. & Xiong, H. (2012a). Climatic and geological factors contributing to the natural water chemistry in an arid environment from watersheds in northern Xinjiang, China. *Geomorphology*, 153-154, 102-114.

Zhu, B., Yu, J., Qin, X., Rioual, P., Zhang, Y., Liu, Z., Mu, Y., Li, H., Ren, X. & Xiong, H. (2013). Identification of rock weathering and environmental control in arid catchments (northern Xinjiang) of Central Asia. *Journal of Asian Earth Sciences*, *66*, 277-294.

Zhu, B., Yu, J., Qin, X., Rioual, P., Liu, Z., Zhang, Y., Jiang, F., Mu, Y., Li, H., Ren, X. & Xiong, H. (2013). The Significance of mid-latitude rivers for weathering rates and chemical fluxes: evidence from northern Xinjiang rivers. *Journal of Hydrology*, *486*, 151-174.

Zhu, B., Yu, J., Rioual, P., Gao, Y., Zhang, Y. & Xiong, H. (2015). Climate effects on recharge and evolution of natural water resources in middle-latitude watersheds under arid climate. In: Ramkumar, M. U.; Kumaraswamy, K.; Mohanraj, R. (Eds). *Environmental Management of River Basin Ecosystems. Springer Earth System Sciences*. Springer-Verlag, Heidelberg, doi: 10.1007/978-3-319-13425-3_5. pp. 91-109.

Zhu, B., Yu, J. & Rioual, P. (2016a). Geochemical signature of natural water recharge in the Jungar Basin and their response to climate. *Water Environment Research*, *88*, 79-86.

Zhu, B. & Wang, Y. (2016b). Statistical study to identify the key factors governing ground water recharge in the watersheds of the arid Central Asia. *Environmental Monitoring and Assessment*, *188*(1), 66, doi: 10.1007/s10661-015-5075-4.

ABOUT THE EDITORS

Saeid Eslamian is a full professor of environmental hydrology and water resources engineering in the Department of Water Engineering at Isfahan University of Technology, where he has been since 1995. His research focuses mainly on statistical and environmental hydrology in a changing climate. In recent years, he has worked on modeling natural hazards, including floods, severe storms, wind, drought, pollution, water reuses, sustainable development and resiliency, etc. Formerly, he was a visiting professor at Princeton University, New Jersey, and the University of ETH Zurich, Switzerland. On the research side, he started a research partnership in 2014 with McGill University, Canada. He has contributed to more than 600 publications in journals, books, and technical reports. He is the founder and chief editor of both the International Journal of Hydrology Science and Technology (IJHST). Eslamian is now associate editor of three important publications: Journal of Hydrology (Elsevier), Eco-Hydrology and Hydrobiology (Elsevier), Journal of Water Reuse and Desalination (IWA) and Journal of the Saudi Society of Agricultural Sciences (Elsevier). Professor Eslamian is the author of more than 40 books and 200 chapter books.

Dr. Eslamian's professional experience includes membership on editorial boards, and he is a reviewer of more than 100 Web of Science (ISI) journals, including the ASCE Journal of Hydrologic Engineering, ASCE Journal of Water Resources Planning and Management, ASCE Journal of Irrigation and Drainage Engineering, Advances in Water Resources, Groundwater, Hydrological Processes, Hydrological Sciences Journal, Global Planetary Changes, Water Resources Management, Water Science and Technology, Eco-Hydrology, Journal of American Water Resources Association, American Water Works Association Journal, etc. UNESCO has also nominated him for a special issue of the Eco-Hydrology and Hydrobiology Journal in 2015.

Professor Eslamian was selected as an outstanding reviewer for the Journal of Hydrologic Engineering in 2009 and received the EWRI/ASCE Visiting International Fellowship in Rhode Island (2010). He was also awarded outstanding prizes from the Iranian Hydraulics Association in 2005 and Iranian Petroleum and Oil Industry in 2011. Professor Eslamian has been chosen as a distinguished researcher of Isfahan University of Technology (IUT) and Isfahan Province in 2012 and 2014, respectively. In 2016, he was a candidate for national distinguished researcher in Iran.

He has also been the referee of many international organizations and universities. Some examples include the U.S. Civilian Research and Development Foundation (USCRDF), the Swiss Network for International Studies, the Majesty Research Trust Fund of Sultan Qaboos University of Oman, the Royal Jordanian Geography Center College, and the Research Department of Swinburne University of Technology of Australia. He is also a member of the following associations: American Society of Civil Engineers (ASCE), International Association of Hydrologic Science (IAHS), World Conservation Union (IUCN), GC Network for Drylands Research and Development (NDRD), International Association for Urban Climate (IAUC), International Society for Agricultural Meteorology (ISAM), Association of Water and Environment Modeling (AWEM), International Hydrological Association (STAHS), and UK Drought National Center (UKDNC).

Professor Eslamian finished Hakimsanaei High School in Isfahan in 1979. After the Islamic Revolution, he was admitted to IUT for a BS in water engineering and graduated in 1986. After graduation, he was offered a scholarship for a master's degree program at Tarbiat Modares University, Tehran. He finished his studies in hydrology and water resources engineering in 1989. In 1991, he was awarded a scholarship for a PhD in civil engineering at the University of New South Wales, Australia. His supervisor was Professor David H. Pilgrim, who encouraged him to work on "Regional Flood Frequency Analysis Using a New Region of Influence Approach." He earned a PhD in 1995 and returned to his home country and IUT. In 2001, he was promoted to associate professor and in 2014 to full professor. For the past 24 years, he has been nominated for different positions at IUT, including university president consultant, faculty deputy of education, and head of department. Eslamian is now director for center of excellence in Risk Management and Natural Hazards (RiMaNaH).

Professor Eslamian has made three scientific visits to the United States, Switzerland, and Canada in 2006, 2008, and 2015, respectively. In the first, he was offered the position of visiting professor by Princeton University and worked jointly with Professor Eric F. Wood at the School of Engineering and Applied Sciences for one year. The outcome was a contribution in hydrological and agricultural drought interaction knowledge by developing multivariate L-moments between soil moisture and low flows for northeastern U.S. streams.

Recently, Professor Eslamian has published the editorship of nine handbooks published by Taylor & Francis (CRC Press): the three-volume Handbook of Engineering

Hydrology in 2014, Urban Water Reuse Handbook in 2016, Underground Aqueducts Handbook (2017), the three-volume Handbook of Drought and Water Scarcity (2017), Constructed Wetlands: Hydraulic Design (2020). An Evaluation of Groundwater Storage Potentials in a Semiarid Climate by Nova Science Publishers is also his joint book publication in 2019.

Faezeh Eslamian is a PhD Holder of Bioresources Engineering in McGill University. Her research focuses on the development of a novel lime-based product to mitigate phosphorus loss from agricultural fields. Faezeh completed her bachelor and master's degrees in Civil and Environmental Engineering from Isfahan University of Technology, Iran where she evaluated natural and low-cost absorbents for the removal of pollutants such as textile dyes and heavy metals. Furthermore, she has conducted research on the worldwide water quality standards and wastewater reuse guidelines. Faezeh is an experienced multidisciplinary researcher with interest in soil and water quality, environmental remediation, water reuse and drought management.

INDEX

#

A

B

C

D

J

K

L

M

N

O

P

Q

R

T

U

V

W

Y

Z